AF855806

APPLIED CALCULUS

4th Edition

Hughes-Hallett | Gleason | Lock | Flath | et al.

Custom Edition for Virginia Tech

WILEY Custom
LEARNING SOLUTIONS

To order books or for customer service, please call 1(800)-CALL-WILEY (225-5945).

Printed in the United States of America.

ISBN 978-0-470-90034-5

10 9 8 7 6 5 4 3 2 1

Printed and bound by ePAC Technologies, Inc.

Contents

APPLIED CALCULUS

Fourth Edition

Produced by the Calculus Consortium and initially funded by a National Science Foundation Grant.

Deborah Hughes-Hallett
University of Arizona

William G. McCallum
University of Arizona

Andrew M. Gleason
Harvard University

Brad G. Osgood
Stanford University

Patti Frazer Lock
St. Lawrence University

Douglas Quinney
University of Keele

Daniel E. Flath
Macalester College

Karen Rhea
University of Michigan

David O. Lomen
University of Arizona

Jeff Tecosky-Feldman
Haverford College

David Lovelock
University of Arizona

Thomas W. Tucker
Colgate University

with the assistance of

Otto K. Bretscher
Colby College

Eric Connally
Harvard Extension School

Richard D. Porter
Northeastern University

Sheldon P. Gordon
SUNY at Farmingdale

Andrew Pasquale
Chelmsford High School

Joe B. Thrash
University of Southern Mississippi

Coordinated by
Elliot J. Marks

WILEY

John Wiley & Sons, Inc.

PUBLISHER	Laurie Rosatone
ACQUISITIONS EDITOR	David Dietz
ASSOCIATE EDITOR	Shannon Corliss
EDITORIAL ASSISTANT	Pamela Lashbrook
DEVELOPMENTAL EDITOR	Anne Scanlan-Rohrer/Two Ravens Editorial
MARKETING MANAGER	Sarah Davis
MEDIA EDITOR	Melissa Edwards
SENIOR PRODUCTION EDITOR	Ken Santor
COVER DESIGNER	Madelyn Lesure
COVER AND CHAPTER OPENING PHOTO	©Patrick Zephyr/Patrick Zephyr Nature Photography Images

Problems from *Calculus: The Analysis of Functions*, by Peter D. Taylor (Toronto: Wall & Emerson, Inc., 1992). Reprinted with permission of the publisher.

This book was set in Times Roman by the Consortium using TeX, Mathematica, and the package AsTeX, which was written by Alex Kasman. It was printed and bound by R.R. Donnelley / Jefferson City. The cover was printed by R.R. Donnelley.

This book is printed on acid-free paper.

This material is based upon work supported by the National Science Foundation under Grant No. DUE-9352905. Opinions expressed are those of the authors and not necessarily those of the Foundation.

ISBN: 978-0-470-17052-6
Binder Ready: 978-0-470-55662-7

Printed in the United States of America

10 9 8 7 6 5 4 3 2 1

Chapter Five

ACCUMULATED CHANGE: THE DEFINITE INTEGRAL

Contents

5.1 DISTANCE AND ACCUMULATED CHANGE

In Chapter 2, we used the derivative to find the rate of change of a function. Here we see how to go in the other direction. If we know the rate of change, can we find the original function? We start by finding the distance traveled from the velocity.

How Do We Measure Distance Traveled?

The rate of change of distance with respect to time is velocity. If we are given the velocity, can we find the distance traveled? Suppose the velocity was 50 miles per hour throughout a four-hour trip. What is the total distance traveled? Since

$$\text{Distance} = \text{Velocity} \times \text{Time},$$

we have

$$\begin{aligned}\text{Distance traveled} &= (50 \text{ miles/hour}) \times (4 \text{ hours})\\ &= 200 \text{ miles.}\end{aligned}$$

The graph of velocity against time is the horizontal line in Figure 5.1. Notice that the distance traveled is represented by the shaded area under the graph.

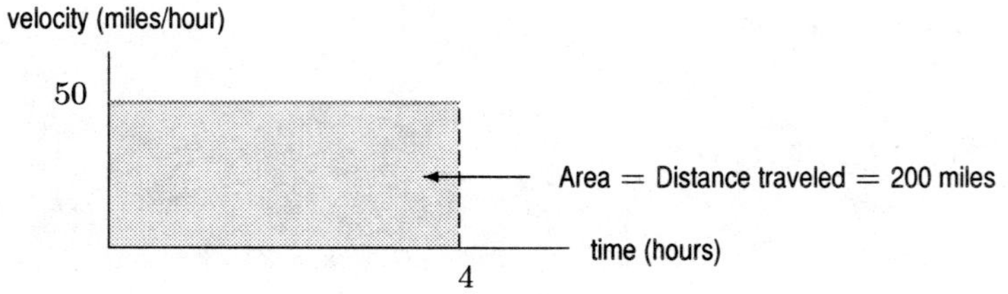

Figure 5.1: Area shaded represents distance traveled in 4 hours at 50 mph

Now let's see what happens if the velocity is not constant.

Example 1 Suppose that you travel 30 miles/hour for 2 hours, then 40 miles/hour for $1/2$ hour, then 20 miles/hour for 4 hours. What is the total distance you traveled?

Solution We compute the distances traveled for each of the three legs of the trip and add them to find the total distance traveled:

$$\begin{aligned}\text{Distance} &= (30 \text{ miles/hour})(2 \text{ hours}) + (40 \text{ miles/hour})(1/2 \text{ hour}) + (20 \text{ miles/hour})(4 \text{ hours})\\ &= 60 \text{ miles} + 20 \text{ miles} + 80 \text{ miles}\\ &= 160 \text{ miles.}\end{aligned}$$

You travel 160 miles on this trip.

A Thought Experiment: How Far Did the Car Go?

In Example 1, the velocity was constant over intervals. Of course, this is not always the case; we now look at an example where the velocity is continually changing.

Velocity Data Every Two Seconds

Suppose a car is moving with increasing velocity and suppose we measure the car's velocity every two seconds, obtaining the data in Table 5.1:

Table 5.1 *Velocity of car every two seconds*

Time (sec)	0	2	4	6	8	10
Velocity (ft/sec)	20	30	38	44	48	50

How far has the car traveled? Since we don't know how fast the car is moving at every moment, we can't calculate the distance exactly, but we can make an estimate. The velocity is increasing, so the car is going at least 20 ft/sec for the first two seconds. Since Distance = Velocity × Time, the car goes at least $20 \cdot 2 = 40$ feet during the first two seconds. Likewise, it goes at least $30 \cdot 2 = 60$ feet during the next two seconds, and so on. During the ten-second period it goes at least

$$20 \cdot 2 + 30 \cdot 2 + 38 \cdot 2 + 44 \cdot 2 + 48 \cdot 2 = 360 \text{ feet.}$$

Thus, 360 feet is an underestimate of the total distance traveled during the ten seconds.

To get an overestimate, we can reason in a similar way: During the first two seconds, the car's velocity is at most 30 ft/sec, so it moves at most $30 \cdot 2 = 60$ feet. In the next two seconds it moves at most $38 \cdot 2 = 76$ feet, and so on. Therefore, over the ten-second period it moves at most

$$30 \cdot 2 + 38 \cdot 2 + 44 \cdot 2 + 48 \cdot 2 + 50 \cdot 2 = 420 \text{ feet.}$$

Therefore,

$$360 \text{ feet} \le \text{Total distance traveled} \le 420 \text{ feet.}$$

There is a difference of 60 feet between the upper and lower estimates.

Velocity Data Every One Second

What if we want a more accurate estimate? We could make more frequent velocity measurements, say every second. The data is in Table 5.2.

As before, we get a lower estimate for each second by using the velocity at the beginning of that second. During the first second the velocity is at least 20 ft/sec, so the car travels at least $20 \cdot 1 = 20$ feet. During the next second the car moves at least 26 feet, and so on. So we say

$$\begin{aligned}\text{New lower estimate} &= 20 \cdot 1 + 26 \cdot 1 + 30 \cdot 1 + 35 \cdot 1 + 38 \cdot 1 \\ &\quad + 42 \cdot 1 + 44 \cdot 1 + 46 \cdot 1 + 48 \cdot 1 + 49 \cdot 1 \\ &= 378 \text{ feet.}\end{aligned}$$

Notice that this is greater than the old lower estimate of 360 feet.

We get a new upper estimate by considering the velocity at the end of each second. During the first second the velocity is at most 26 ft/sec, and so the car moves at most $26 \cdot 1 = 26$ feet; in the next second it moves at most 30 feet, and so on.

$$\begin{aligned}\text{New upper estimate} &= 26 \cdot 1 + 30 \cdot 1 + 35 \cdot 1 + 38 \cdot 1 + 42 \cdot 1 \\ &\quad + 44 \cdot 1 + 46 \cdot 1 + 48 \cdot 1 + 49 \cdot 1 + 50 \cdot 1 \\ &= 408 \text{ feet.}\end{aligned}$$

This is less than the old upper estimate of 420 feet. Now we know that

$$378 \text{ feet} \le \text{Total distance traveled} \le 408 \text{ feet.}$$

Notice that the difference between the new upper and lower estimates is now 30 feet, half of what it was before. By halving the interval of measurement, we have halved the difference between the upper and lower estimates.

Table 5.2 *Velocity of car every second*

Time (sec)	0	1	2	3	4	5	6	7	8	9	10
Velocity (ft/sec)	20	26	30	35	38	42	44	46	48	49	50

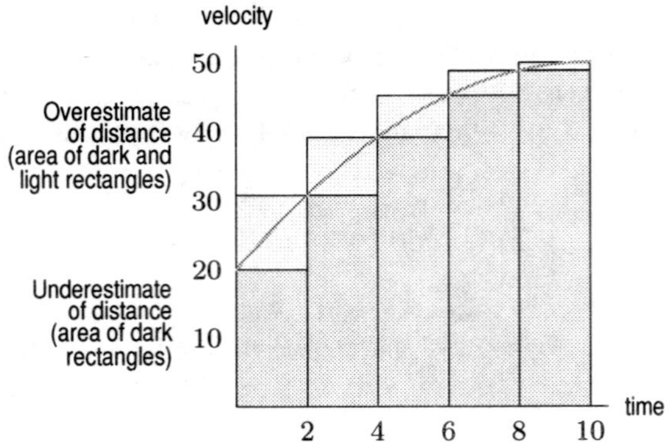

Figure 5.2: Shaded area estimates distance traveled. Velocity measured every 2 seconds

Visualizing Distance on the Velocity Graph

Consider the two-second data in Table 5.1. We can represent both upper and lower estimates on a graph of the velocity against time. The velocity can be graphed by plotting these data and drawing a smooth curve through the data points. (See Figure 5.2.)

We use the fact that for a rectangle, Area $=$ Height $\times$ Width. The area of the first dark rectangle is $20 \cdot 2 = 40$, the lower estimate of the distance moved during the first two seconds. The area of the second dark rectangle is $30 \cdot 2 = 60$, the lower estimate for the distance moved in the next two seconds. The total area of the dark rectangles represents the lower estimate for the total distance moved during the ten seconds.

If the dark and light rectangles are considered together, the first area is $30 \cdot 2 = 60$, the upper estimate for the distance moved in the first two seconds. The second area is $38 \cdot 2 = 76$, the upper estimate for the next two seconds. Continuing this calculation suggests that the upper estimate for the total distance is represented by the sum of the areas of the dark and light rectangles. Therefore, the area of the light rectangles alone represents the difference between the two estimates.

Figure 5.3 shows a graph of the one-second data. The area of the dark rectangles again represents the lower estimate, and the area of the dark and light rectangles together represents the upper estimate. The total area of the light rectangles is smaller in Figure 5.3 than in Figure 5.2, so the underestimate and overestimate are closer for the one-second data than for the two-second data.

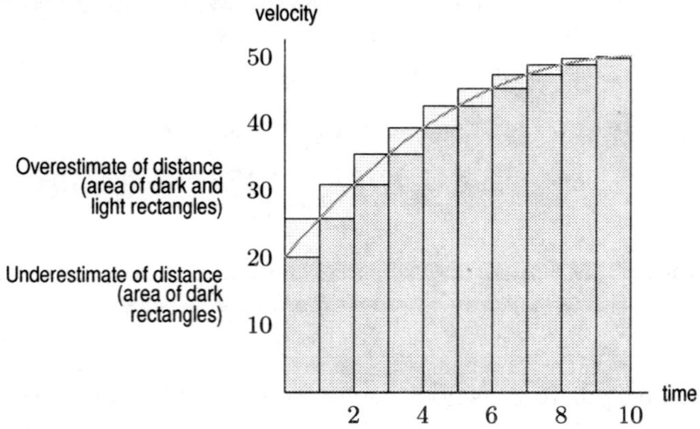

Figure 5.3: Shaded area estimates distance traveled. Velocity measured every second

Visualizing Distance on the Velocity Graph: Area Under Curve

As we make more frequent velocity measurements, the rectangles used to estimate the distance traveled fit the curve more closely. See Figures 5.4 and 5.5. In the limit, as the number of subdivisions increases, we see that the distance traveled is given by the area between the velocity curve and the horizontal axis. See Figure 5.6. In general:

If the velocity is positive, the total distance traveled is the area under the velocity curve.

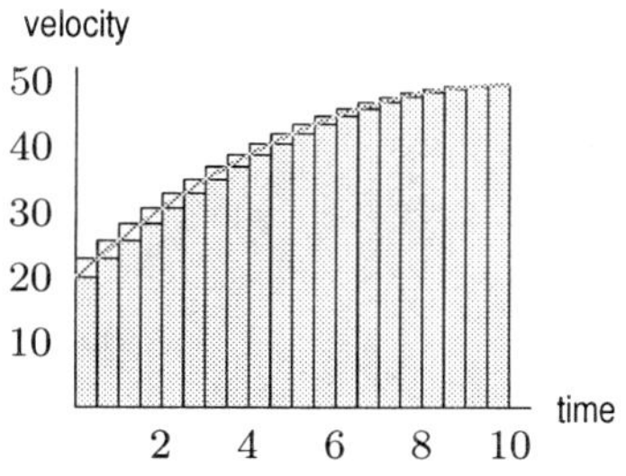

Figure 5.4: Velocity measured every 1/2 second

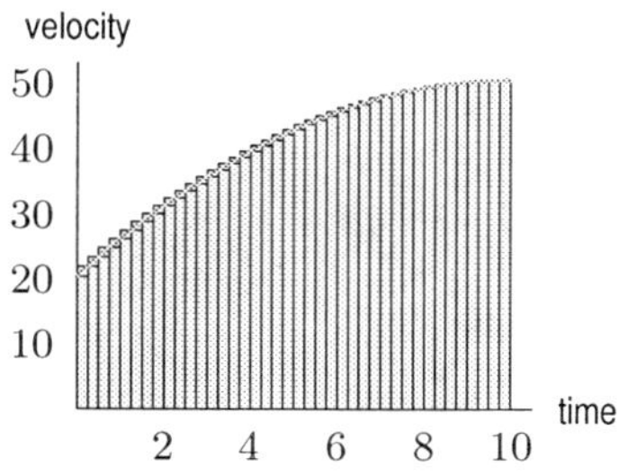

Figure 5.5: Velocity measured every 1/4 second

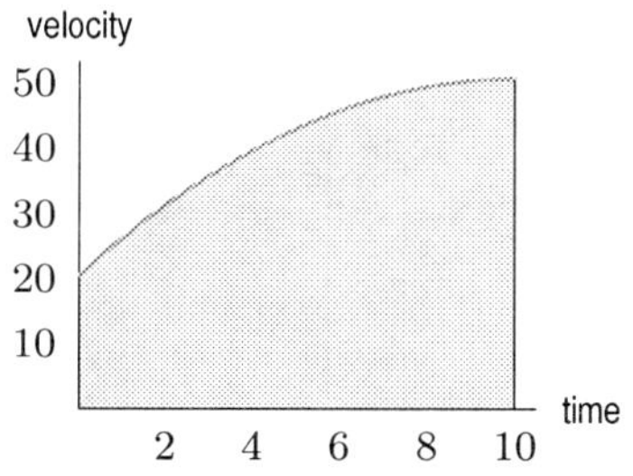

Figure 5.6: Distance traveled is area under curve

Example 2 With time t in seconds, the velocity of a bicycle, in feet per second, is given by $v(t) = 5t$. How far does the bicycle travel in 3 seconds?

Solution The velocity is linear. See Figure 5.7. The distance traveled is the area between the line $v(t) = 5t$ and the t-axis. Since this region is a triangle of height 15 and base 3

$$\text{Distance traveled} = \text{Area of triangle} = \frac{1}{2} \cdot 15 \cdot 3 = 22.5 \text{ feet.}$$

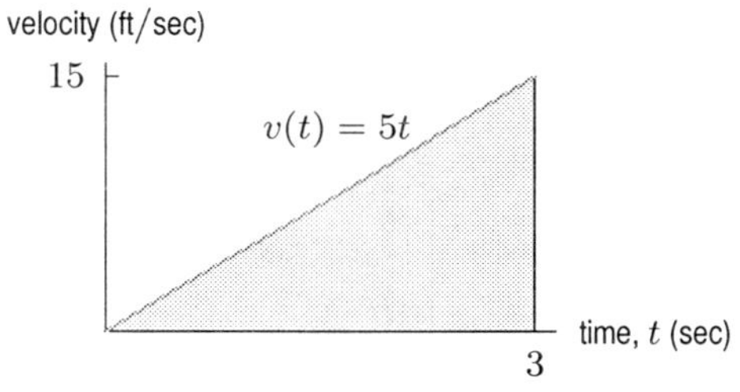

Figure 5.7: Shaded area represents distance traveled

Approximating Total Change from Rate of Change

We have seen how to use the rate of change of distance (the velocity) to calculate the total distance traveled. We can use the same method to find total change from rate of change of other quantities.

Example 3 A city's population grows at the rate of 5000 people/year for 3 years and then grows at the rate of 3000 people/year for the next 4 years. What is the total change in the population of the city during this 7-year period?

Solution The units of people/year remind us that we are given a rate of change of the population (people) with respect to time (years). If the rate of change is constant, we know that

$$\text{Total change in population} = \text{Rate of change per year} \times \text{Number of years.}$$

Thus, the total change in this population is

$$\begin{aligned}\text{Total change} &= (5000 \text{ people/year})(3 \text{ years}) + (3000 \text{ people/year})(4 \text{ years}) \\ &= 15{,}000 \text{ people} + 12{,}000 \text{ people} \\ &= 27{,}000 \text{ people.}\end{aligned}$$

The change in population is 27,000 people. Notice that this does not tell us the population of the city at the end of 7 years; it tells us the change in the population. For example, if the population were initially 100,000, it would be 127,000 at the end of the 7-year period.

Example 4 The rate of sales (in games per week) of a new video game is shown in Table 5.3. Assuming that the rate of sales increased throughout the 20-week period, estimate the total number of games sold during this period.

Table 5.3 *Weekly sales of a video game*

Time (weeks)	0	5	10	15	20
Rate of sales (games per week)	0	585	892	1875	2350

Solution If the rate of sales is constant, we have

$$\text{Total sales} = \text{Rate of sales per week} \times \text{Number of weeks.}$$

How many games were sold during the first five weeks? During this time, sales went from 0 to 585 games per week. If we assume that 585 games were sold every week, we get an overestimate for the sales in the first five weeks of (585 games/week)(5 weeks) $=$ 2925 games. Similar overestimates for each of the five-week periods gives an overestimate for the entire 20-week period:

$$\text{Overestimate for total sales} = 585 \cdot 5 + 892 \cdot 5 + 1875 \cdot 5 + 2350 \cdot 5 = 28{,}510 \text{ games.}$$

We underestimate the total sales by taking the lower value for rate of sales during each of the five-week periods:

$$\text{Underestimate for total sales} = 0 \cdot 5 + 585 \cdot 5 + 892 \cdot 5 + 1875 \cdot 5 = 16{,}760 \text{ games.}$$

Thus, the total sales of the game during the 20-week period is between 16,760 and 28,510 games. A good single estimate of total sales is the average of these two numbers:

$$\text{Total sales} \approx \frac{16{,}760 + 28{,}510}{2} = 22{,}635 \text{ games.}$$

Problems for Section 5.1

1. (a) Sketch a graph of the velocity function for the trip described in Example 1 on page 234.
   (b) Represent the total distance traveled on this graph.

2. Graph the rate of sales against time for the video game data in Example 4. Represent graphically the overestimate and the underestimate calculated in that example.

3. A car comes to a stop six seconds after the driver applies the brakes. While the brakes are on, the velocities recorded are in Table 5.4.

Table 5.4

Time since brakes applied (sec)	0	2	4	6
Velocity (ft/sec)	88	45	16	0

   (a) Give lower and upper estimates for the distance the car traveled after the brakes were applied.
   (b) On a sketch of velocity against time, show the lower and upper estimates of part (a).

4. A car starts moving at time $t = 0$ and goes faster and faster. Its velocity is shown in the following table. Estimate how far the car travels during the 12 seconds.

t (seconds)	0	3	6	9	12
Velocity (ft/sec)	0	10	25	45	75

5. The velocity of a car is $f(t) = 5t$ meters/sec. Use a graph of $f(t)$ to find the exact distance traveled by the car, in meters, from $t = 0$ to $t = 10$ seconds.

6. Figure 5.8 shows the velocity, v, of an object (in meters/sec). Estimate the total distance the object traveled between $t = 0$ and $t = 6$.

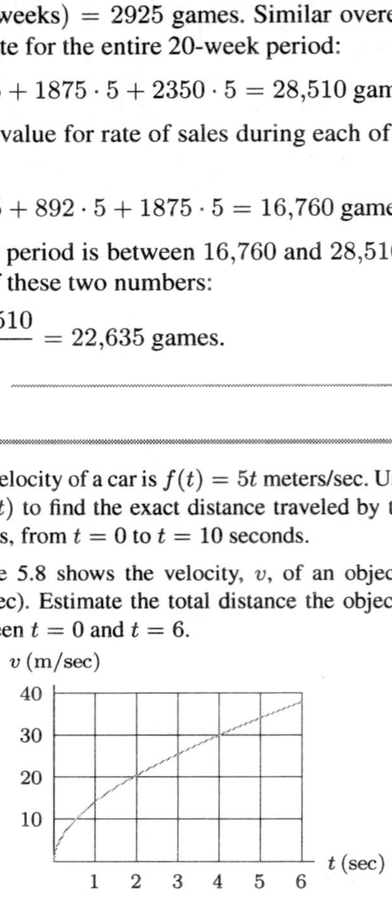

Figure 5.8

7. A car accelerates smoothly from 0 to 60 mph in 10 seconds with the velocity given in Figure 5.9. Estimate how far the car travels during the 10-second period.

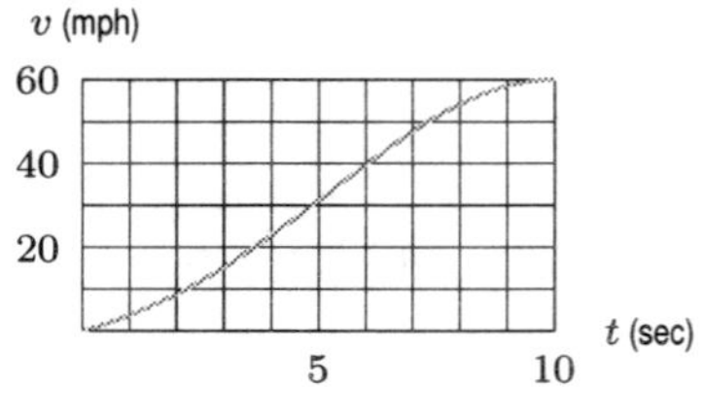

Figure 5.9

8. The following table gives world oil consumption, in billions of barrels per year.[1] Estimate total oil consumption during this 25-year period.

Year	1980	1985	1990	1995	2000	2005
Oil (bn barrels/yr)	22.3	21.3	23.9	24.9	27.0	29.3

9. Filters at a water treatment plant become less effective over time. The rate at which pollution passes through the filters into a nearby lake is given in the following table.

(a) Estimate the total quantity of pollution entering the lake during the 30-day period.
(b) Your answer to part (a) is only an estimate. Give bounds (lower and upper estimates) between which the true quantity of pollution must lie. (Assume the rate of pollution is continually increasing.)

Day	0	6	12	18	24	30
Rate (kg/day)	7	8	10	13	18	35

10. The rate of change of the world's population, in millions of people per year, is given in the following table.

(a) Use this data to estimate the total change in the world's population between 1950 and 2000.
(b) The world population was 2555 million people in 1950 and 6085 million people in 2000. Calculate the true value of the total change in the population. How does this compare with your estimate in part (a)?

Year	1950	1960	1970	1980	1990	2000
Rate of change	37	41	78	77	86	79

11. A village wishes to measure the quantity of water that is piped to a factory during a typical morning. A gauge on the water line gives the flow rate (in cubic meters per hour) at any instant. The flow rate is about 100 m^3/hr at 6 am and increases steadily to about 280 m^3/hr at 9 am. Using only this information, give your best estimate of the total volume of water used by the factory between 6 am and 9 am.

12. Roger runs a marathon. His friend Jeff rides behind him on a bicycle and clocks his speed every 15 minutes. Roger starts out strong, but after an hour and a half he is so exhausted that he has to stop. Jeff's data follow:

Time since start (min)	0	15	30	45	60	75	90
Speed (mph)	12	11	10	10	8	7	0

(a) Assuming that Roger's speed is never increasing, give upper and lower estimates for the distance Roger ran during the first half hour.
(b) Give upper and lower estimates for the distance Roger ran in total during the entire hour and a half.

13. A car initially going 50 ft/sec brakes at a constant rate (constant negative acceleration), coming to a stop in 5 seconds.

(a) Graph the velocity from $t = 0$ to $t = 5$.
(b) How far does the car travel?
(c) How far does the car travel if its initial velocity is doubled, but it brakes at the same constant rate?

14. Figure 5.10 shows the rate of change of a fish population. Estimate the total change in the population during this 12-month period.

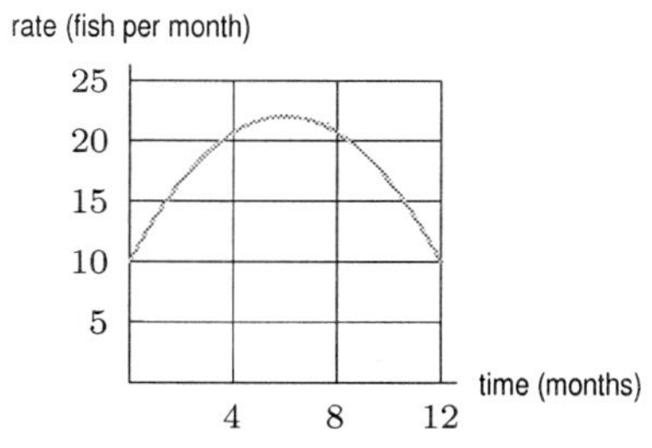

Figure 5.10

15. Two cars start at the same time and travel in the same direction along a straight road. Figure 5.11 gives the velocity, v, of each car as a function of time, t. Which car:

(a) Attains the larger maximum velocity?
(b) Stops first?
(c) Travels farther?

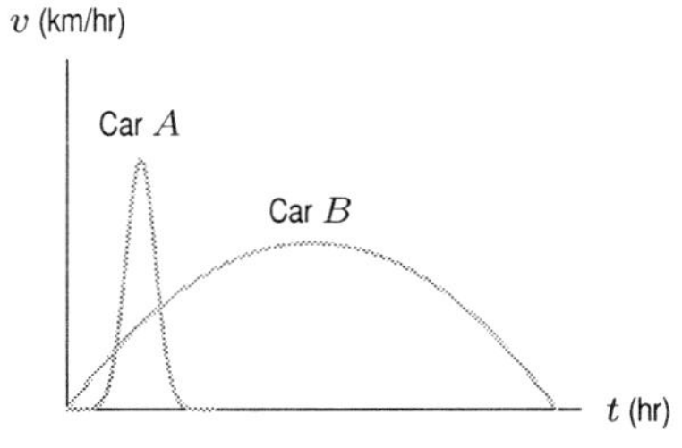

Figure 5.11

[1] www.bp.com/centres/energy/world_stat_rev/oil/reserves.asp. and https://www.cia.gov/library/publications/the-world-factbook/print/xx.html.

16. Two cars travel in the same direction along a straight road. Figure 5.12 shows the velocity, v, of each car at time t. Car B starts 2 hours after car A and car B reaches a maximum velocity of 50 km/hr.

(a) For approximately how long does each car travel?
(b) Estimate car A's maximum velocity.
(c) Approximately how far does each car travel?

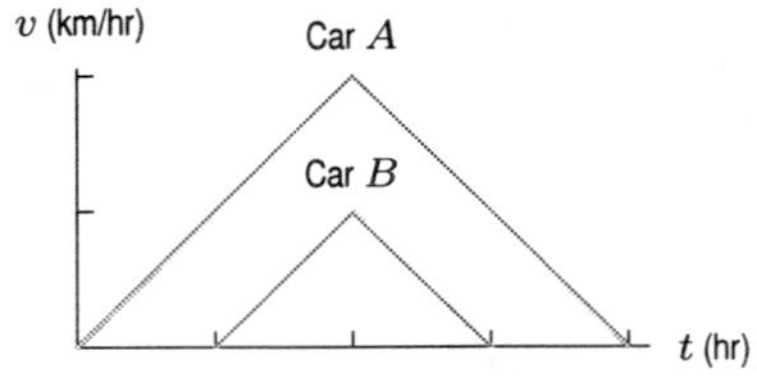

Figure 5.12

17. Your velocity is given by $v(t) = t^2+1$ in m/sec, with t in seconds. Estimate the distance, s, traveled between $t = 0$ and $t = 5$. Explain how you arrived at your estimate.

18. An old rowboat has sprung a leak. Water is flowing into the boat at a rate, $r(t)$, given in the following table.

t minutes	0	5	10	15
$r(t)$ liters/min	12	20	24	16

(a) Compute upper and lower estimates for the volume of water that has flowed into the boat during the 15 minutes.
(b) Draw a graph to illustrate the lower estimate.

19. The value of a mutual fund increases at a rate of $R = 500e^{0.04t}$ dollars per year, with t in years since 2010.

(a) Using $t = 0, 2, 4, 6, 8, 10$, make a table of values for R.
(b) Use the table to estimate the total change in the value of the mutual fund between 2010 and 2020.

20. A car speeds up at a constant rate from 10 to 70 mph over a period of half an hour. Its fuel efficiency (in miles per gallon) increases with speed; values are in the table. Make lower and upper estimates of the quantity of fuel used during the half hour.

Speed (mph)	10	20	30	40	50	60	70
Fuel efficiency (mpg)	15	18	21	23	24	25	26

5.2 THE DEFINITE INTEGRAL

In Section 5.1 we saw how to approximate total change given the rate of change. We now see how to make the approximation more accurate.

Improving the Approximation: The Role of n and Δt

To approximate total change, we construct a sum. We use the notation Δt for the size of the t-intervals used. We use n to represent the number of subintervals of length Δt. In the following example, we see how decreasing Δt (and increasing n) improves the accuracy of the approximation.

Example 1 If t is in hours since the start of a 20-hour period, a bacteria population increases at a rate given by

$$f(t) = 3 + 0.1t^2 \text{ millions of bacteria per hour.}$$

Make an underestimate of the total change in the number of bacteria over this period using

(a) $\Delta t = 4$ hours (b) $\Delta t = 2$ hours (c) $\Delta t = 1$ hour

Solution (a) The rate of change is $f(t) = 3 + 0.1t^2$. If we use $\Delta t = 4$, we measure the rate every 4 hours and $n = 20/4 = 5$. See Table 5.5. An underestimate for the population change during the first 4 hours is (3.0 million/hour)(4 hours) $= 12$ million. Combining the contributions from all the subintervals gives the underestimate:

$$\text{Total change} \approx 3.0 \cdot 4 + 4.6 \cdot 4 + 9.4 \cdot 4 + 17.4 \cdot 4 + 28.6 \cdot 4 = 252.0 \text{ million bacteria.}$$

The rate of change is graphed in Figure 5.13 (a); the area of the shaded rectangles represents this underestimate. Notice that $n = 5$ is the number of rectangles in the graph.

Table 5.5 *Rate of change with $\Delta t = 4$ using $f(t) = 3 + 0.1t^2$ million bacteria/hour*

t (hours)	0	4	8	12	16	20
$f(t)$	3.0	4.6	9.4	17.4	28.6	43.0

(b) If we use $\Delta t = 2$, we measure $f(t)$ every 2 hours and $n = 20/2 = 10$. See Table 5.6. The underestimate is

$$\text{Total change} \approx 3.0 \cdot 2 + 3.4 \cdot 2 + 4.6 \cdot 2 + \cdots + 35.4 \cdot 2 = 288.0 \text{ million bacteria.}$$

Figure 5.13 (b) suggests that this estimate is more accurate than the estimate made in part (a).

Table 5.6 *Rate of change with $\Delta t = 2$ using $f(t) = 3 + 0.1t^2$ million bacteria/hour*

t (hours)	0	2	4	6	8	10	12	14	16	18	20
$f(t)$	3.0	3.4	4.6	6.6	9.4	13.0	17.4	22.6	28.6	35.4	43.0

(c) If we use $\Delta t = 1$, then $n = 20$ and a similar calculation shows that we have

$$\text{Total change} \approx 307.0 \text{ million bacteria.}$$

The shaded area in Figure 5.13 (c) represents this estimate; it is the most accurate of the three.

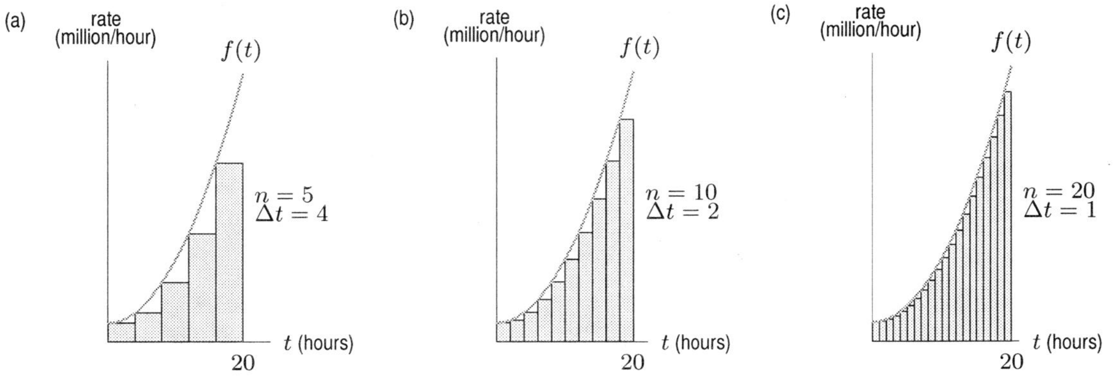

Figure 5.13: More and more accurate estimates of total change from rate of change. In each case, $f(t)$ is the rate of change, and the shaded area approximates total change. Largest n and smallest Δt give the best estimate.

Notice that as n gets larger, the estimate improves and the area of the shaded rectangles approaches the area under the curve.

Left- and Right-Hand Sums

Suppose we have a function $f(t)$ that is continuous for $a \le t \le b$. We divide the interval from a to b into n equal subdivisions, each of width Δt, so

$$\Delta t = \frac{b-a}{n}.$$

We let $t_0, t_1, t_2, \ldots, t_n$ be endpoints of the subdivisions, as in Figures 5.14 and 5.15. We construct two sums, similar to the overestimates and underestimates in Section 3.1. For a *left-hand sum*, we use the values of the function from the left end of the interval. For a *right-hand sum*, we use the values of the function from the right end of the interval. We have:

$$\text{Left-hand sum} = f(t_0)\Delta t + f(t_1)\Delta t + \cdots + f(t_{n-1})\Delta t$$

and

$$\text{Right-hand sum} = f(t_1)\Delta t + f(t_2)\Delta t + \cdots + f(t_n)\Delta t.$$

These sums represent the shaded areas in Figures 5.14 and 5.15, provided $f(t) \geq 0$. In Figure 5.14, the first rectangle has width Δt and height $f(t_0)$, since the top of its left edge just touches the curve, and hence it has area $f(t_0)\Delta t$. The second rectangle has width Δt and height $f(t_1)$, and hence has area $f(t_1)\Delta t$, and so on. The sum of all these areas is the left-hand sum. The right-hand sum, shown in Figure 5.15, is constructed in the same way, except that each rectangle touches the curve on its right edge instead of its left.

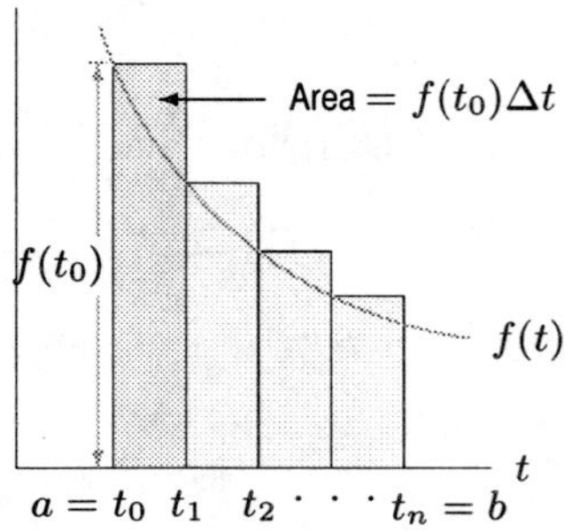

Figure 5.14: Left-hand sum: Area of rectangles

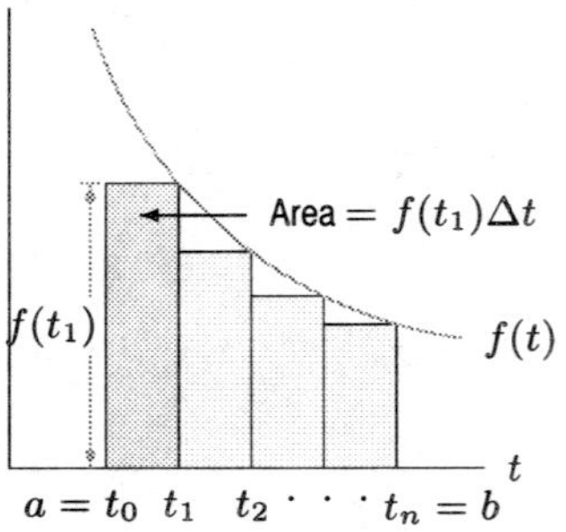

Figure 5.15: Right-hand sum: Area of rectangles

Writing Left- and Right-Hand Sums Using Sigma Notation

Both the left-hand and right-hand sums can be written more compactly using *sigma*, or summation, notation. The symbol $\sum$ is a capital sigma, or Greek letter "S." We write

$$\text{Right-hand sum} = \sum_{i=1}^{n} f(t_i)\Delta t = f(t_1)\Delta t + f(t_2)\Delta t + \cdots + f(t_n)\Delta t.$$

The $\sum$ tells us to add terms of the form $f(t_i)\Delta t$. The "$i = 1$" at the base of the sigma sign tells us to start at $i = 1$, and the "n" at the top tells us to stop at $i = n$.

In the left-hand sum we start at $i = 0$ and stop at $i = n - 1$, so we write

$$\text{Left-hand sum} = \sum_{i=0}^{n-1} f(t_i)\Delta t = f(t_0)\Delta t + f(t_1)\Delta t + \cdots + f(t_{n-1})\Delta t.$$

Taking the Limit to Obtain the Definite Integral

If f is a rate of change of some quantity, then the left-hand sum and the right-hand sum approximate the total change in the quantity. For most functions f, the approximation is improved by increasing the value of n. To find the total change exactly, we take larger and larger values of n and look at the values approached by the left and right sums. This is called taking the *limit* of these sums as n goes to infinity and is written $\lim_{n\to\infty}$. If f is continuous for $a \leq t \leq b$, the limits of the left- and right-hand sums exist and are equal. The *definite integral* is the common limit of these sums.

Suppose f is continuous for $a \le t \le b$. The **definite integral** of f from a to b, written

$$\int_a^b f(t)\,dt,$$

is the limit of the left-hand or right-hand sums with n subdivisions of $[a, b]$ as n gets arbitrarily large. In other words, if $t_0, t_1, \dots t_n$ are the endpoints of the subdivisions,

$$\int_a^b f(t)\,dt = \lim_{n\to\infty} (\text{Left-hand sum}) = \lim_{n\to\infty} \left(\sum_{i=0}^{n-1} f(t_i)\Delta t \right)$$

and

$$\int_a^b f(t)\,dt = \lim_{n\to\infty} (\text{Right-hand sum}) = \lim_{n\to\infty} \left(\sum_{i=1}^{n} f(t_i)\Delta t \right).$$

Each of these sums is called a *Riemann sum*, f is called the *integrand*, and a and b are called the *limits of integration*.

The "$\int$" notation comes from an old-fashioned "S," which stands for "sum" in the same way that $\sum$ does. The "dt" in the integral comes from the factor Δt. Notice that the limits on the $\sum$ symbol are 0 and $n-1$ for the left-hand sum, and 1 and n for the right-hand sum, whereas the limits on the $\int$ sign are a and b.

When $f(t)$ is positive, the left- and right-hand sums are represented by the sums of areas of rectangles, so the definite integral is represented graphically by an area.

Computing a Definite Integral

In practice, we often approximate definite integrals numerically using a calculator or computer. They compute sums for larger and larger values of n, and eventually give a value for the integral. Different calculators and computers may give slightly different estimates, owing to round-off error and the fact that they may use different approximation methods.

Example 2 Compute $\int_1^3 t^2\,dt$ and represent this integral as an area.

Solution Using a calculator, we find

$$\int_1^3 t^2\,dt = 8.667.$$

The integral represents the area between $t = 1$ and $t = 3$ under the curve $f(t) = t^2$. See Figure 5.16.

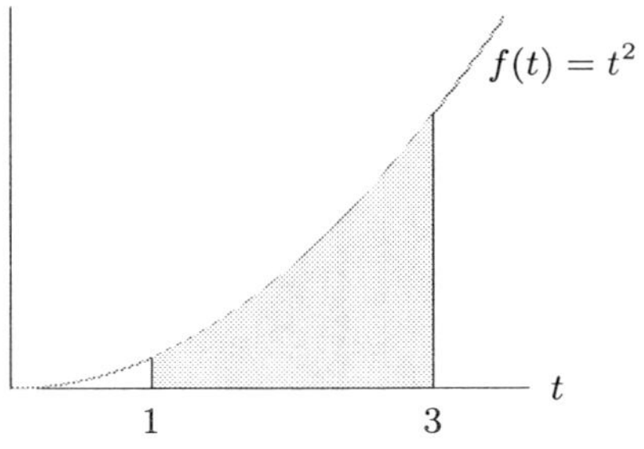

Figure 5.16: Shaded area $= \int_1^3 t^2\,dt$

Estimating a Definite Integral from a Table or Graph

If we have a formula for the integrand, $f(x)$, we can calculate the integral $\int_a^b f(x)\,dx$ using a calculator or computer. If, however, we have only a table of values or a graph of $f(x)$, we can still estimate the integral.

Example 3 Values for a function $f(t)$ are in the following table. Estimate $\displaystyle\int_{20}^{30} f(t)dt$.

t	20	22	24	26	28	30
$f(t)$	5	7	11	18	29	45

Solution Since we have only a table of values, we use left- and right-hand sums to approximate the integral. The values of $f(t)$ are spaced 2 units apart, so $\Delta t = 2$ and $n = (30 - 20)/2 = 5$. Calculating the left-hand and right-hand sums gives

$$\begin{aligned}\text{Left-hand sum} &= f(20)\cdot 2 + f(22)\cdot 2 + f(24)\cdot 2 + f(26)\cdot 2 + f(28)\cdot 2\\ &= 5\cdot 2 + 7\cdot 2 + 11\cdot 2 + 18\cdot 2 + 29\cdot 2\\ &= 10 + 14 + 22 + 36 + 58\\ &= 140.\end{aligned}$$

$$\begin{aligned}\text{Right-hand sum} &= f(22)\cdot 2 + f(24)\cdot 2 + f(26)\cdot 2 + f(28)\cdot 2 + f(30)\cdot 2\\ &= 7\cdot 2 + 11\cdot 2 + 18\cdot 2 + 29\cdot 2 + 45\cdot 2\\ &= 14 + 22 + 36 + 58 + 90\\ &= 220.\end{aligned}$$

Both left- and right-hand sums approximate the integral. We generally get a better estimate by averaging the two:

$$\int_{20}^{30} f(t)dt \approx \frac{140 + 220}{2} = 180.$$

Example 4 The function $f(x)$ is graphed in Figure 5.17. Estimate $\displaystyle\int_0^6 f(x)\,dx$.

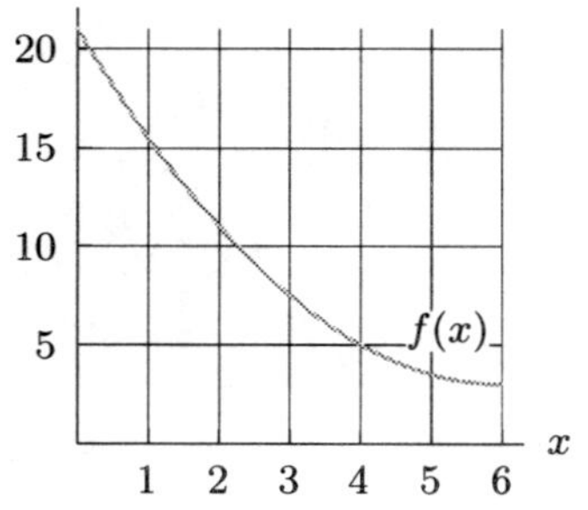

Figure 5.17: Estimate $\int_0^6 f(x)\,dx$

Solution We approximate the integral using left- and right-hand sums with $n = 3$, so $\Delta x = 2$. Figures 5.18 and 5.19 give

$$\text{Left-hand sum} = f(0)\cdot 2 + f(2)\cdot 2 + f(4)\cdot 2 = 21\cdot 2 + 11\cdot 2 + 5\cdot 2 = 74,$$
$$\text{Right-hand sum} = f(2)\cdot 2 + f(4)\cdot 2 + f(6)\cdot 2 = 11\cdot 2 + 5\cdot 2 + 3\cdot 2 = 38.$$

We estimate the integral by taking the average:

$$\int_0^6 f(x)\,dx \approx \frac{74+38}{2} = 56.$$

Alternatively, since the integral equals the area under the curve between $x = 0$ and $x = 6$, we can estimate it by counting grid squares. Each grid square has area $5 \cdot 1 = 5$, and the region under $f(x)$ includes about 10.5 grid squares, so the area is about $10.5 \cdot 5 = 52.5$.

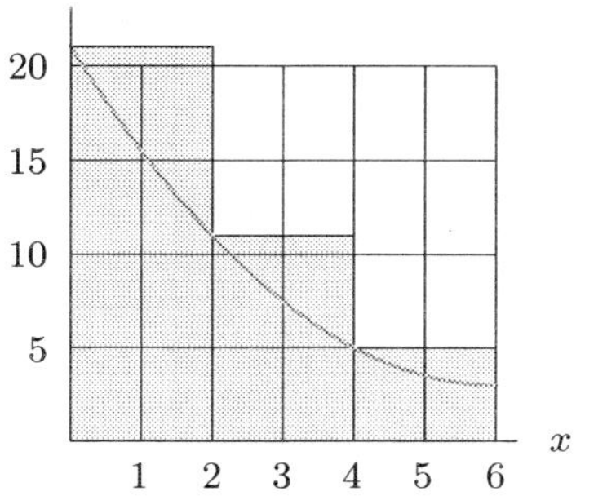

Figure 5.18: Area of shaded region is left-hand sum with $n = 3$

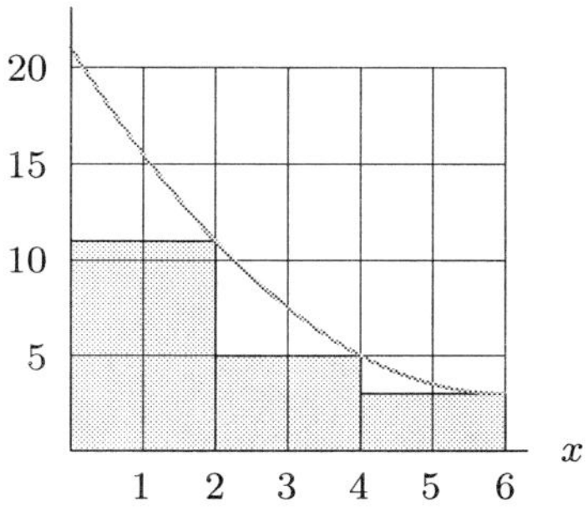

Figure 5.19: Area of shaded region is right-hand sum with $n = 3$

Rough Estimates of a Definite Integral

When calculating an integral using a calculator or computer, it is useful to have a rough idea of the value you expect. This helps detect errors in entering the integral.

Example 5 Three people calculated $\displaystyle\int_1^3 \frac{1}{t}\,dt$ on a calculator and got values 0.023, 11.984, and 1.526. Explain how you can be sure that none of these values is correct.

Solution Figure 5.20 shows left- and right-hand approximations to $\int_1^3 \frac{1}{t}\,dt$ with $n = 1$. We see that the right-hand sum $2(1/3)$ is an underestimate of $\int_1^3 \frac{1}{t}\,dt$. Since 0.023 is less than $2/3$, this value must be wrong. Similarly, we see that the left-hand sum $2(1)$ is an overestimate of the integral. Since 11.984 is larger than 2, this value is wrong. Since the graph is concave up, the value of the integral is closer to the smaller value of the two sums, $2/3$, than to the larger value, 2. Thus, the integral is less than the average of the two sums, $4/3$. Since $1.526 > 4/3$, this value is wrong too.

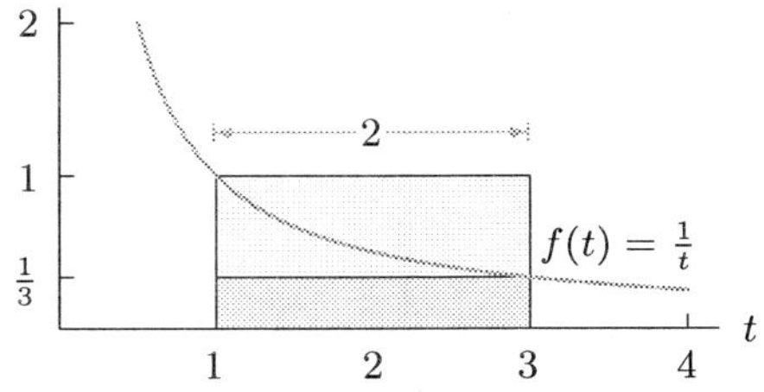

Figure 5.20: Left- and right-hand sums with $n = 1$

Problems for Section 5.2

1. Estimate $\int_0^6 2^x\,dx$ using a left-hand sum with $n = 2$.

2. Estimate $\int_0^{12} \frac{1}{x+1}\,dx$ using a left-hand sum with $n = 3$.

3. Use the following table to estimate $\int_0^{25} f(x)dx$.

x	0	5	10	15	20	25
$f(x)$	100	82	69	60	53	49

4. Use the following table to estimate $\int_3^4 W(t)\,dt$. What are n and Δt?

t	3.0	3.2	3.4	3.6	3.8	4.0
$W(t)$	25	23	20	15	9	2

5. Use the following table to estimate $\int_0^{15} f(x)\,dx$.

x	0	3	6	9	12	15
$f(x)$	50	48	44	36	24	8

6. Use the table to estimate $\int_0^{40} f(x)dx$. What values of n and Δx did you use?

x	0	10	20	30	40
$f(x)$	350	410	435	450	460

7. Using Figure 5.21, draw rectangles representing each of the following Riemann sums for the function f on the interval $0 \le t \le 8$. Calculate the value of each sum.

(a) Left-hand sum with $\Delta t = 4$
(b) Right-hand sum with $\Delta t = 4$
(c) Left-hand sum with $\Delta t = 2$
(d) Right-hand sum with $\Delta t = 2$

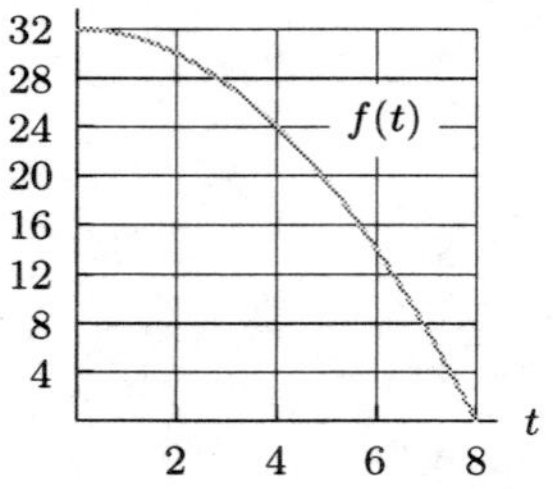

Figure 5.21

8. Use Figure 5.22 to estimate $\int_0^{20} f(x)\,dx$.

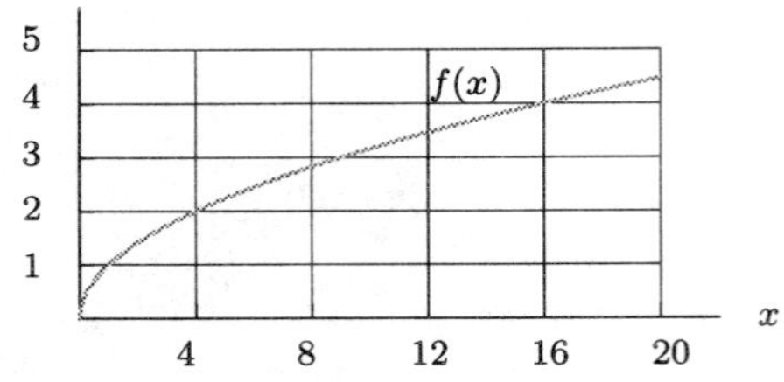

Figure 5.22

9. Use Figure 5.23 to estimate $\int_{-10}^{15} f(x)dx$.

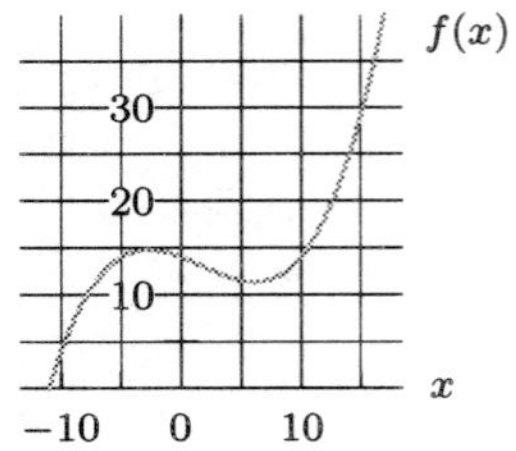

Figure 5.23

Use the graphs in Problems 10–11 to estimate $\int_0^3 f(x)\,dx$.

10.

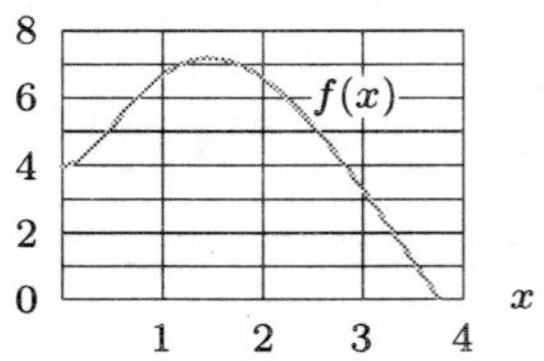

11.

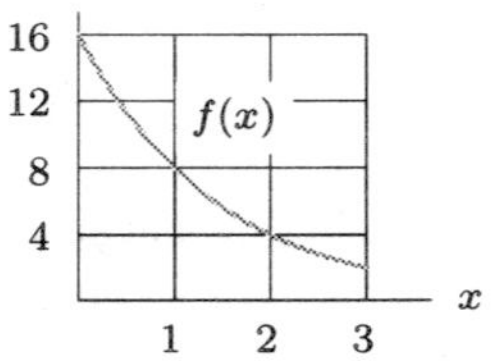

For Problems 12–15:

(a) Use a graph of the integrand to make a rough estimate of the integral. Explain your reasoning.
(b) Use a computer or calculator to find the value of the definite integral.

12. $\int_0^1 x^3\,dx$

13. $\int_0^3 \sqrt{x}\,dx$

14. $\int_0^1 3^t\,dt$

15. $\int_1^2 x^x\,dx$

16. The rate of change of a quantity is given by $f(t) = t^2 + 1$. Make an underestimate and an overestimate of the total change in the quantity between $t = 0$ and $t = 8$ using

(a) $\Delta t = 4$ **(b)** $\Delta t = 2$ **(c)** $\Delta t = 1$

What is n in each case? Graph $f(t)$ and shade rectangles to represent each of your six answers.

17. **(a)** Use a calculator or computer to find $\int_0^6 (x^2+1)\,dx$. Represent this value as the area under a curve.

(b) Estimate $\int_0^6 (x^2+1)\,dx$ using a left-hand sum with $n = 3$. Represent this sum graphically on a sketch of $f(x) = x^2 + 1$. Is this sum an overestimate or underestimate of the true value found in part (a)?

(c) Estimate $\int_0^6 (x^2+1)\,dx$ using a right-hand sum with $n = 3$. Represent this sum on your sketch. Is this sum an overestimate or underestimate?

18. Using Figure 5.24, find the value of $\int_1^6 f(x)\,dx$.

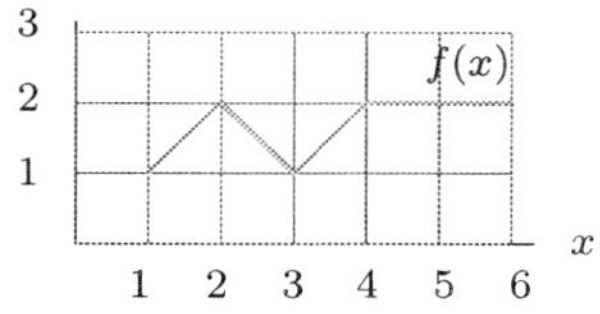

Figure 5.24

19. The graph of a function $f(t)$ is given in Figure 5.25. Which of the following four numbers could be an estimate of $\int_0^1 f(t)dt$ accurate to two decimal places? Explain how you chose your answer.

(a) -98.35 **(b)** 71.84
(c) 100.12 **(d)** 93.47

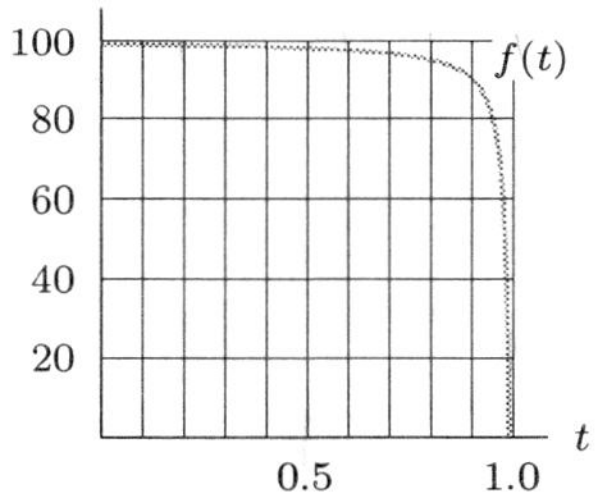

Figure 5.25

In Problems 20–29, use a calculator or computer to evaluate the integral.

20. $\displaystyle\int_0^5 x^2\,dx$ **21.** $\displaystyle\int_1^5 (3x+1)^2\,dx$

22. $\displaystyle\int_1^4 \frac{1}{\sqrt{1+x^2}}\,dx$ **23.** $\displaystyle\int_{-1}^1 \frac{1}{e^t}\,dt$

24. $\displaystyle\int_{1.1}^{1.7} 10(0.85)^t\,dt$ **25.** $\displaystyle\int_1^2 2^x\,dx$

26. $\displaystyle\int_1^2 (1.03)^t\,dt$ **27.** $\displaystyle\int_1^3 \ln x\,dx$

28. $\displaystyle\int_{1.1}^{1.7} e^t \ln t\,dt$ **29.** $\displaystyle\int_{-3}^3 e^{-t^2}\,dt$

30. Use the expressions for left and right sums on page 242 and Table 5.7.

(a) If $n = 4$, what is Δt? What are t_0, t_1, t_2, t_3, t_4? What are $f(t_0), f(t_1), f(t_2), f(t_3), f(t_4)$?
(b) Find the left and right sums using $n = 4$.
(c) If $n = 2$, what is Δt? What are t_0, t_1, t_2? What are $f(t_0), f(t_1), f(t_2)$?
(d) Find the left and right sums using $n = 2$.

Table 5.7

t	15	17	19	21	23
$f(t)$	10	13	18	20	30

31. Use the expressions for left and right sums on page 242 and Table 5.8.

(a) If $n = 4$, what is Δt? What are t_0, t_1, t_2, t_3, t_4? What are $f(t_0), f(t_1), f(t_2), f(t_3), f(t_4)$?
(b) Find the left and right sums using $n = 4$.
(c) If $n = 2$, what is Δt? What are t_0, t_1, t_2? What are $f(t_0), f(t_1), f(t_2)$?
(d) Find the left and right sums using $n = 2$.

Table 5.8

t	0	4	8	12	16
$f(t)$	25	23	22	20	17

5.3 THE DEFINITE INTEGRAL AS AREA

The Definite Integral as an Area: When $f(x)$ is Positive

If $f(x)$ is continuous and positive, each term $f(x_0)\Delta x, f(x_1)\Delta x, \ldots$ in a left- or right-hand Riemann sum represents the area of a rectangle. See Figure 5.26. As the width Δx of the rectangles approaches zero, the rectangles fit the curve of the graph more exactly, and the sum of their areas gets closer to the area under the curve shaded in Figure 5.27. In other words:

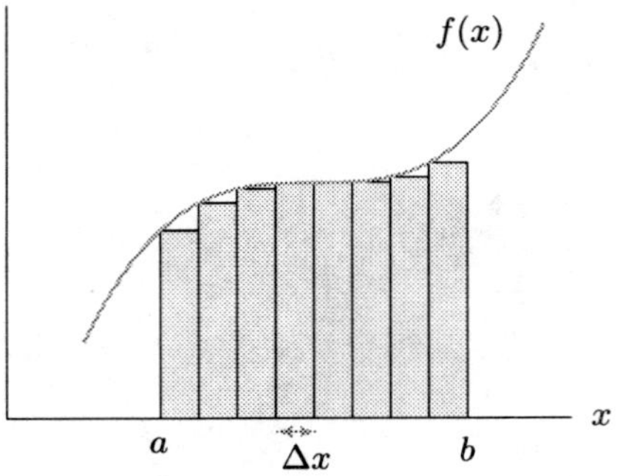

Figure 5.26: Area of rectangles approximating the area under the curve

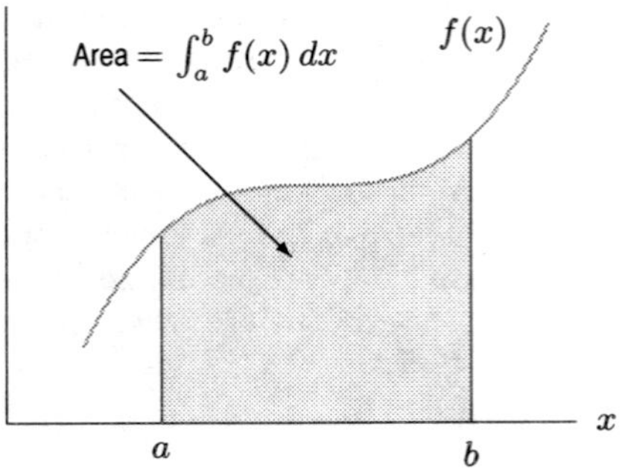

Figure 5.27: Shaded area is the definite integral $\int_a^b f(x)\,dx$

> When $f(x)$ is positive and $a < b$:
>
> $$\begin{array}{c}\text{Area under graph of } f \\ \text{between } a \text{ and } b\end{array} = \int_a^b f(x)\,dx.$$

Example 1 Find the area under the graph of $y = 10x(3^{-x})$ between $x = 0$ and $x = 3$.

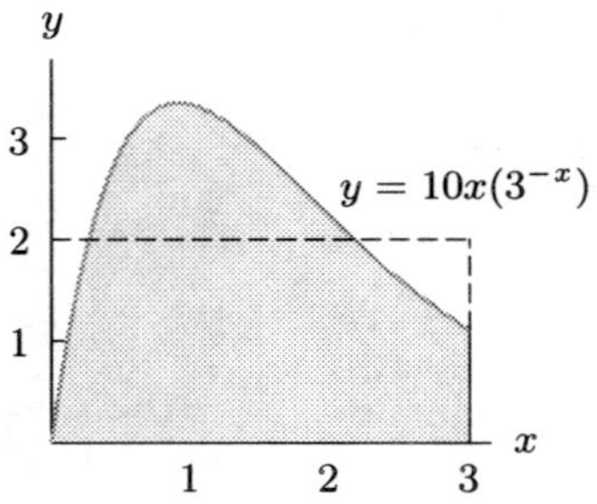

Figure 5.28: Area shaded $= \int_0^3 10x(3^{-x})\,dx$

Solution The area we want is shaded in Figure 5.28. A rough estimate of this area is 6, since it has about the same area as a rectangle of width 3 and height 2. To find the area more accurately, we say

$$\text{Area shaded} = \int_0^3 10x(3^{-x})dx.$$

Using a calculator or computer to evaluate the integral, we obtain

$$\text{Area shaded} = \int_0^3 10x(3^{-x})\,dx = 6.967 \approx 7 \text{ square units.}$$

Relationship Between Definite Integral and Area: When $f(x)$ is Not Positive

We assumed in drawing Figure 5.27 that the graph of $f(x)$ lies above the x-axis. If the graph lies below the x-axis, then each value of $f(x)$ is negative, so each $f(x)\Delta x$ is negative, and the area gets counted negatively. In that case, the definite integral is the negative of the area between the graph of f and the horizontal axis.

Example 2 What is the relation between the definite integral $\int_{-1}^{1} (x^2 - 1)\,dx$ and the area between the parabola $y = x^2 - 1$ and the x-axis?

Solution The parabola lies below the x-axis between $x = -1$ and $x = 1$. (See Figure 5.29.) So,

$$\int_{-1}^{1} (x^2 - 1)\, dx = -\text{Area} = -1.33.$$

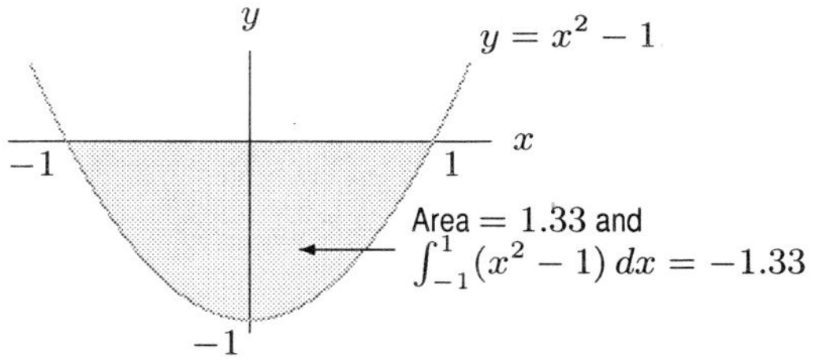

Figure 5.29: Integral $\int_{-1}^{1} (x^2 - 1)\, dx$ is negative of shaded area

Summarizing, assuming $f(x)$ is continuous, we have:

> **When $f(x)$ is positive for some x-values and negative for others**, and $a < b$:
> $\int_a^b f(x)\, dx$ is the sum of the areas above the x-axis, counted positively, and the areas below the x-axis, counted negatively.

In the following example, we break up the integral. The properties that allow us to do this are discussed in the Focus on Theory section at the end of this chapter.

Example 3 Interpret the definite integral $\int_0^4 (x^3 - 7x^2 + 11x)\, dx$ in terms of areas.

Solution Figure 5.30 shows the graph of $f(x) = x^3 - 7x^2 + 11x$ crossing below the x-axis at about $x = 2.38$. The integral is the area above the x-axis, A_1, minus the area below the x-axis, A_2. Computing the integral with a calculator or computer shows

$$\int_0^4 (x^3 - 7x^2 + 11x)\, dx = 2.67.$$

Breaking the integral into two parts and calculating each one separately gives

$$\int_0^{2.38} (x^3 - 7x^2 + 11x)\, dx = 7.72 \quad \text{and} \quad \int_{2.38}^{4} (x^3 - 7x^2 + 11x)\, dx = -5.05,$$

so $A_1 = 7.72$ and $A_2 = 5.05$. Then, as we would expect,

$$\int_0^4 (x^3 - 7x^2 + 11x)\, dx = A_1 - A_2 = 7.72 - 5.05 = 2.67.$$

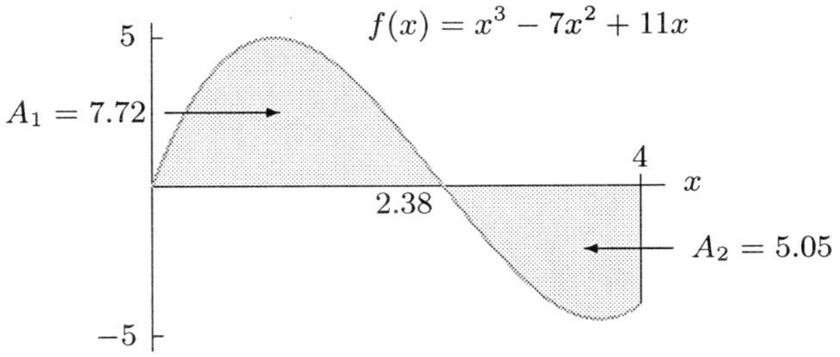

Figure 5.30: Integral $\int_0^4 (x^3 - 7x^2 + 11x)\, dx = A_1 - A_2$

Example 4 Find the total area of the shaded regions in Figure 5.30.

Solution We saw in Example 3 that $A_1 = 7.72$ and $A_2 = 5.05$. Thus we have

$$\text{Total shaded area} = A_1 + A_2 = 7.72 + 5.05 = 12.77.$$

Example 5 For each of the functions graphed in Figure 5.31, decide whether $\int_0^5 f(x)\,dx$ is positive, negative or approximately zero.

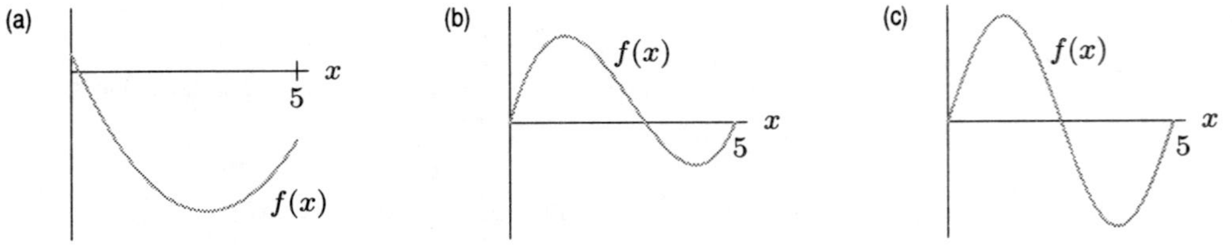

Figure 5.31: Is $\int_0^5 f(x)dx$ positive, negative or zero?

Solution
(a) The graph lies almost entirely below the x-axis, so the integral is negative.
(b) The graph lies partly below the x-axis and partly above the x-axis. However, the area above the x-axis is larger than the area below the x-axis, so the integral is positive.
(c) The graph lies partly below the x-axis and partly above the x-axis. Since the areas above and below the x-axis appear to be approximately equal in size, the integral is approximately zero.

Area Between Two Curves

We can use rectangles to approximate the area between two curves. If $g(x) \leq f(x)$, as in Figure 5.32, the height of a rectangle is $f(x) - g(x)$. The area of the rectangle is $(f(x) - g(x))\Delta x$, and we have the following result:

If $g(x) \leq f(x)$ for $a \leq x \leq b$:

$$\begin{array}{c}\text{Area between graphs of } f(x) \text{ and } g(x) \\ \text{for } a \leq x \leq b\end{array} = \int_a^b (f(x) - g(x))\,dx.$$

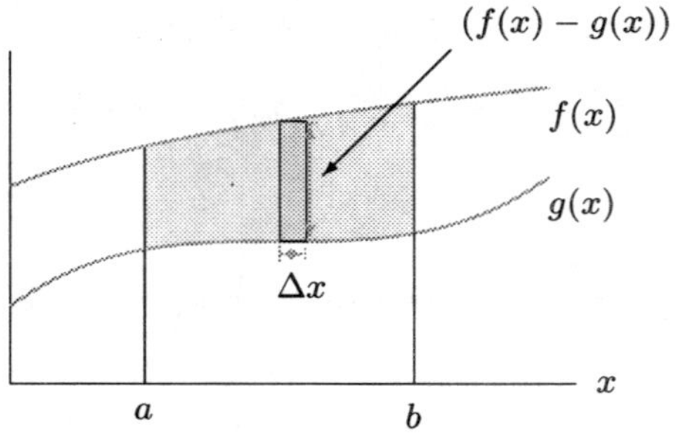

Figure 5.32: Area between two curves $= \int_a^b (f(x) - g(x))\,dx$

Example 6 Graphs of $f(x) = 4x - x^2$ and $g(x) = \frac{1}{2}x^{3/2}$ for $x \geq 0$ are shown in Figure 5.33. Use a definite integral to estimate the area enclosed by the graphs of these two functions.

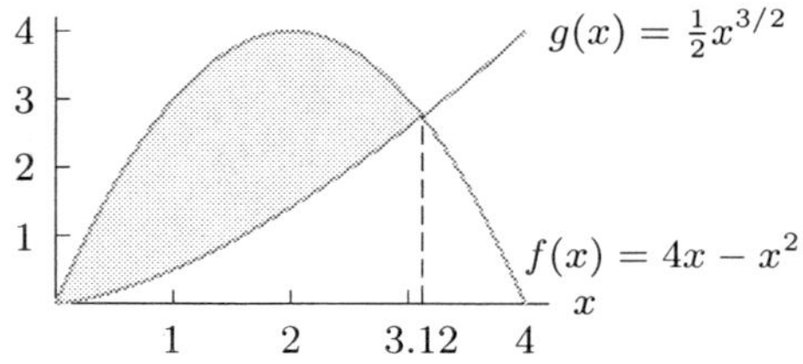

Figure 5.33: Find the area between $f(x) = 4x - x^2$ and $g(x) = \frac{1}{2}x^{3/2}$ using an integral

Solution The region enclosed by the graphs of the two functions is shaded in Figure 5.33. The two graphs cross at $x = 0$ and at $x \approx 3.12$. Between these values, the graph of $f(x) = 4x - x^2$ lies above the graph of $g(x) = \frac{1}{2}x^{3/2}$. Using a calculator or computer to evaluate the integral, we get

$$\text{Area between graphs} = \int_0^{3.12} \left((4x - x^2) - \tfrac{1}{2}x^{3/2}\right) dx = 5.906.$$

Problems for Section 5.3

1. Find the area under $y = x^3 + 2$ between $x = 0$ and $x = 2$. Sketch this area.

2. Find the area under $P = 100(0.6)^t$ between $t = 0$ and $t = 8$.

3. Find the area between $y = x+5$ and $y = 2x+1$ between $x = 0$ and $x = 2$.

4. Find the area enclosed by $y = 3x$ and $y = x^2$.

5. **(a)** What is the area between the graph of $f(x)$ in Figure 5.34 and the x-axis, between $x = 0$ and $x = 5$?
(b) What is $\int_0^5 f(x)\,dx$?

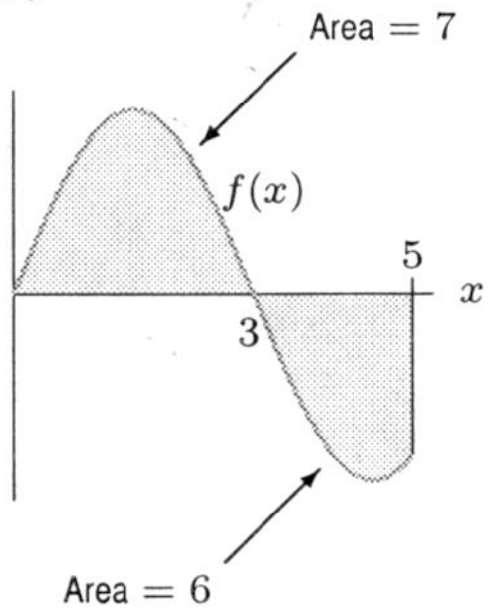

Figure 5.34

In Problems 6–9, decide whether $\int_{-3}^{3} f(x)\,dx$ is positive, negative, or approximately zero.

6.

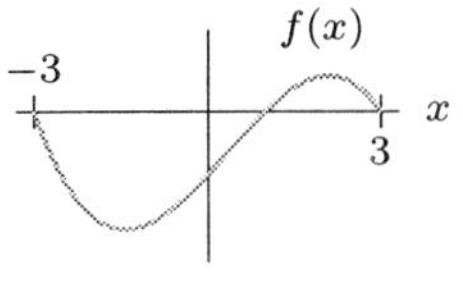

7.

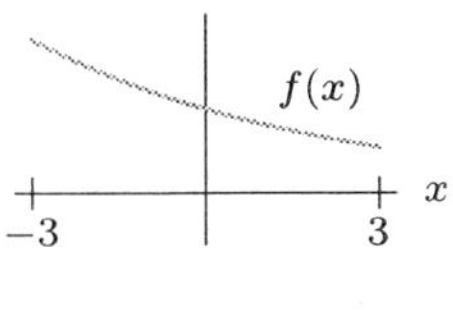

8.

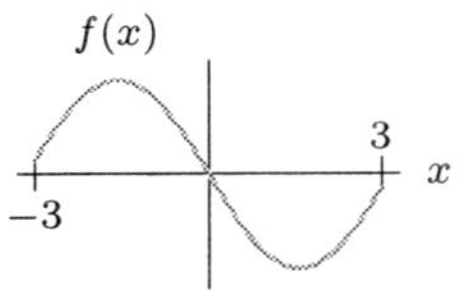

9.

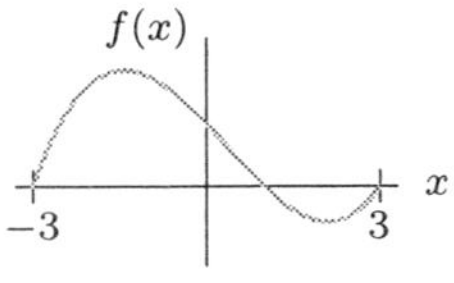

10. **(a)** Estimate (by counting the squares) the total area shaded in Figure 5.35.
(b) Using Figure 5.35, estimate $\int_0^8 f(x)\,dx$.
(c) Why are your answers to parts (a) and (b) different?

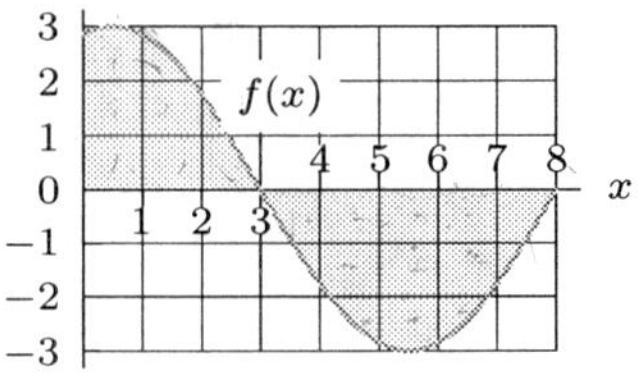

Figure 5.35

11. Using Figure 5.36, estimate $\int_{-3}^{5} f(x)dx$.

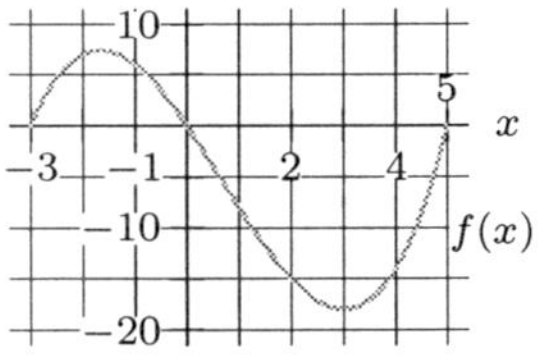

Figure 5.36

12. Given $\int_{-1}^{0} f(x)\,dx = 0.25$ and Figure 5.37, estimate:

(a) $\int_0^1 f(x)\,dx$ **(b)** $\int_{-1}^{1} f(x)\,dx$
(c) The total shaded area.

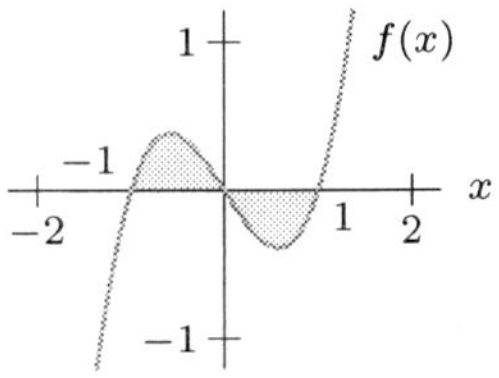

Figure 5.37

13. Given $\int_{-2}^{0} f(x)dx = 4$ and Figure 5.38, estimate:

(a) $\int_{0}^{2} f(x)dx$ (b) $\int_{-2}^{2} f(x)dx$

(c) The total shaded area.

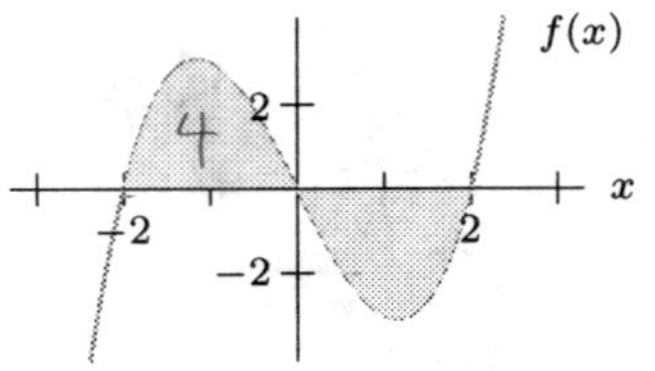

Figure 5.38

In Problems 14–17, match the graph with one of the following possible values for the integral $\int_{0}^{5} f(x)\,dx$:

I. -10.4 II. -2.1 III. 5.2 IV. 10.4

14.

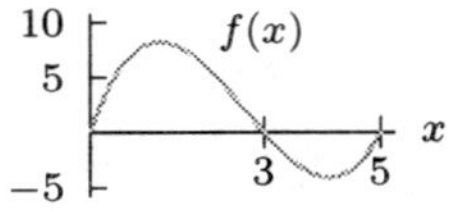

15.

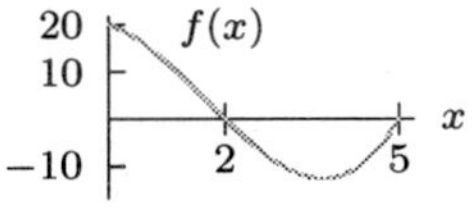

16.

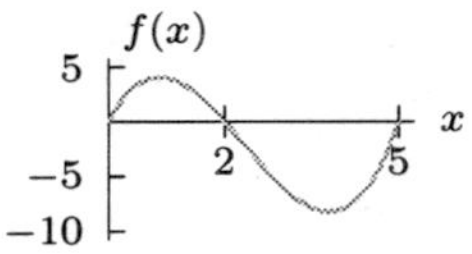

17.

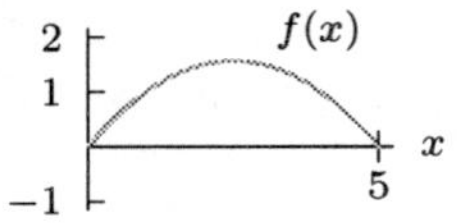

18. Use Figure 5.39 to find the values of

(a) $\int_{a}^{b} f(x)\,dx$ (b) $\int_{b}^{c} f(x)\,dx$

(c) $\int_{a}^{c} f(x)\,dx$ (d) $\int_{a}^{c} |f(x)|\,dx$

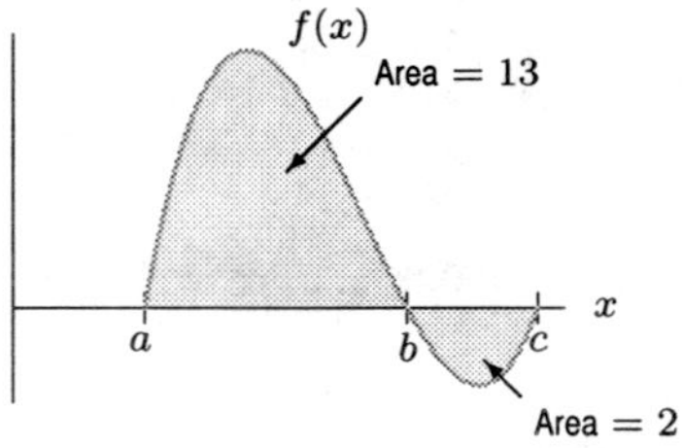

Figure 5.39

19. Using Figure 5.40, list the following integrals in increasing order (from smallest to largest). Which integrals are negative, which are positive? Give reasons.

I. $\int_{a}^{b} f(x)\,dx$ II. $\int_{a}^{c} f(x)\,dx$ III. $\int_{a}^{e} f(x)\,dx$

IV. $\int_{b}^{e} f(x)\,dx$ V. $\int_{b}^{c} f(x)\,dx$

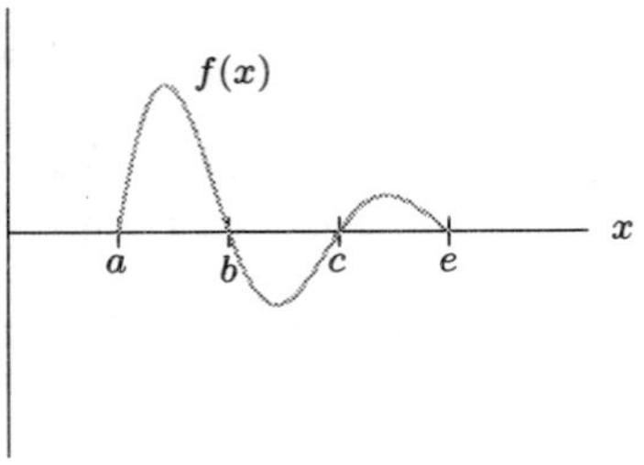

Figure 5.40

20. (a) Graph $f(x) = x(x+2)(x-1)$.

(b) Find the total area between the graph and the x-axis between $x = -2$ and $x = 1$.

(c) Find $\int_{-2}^{1} f(x)\,dx$ and interpret it in terms of areas.

21. (a) Using Figure 5.41, find $\int_{-3}^{0} f(x)\,dx$.

(b) If the area of the shaded region is A, estimate $\int_{-3}^{4} f(x)\,dx$.

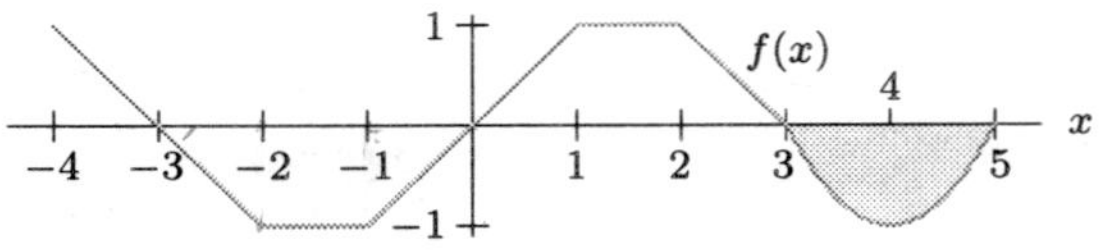

Figure 5.41

22. Use the following table to estimate the area between $f(x)$ and the x-axis on the interval $0 \le x \le 20$.

x	0	5	10	15	20
$f(x)$	15	18	20	16	12

For Problems 23–24, compute the definite integral and interpret the result in terms of areas.

23. $\displaystyle\int_{1}^{4} \frac{x^2-3}{x}\,dx.$ 24. $\displaystyle\int_{1}^{4} (x - 3\ln x)\,dx.$

25. Compute the definite integral $\int_{0}^{4} \cos\sqrt{x}\,dx$ and interpret the result in terms of areas.

26. Find the area between the graph of $y = x^2 - 2$ and the x-axis, between $x = 0$ and $x = 3$.

In Problems 27–34, use an integral to find the specified area.

27. Under $y = 6x^3 - 2$ for $5 \le x \le 10$.

28. Under $y = 2\cos(t/10)$ for $1 \le t \le 2$.

29. Under $y = 5\ln(2x)$ and above $y = 3$ for $3 \le x \le 5$.

30. Between $y = \sin x + 2$ and $y = 0.5$ for $6 \le x \le 10$.

31. Between $y = \cos x + 7$ and $y = \ln(x-3)$, $5 \le x \le 7$.

32. Above the curve $y = x^4 - 8$ and below the x-axis.

33. Above the curve $y = -e^x + e^{2(x-1)}$ and below the x-axis, for $x \ge 0$.

34. Between $y = \cos t$ and $y = \sin t$ for $0 \le t \le \pi$.

5.4 INTERPRETATIONS OF THE DEFINITE INTEGRAL

The Notation and Units for the Definite Integral

Just as the Leibniz notation dy/dx for the derivative reminds us that the derivative is the limit of a quotient of differences, the notation for the definite integral,

$$\int_a^b f(x)\,dx,$$

reminds us that an integral is a limit of a sum. The integral sign is a misshapen S. Since the terms being added are products of the form "$f(x)$ times a difference in x," we have the following result:

> The unit of measurement for $\int_a^b f(x)\,dx$ is the product of the units for $f(x)$ and the units for x.

For example, if x and $f(x)$ have the same units, then the integral $\int_a^b f(x)\,dx$ is measured in square units, say cm $\times$ cm $=$ cm^2. This is what we would expect, since the integral represents an area.

Similarly, if $f(t)$ is velocity in meters/second and t is time in seconds, then the integral

$$\int_a^b f(t)\,dt$$

has units of (meters/sec) $\times$ (sec) $=$ meters, which is what we expect since the integral represents change in position. We saw in Section 5.1 that total change could be approximated by a Riemann sum formed using the rate of change. In the limit, we have:

> If $f(t)$ is a rate of change of a quantity, then
>
> $$\begin{matrix}\text{Total change in quantity between}\\ t=a \text{ and } t=b\end{matrix} = \int_a^b f(t)\,dt$$

The units correspond: If $f(t)$ is a rate of change, with units of quantity/time, then $f(t)\Delta t$ and the definite integral have units of (quantity/time) $\times$ (time) $=$ quantity.

Example 1 A bacteria colony initially has a population of 14 million bacteria. Suppose that t hours later the population is growing at a rate of $f(t) = 2^t$ million bacteria per hour.

(a) Give a definite integral that represents the total change in the bacteria population during the time from $t = 0$ to $t = 2$.

(b) Find the population at time $t = 2$.

Solution

(a) Since $f(t) = 2^t$ gives the rate of change of population, we have

$$\begin{matrix}\text{Change in population between}\\ t=0 \text{ and } t=2\end{matrix} = \int_0^2 2^t\,dt.$$

(b) Using a calculator, we find $\int_0^2 2^t\,dt = 4.328$. The bacteria population was 14 million at time $t = 0$ and increased 4.328 million between $t = 0$ and $t = 2$. Therefore, at time $t = 2$,

$$\text{Population} = 14 + 4.328 = 18.328 \text{ million bacteria.}$$

Example 2 Suppose that $C(t)$ represents the cost per day to heat your home in dollars per day, where t is time measured in days and $t = 0$ corresponds to January 1, 2010. Interpret $\int_0^{90} C(t)\,dt$.

Solution The units for the integral $\int_0^{90} C(t)\,dt$ are (dollars/day) $\times$ (days) $=$ dollars. The integral represents the cost in dollars to heat your house for the first 90 days of 2010, namely the months of January, February, and March.

Example 3 A man starts 50 miles away from his home and takes a trip in his car. He moves on a straight line, and his home lies on this line. His velocity is given in Figure 5.42.

(a) When is the man closest to his home? Approximately how far away is he then?
(b) When is the man farthest from his home? How far away is he then?

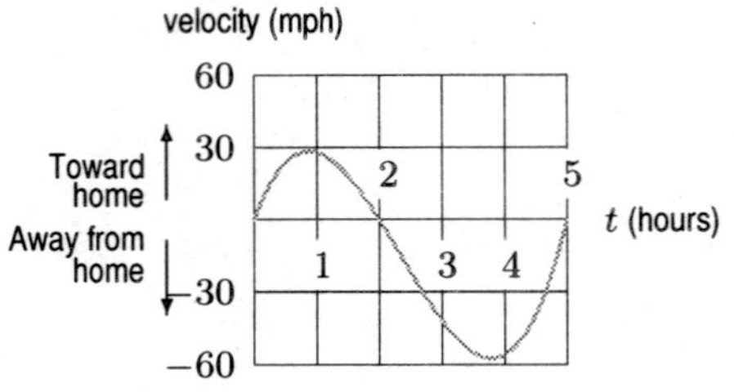

Figure 5.42: Velocity of trip starting 50 miles from home

Solution What happens on this trip? The velocity function is positive the first two hours and negative between $t = 2$ and $t = 5$. So the man moves toward his home during the first two hours, then turns around at $t = 2$ and moves away from his home. The distance he travels is represented by the area between the graph of velocity and the t-axis; since the area below the axis is greater than the area above the axis, we see that he ends up farther away from home than when he started. Thus he is closest to home at $t = 2$ and farthest from home at $t = 5$. We can estimate how far he went in each direction by estimating areas.

(a) The man starts out 50 miles from home. The distance the man travels during the first two hours is the area under the curve between $t = 0$ and $t = 2$. This area corresponds to about one grid square. Since each grid square has area (30 miles/hour)(1 hour) = 30 miles, the man travels about 30 miles toward home. He is closest to home after 2 hours, and he is about 20 miles away at that time.
(b) Between $t = 2$ and $t = 5$, the man moves away from his home. Since this area is equal to about 3.5 grid squares, which is $(3.5)(30) = 105$ miles, he has moved 105 miles farther from home. He was already 20 miles from home, so at $t = 5$ he is about 125 miles from home. He is farthest from home at $t = 5$.

Notice that the man has covered a total distance of $30 + 105 = 135$ miles. However, he went toward his home for 30 miles and away from his home for 105 miles. His *net* change in position is 75 miles.

Example 4 The rates of growth of the populations of two species of plants (measured in new plants per year) are shown in Figure 5.43. Assume that the populations of the two species are equal at time $t = 0$.

(a) Which population is larger after one year? After two years?
(b) How much does the population of species 1 increase during the first two years?

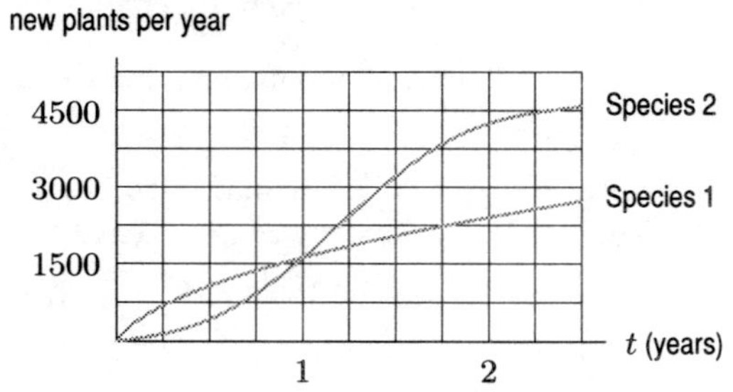

Figure 5.43: Population growth rates for two species of plants

Solution (a) The rate of growth of the population of species 1 is higher than that of species 2 throughout the first year, so the population of species 1 is larger after one year. After two years, the situation is less clear, since the population of species 1 increased faster for the first year and that of species 2 for the second. However, if $r(t)$ is the rate of growth of a population, we have

$$\text{Total change in population during first two years} = \int_0^2 r(t)dt.$$

This integral is the area under the graph of $r(t)$. For $t = 0$ to $t = 2$, the area under the species 1 graph in Figure 5.43 is smaller than the area under the species 2 graph, so the population of species 2 is larger after two years.

(b) The population change for species 1 is the area of the region under the graph of $r(t)$ between $t = 0$ and $t = 2$ in Figure 5.43. The region consists of about 16.5 grid squares, each of area (750 plants/year)(0.25 year) = 187.5 plants, giving a total of (16.5)(187.5) = 3093.75 plants. The population of species 1 increases by about 3100 plants during the two years.

Bioavailability of Drugs

In pharmacology, the definite integral is used to measure *bioavailability*; that is, the overall presence of a drug in the bloodstream during the course of a treatment. Unit bioavailability represents 1 unit concentration of the drug in the bloodstream for 1 hour. For example, a concentration of 3 μg/cm^3 in the blood for 2 hours has bioavailability of $3 \cdot 2 = 6$ (μg/cm^3)-hours.

Ordinarily the concentration of a drug in the blood is not constant. Typically, the concentration in the blood increases as the drug is absorbed into the bloodstream, and then decreases as the drug is broken down and excreted.[2] (See Figure 5.44.)

Suppose that we want to calculate the bioavailability of a drug that is in the blood with concentration $C(t)\mu$g/cm^3 at time t for the time period $0 \le t \le T$. Over a small interval Δt, we estimate

$$\text{Bioavailability} \approx \text{Concentration} \times \text{Time} = C(t)\Delta t.$$

Summing over all subintervals gives

$$\text{Total bioavailability} \approx \sum C(t)\Delta t.$$

In the limit as $n \to \infty$, where n is the number of intervals of width Δt, the sum becomes an integral. So for $0 \le t \le T$, we have

$$\text{Bioavailability} = \int_0^T C(t)dt.$$

That is, the total bioavailability of a drug is equal to the area under the drug concentration curve.

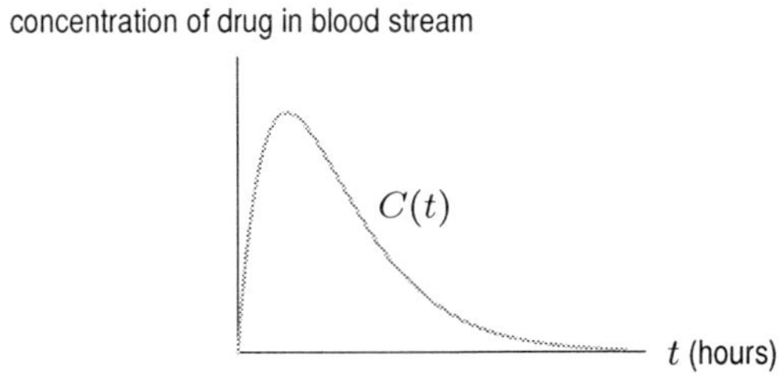

Figure 5.44: Curve showing drug concentration as a function of time

[2] *Drug Treatment*, Graeme S. Avery (Ed.) (Sydney: Adis Press, 1976).

Example 5 Blood concentration curves[3] of two drugs are shown in Figure 5.45. Describe the differences and similarities between the two drugs in terms of peak concentration, speed of absorption into the bloodstream, and total bioavailability.

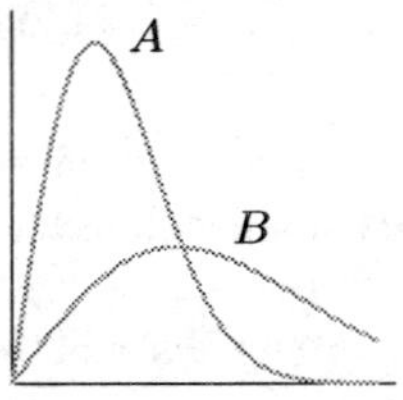

Figure 5.45: Concentration curves of two drugs

Solution Drug A has a peak concentration more than twice as high as that of drug B. Because drug A achieves peak concentration sooner than drug B, drug A appears to be absorbed more rapidly into the blood stream than drug B. Finally, drug A has greater total bioavailability, since the area under the graph of the concentration function for drug A is greater than the area under the graph for drug B.

Problems for Section 5.4

1. The following table gives the emissions, E, of nitrogen oxides in millions of metric tons per year in the US.[4] Let t be the number of years since 1970 and $E = f(t)$.

(a) What are the units and meaning of $\int_0^{30} f(t)dt$?

(b) Estimate $\int_0^{30} f(t)dt$.

Year	1970	1975	1980	1985	1990	1995	2000
E	26.9	26.4	27.1	25.8	25.5	25.0	22.6

2. Annual coal production in the US (in quadrillion BTU per year) is given in the table.[5] Estimate the total amount of coal produced in the US between 1960 and 1990. If $r = f(t)$ is the rate of coal production t years since 1960, write an integral to represent the 1960–1990 coal production.

Year	1960	1965	1970	1975	1980	1985	1990
Rate	10.82	13.06	14.61	14.99	18.60	19.33	22.46

In Problems 3–6, explain in words what the integral represents and give units.

3. $\int_1^3 v(t)\,dt$, where $v(t)$ is velocity in meters/sec and t is time in seconds.

4. $\int_0^6 a(t)\,dt$, where $a(t)$ is acceleration in km/hr^2 and t is time in hours.

5. $\int_{2000}^{2004} f(t)\,dt$, where $f(t)$ is the rate at which the world's population is growing in year t, in billion people per year.

6. $\int_0^5 s(x)\,dx$, where $s(x)$ is rate of change of salinity (salt concentration) in gm/liter per cm in sea water, and where x is depth below the surface of the water in cm.

7. Oil leaks out of a tanker at a rate of $r = f(t)$ gallons per minute, where t is in minutes. Write a definite integral expressing the total quantity of oil which leaks out of the tanker in the first hour.

8. A cup of coffee at 90°C is put into a 20°C room when $t = 0$. The coffee's temperature is changing at a rate of $r(t) = -7(0.9^t)$ °C per minute, with t in minutes. Estimate the coffee's temperature when $t = 10$.

9. After a foreign substance is introduced into the blood, the rate at which antibodies are made is given by

$$r(t) = \frac{t}{t^2+1} \text{ thousands of antibodies per minute,}$$

where time, t, is in minutes. Assuming there are no antibodies present at time $t = 0$, find the total quantity of antibodies in the blood at the end of 4 minutes.

[3] *Drug Treatment*, Graeme S. Avery (Ed.) (Sydney: Adis Press, 1976).
[4] *The World Almanac and Book of Facts 2005*, p. 177 (New York: World Almanac Books).
[5] *World Almanac*, 1995.

10. World annual natural gas[6] consumption, N, in millions of metric tons of oil equivalent, is approximated by $N = 1770 + 53t$, where t is in years since 1990.

(a) How much natural gas was consumed in 1990? In 2010?

(b) Estimate the total amount of natural gas consumed during the 20-year period from 1990 to 2010.

11. Solar photovoltaic (PV) cells are the world's fastest-growing energy source.[7] Annual solar PV production, S, in megawatts, is approximated by $S = 277e^{0.368t}$, where t is in years since 2000. Estimate the total solar PV production between 2000 and 2010.

12. The velocity of a car (in miles per hour) is given by $v(t) = 40t - 10t^2$, where t is in hours.

(a) Write a definite integral for the distance the car travels during the first three hours.

(b) Sketch a graph of velocity against time and represent the distance traveled during the first three hours as an area on your graph.

(c) Use a computer or calculator to find this distance.

13. Your velocity is $v(t) = \ln(t^2 + 1)$ ft/sec for t in seconds, $0 \leq t \leq 3$. Estimate the distance traveled during this time.

14. Figure 5.46 shows the rate of change of the quantity of water in a water tower, in liters per day, during the month of April. If the tower had 12,000 liters of water in it on April 1, estimate the quantity of water in the tower on April 30.

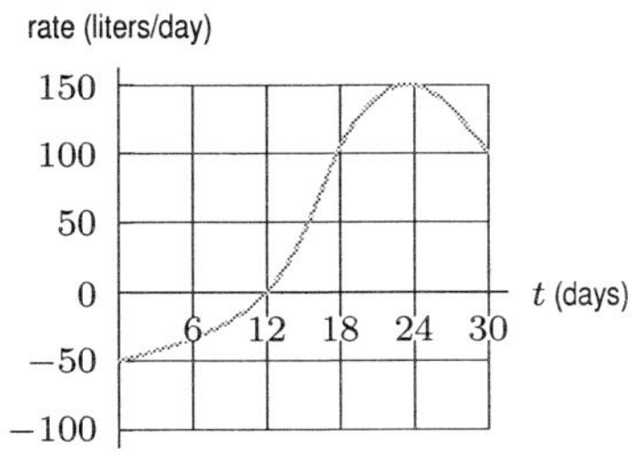

Figure 5.46

15. Figure 5.47 shows the weight growth rate of a human fetus.

(a) What property of a graph of weight as a function of age corresponds to the fact that the function in Figure 5.47 is increasing?

(b) Estimate the weight of a baby born in week 40.

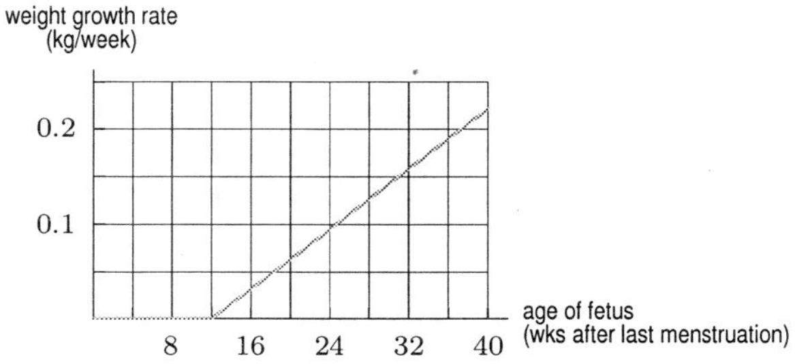

Figure 5.47: Rate of increase of fetal weight

Problems 16–18 concern the future of the US Social Security Trust Fund, out of which pensions are paid. Figure 5.48 shows the rates (billions of dollars per year) at which income, $I(t)$, from taxes and interest is projected to flow into the fund and at which expenditures, $E(t)$, flow out of the fund. Figure 5.49 shows the value of the fund as a function of time.[8]

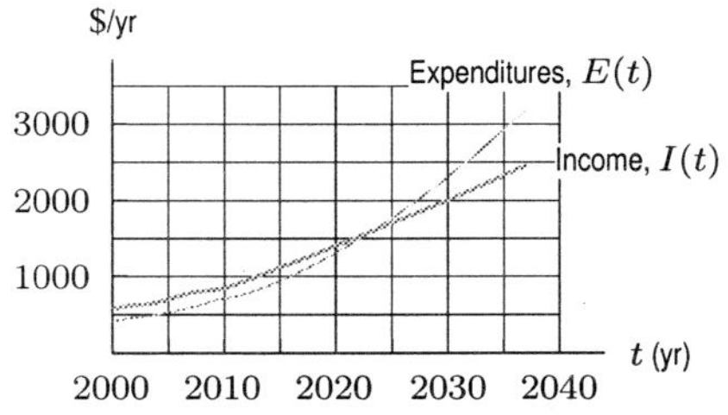

Figure 5.48

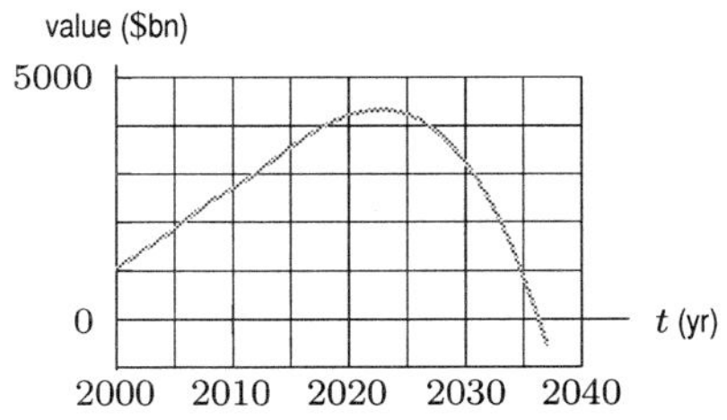

Figure 5.49

16. **(a)** Write each of the following areas in Figure 5.48 from 2000 to 2015 as an integral and explain its significance for the fund.

(i) Under the income curve

(ii) Under the expenditure curve

(iii) Between the income and expenditure curves

(b) Use Figure 5.49 to estimate the area between the two curves from 2000 to 2015.

17. Decide when the value of the fund is projected to be a maximum using

(a) Figure 5.48 **(b)** Figure 5.49

[6] *BP Statistical Review of World Energy 2009*, http://www.bp.com/ (accessed 9/6/2009).
[7] *Vital Signs 2007-2008*, The Worldwatch Institute, W.W. Norton & Company, 2007, p. 38.
[8] The 2009 OASDI Trustees Report, www.ssa.gov/OACT/TR/2009.

18. Express the projected increase in value of the fund from 2000 to 2030 as an integral.

19. A forest fire covers 2000 acres at time $t = 0$. The fire is growing at a rate of $8\sqrt{t}$ acres per hour, where t is in hours. How many acres are covered 24 hours later?

20. Water is pumped out of a holding tank at a rate of $5 - 5e^{-0.12t}$ liters/minute, where t is in minutes since the pump is started. If the holding tank contains 1000 liters of water when the pump is started, how much water does it hold one hour later?

21. Figure 5.50 shows the rate of growth of two trees. If the two trees are the same height at time $t = 0$, which tree is taller after 5 years? After 10 years?

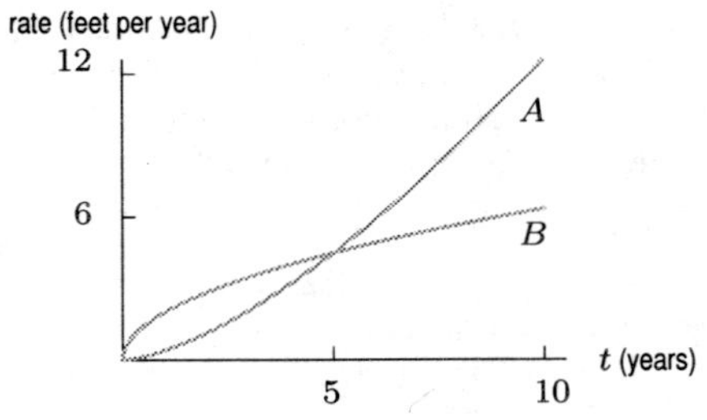

Figure 5.50

22. Figure 5.51 shows the number of sales per month made by two salespeople. Which person has the most total sales after 6 months? After the first year? At approximately what times (if any) have they sold roughly equal total amounts? Approximately how many total sales has each person made at the end of the first year?

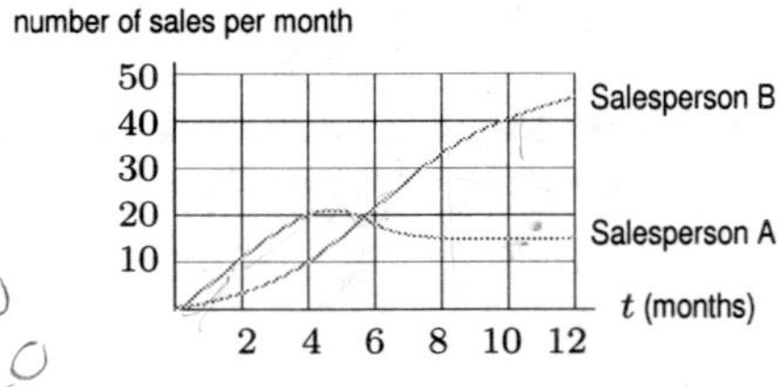

Figure 5.51

23. Height velocity graphs are used by endocrinologists to follow the progress of children with growth deficiencies. Figure 5.52 shows the height velocity curves of an average boy and an average girl between ages 3 and 18.

(a) Which curve is for girls and which is for boys? Explain how you can tell.

(b) About how much does the average boy grow between ages 3 and 10?

(c) The growth spurt associated with adolescence and the onset of puberty occurs between ages 12 and 15 for the average boy and between ages 10 and 12.5 for the average girl. Estimate the height gained by each average child during this growth spurt.

(d) When fully grown, about how much taller is the average man than the average woman? (The average boy and girl are about the same height at age 3.)

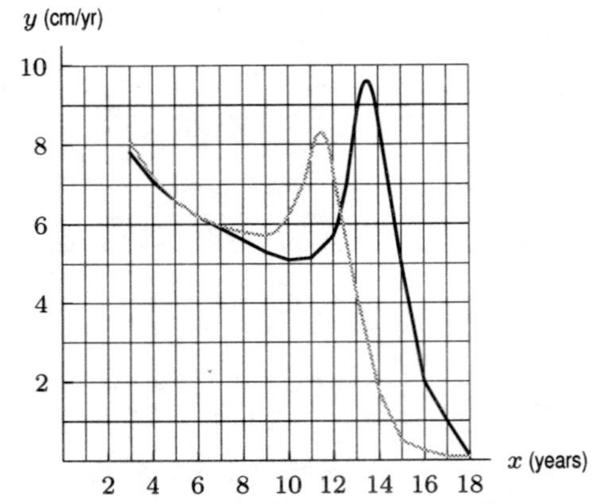

Figure 5.52

24. The birth rate, B, in births per hour, of a bacteria population is given in Figure 5.53. The curve marked D gives the death rate, in deaths per hour, of the same population.

(a) Explain what the shape of each of these graphs tells you about the population.

(b) Use the graphs to find the time at which the net rate of increase of the population is at a maximum.

(c) At time $t = 0$ the population has size N. Sketch the graph of the total number born by time t. Also sketch the graph of the number alive at time t. Estimate the time at which the population is a maximum.

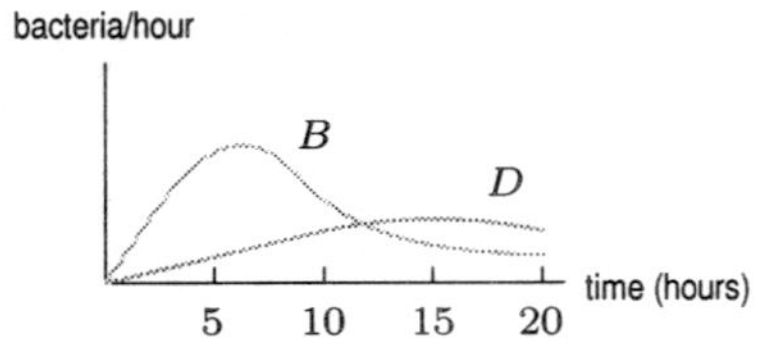

Figure 5.53

25. The rates of consumption of stores of protein and fat in the human body during 8 weeks of starvation are shown in Figure 5.54. Does the body burn more fat or more protein during this period?

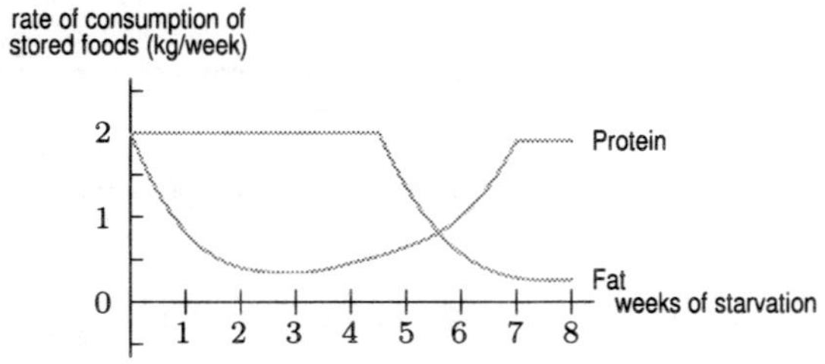

Figure 5.54

A healthy human heart pumps about 5 liters of blood per minute. Problems 26–27 refer to Figure 5.55, which shows the response of the heart to bleeding. The pumping rate drops and then returns to normal if the person recovers fully, or drops to zero if the person dies.

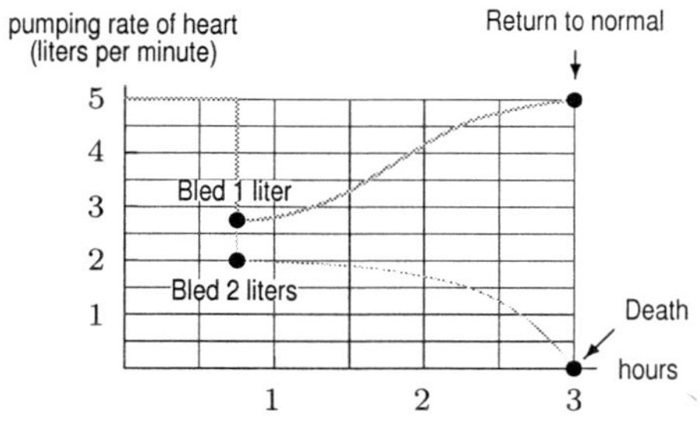

Figure 5.55

26. **(a)** If the body is bled 2 liters, how much blood is pumped during the three hours leading to death?
(b) If $f(t)$ is the pumping rate in liters per minute at time t hours, express your answer to part (a) as a definite integral.
(c) How much more blood would have been pumped during the same time period if there had been no bleeding? Illustrate your answer on the graph.

27. **(a)** If the body is bled 1 liter, how much blood is pumped during the three hours leading to full recovery?
(b) If $g(t)$ is the pumping rate in liters per minute at time t hours, express your answer to part (a) as a definite integral.
(c) How much more blood would have been pumped during the same time period if there had been no bleeding? Show your answer as an area on the graph.

28. The amount of waste a company produces, W, in tons per week, is approximated by $W = 3.75e^{-0.008t}$, where t is in weeks since January 1, 2005. Waste removal for the company costs \$15/ton. How much does the company pay for waste removal during the year 2005?

Problems 29–32 show the velocity, in cm/sec, of a particle moving along the x-axis. Compute the particle's change in position, left (negative) or right (positive), between times $t = 0$ and $t = 5$ seconds.

29.

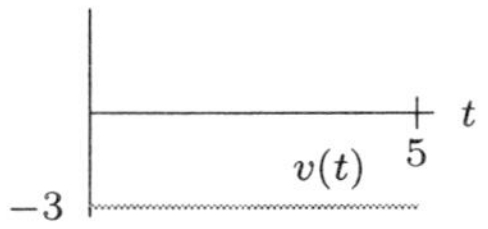

30.

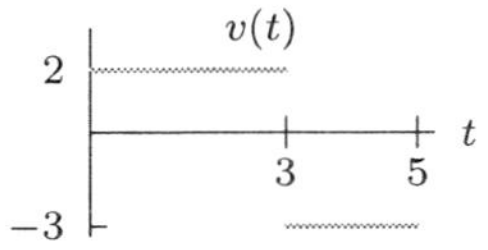

31.

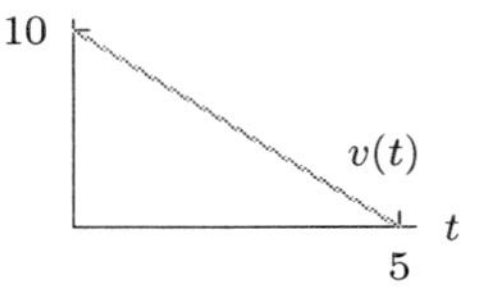

32.

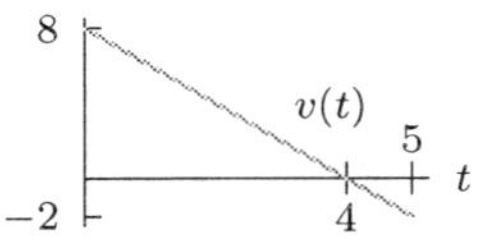

33. Figure 5.56 gives your velocity during a trip starting from home. Positive velocities take you away from home and negative velocities take you toward home. Where are you at the end of the 5 hours? When are you farthest from home? How far away are you at that time?

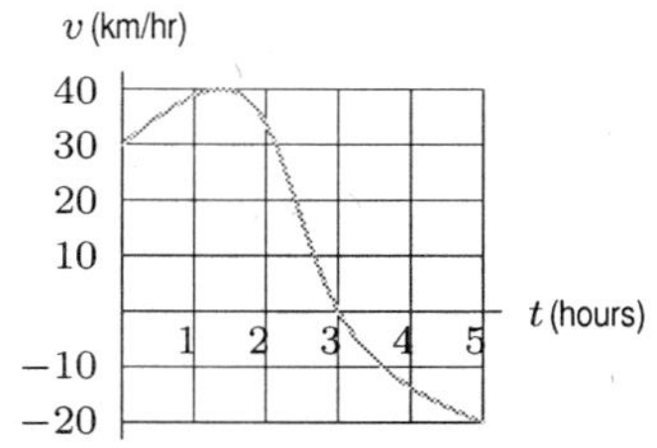

Figure 5.56

34. A bicyclist is pedaling along a straight road for one hour with a velocity v shown in Figure 5.57. She starts out five kilometers from the lake and positive velocities take her toward the lake. [Note: The vertical lines on the graph are at 10 minute (1/6 hour) intervals.]

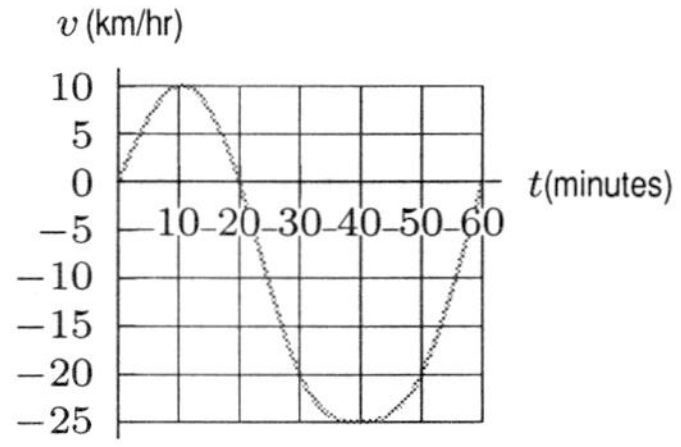

Figure 5.57

(a) Does the cyclist ever turn around? If so, at what time(s)?
(b) When is she going the fastest? How fast is she going then? Toward the lake or away?
(c) When is she closest to the lake? Approximately how close to the lake does she get?
(d) When is she farthest from the lake? Approximately how far from the lake is she then?

35. Figure 5.58 shows plasma concentration curves for two drugs used to slow a rapid heart rate. Compare the two products in terms of level of peak concentration, time until peak concentration, and overall bioavailability.

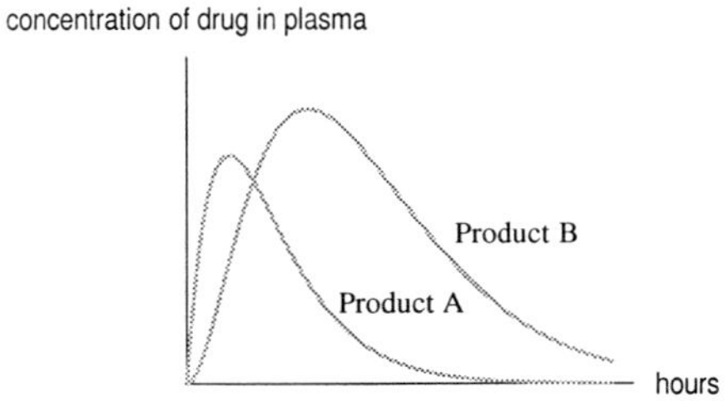

Figure 5.58

36. Figure 5.59 compares the concentration in blood plasma for two pain relievers. Compare the two products in terms of level of peak concentration, time until peak concentration, and overall bioavailability.

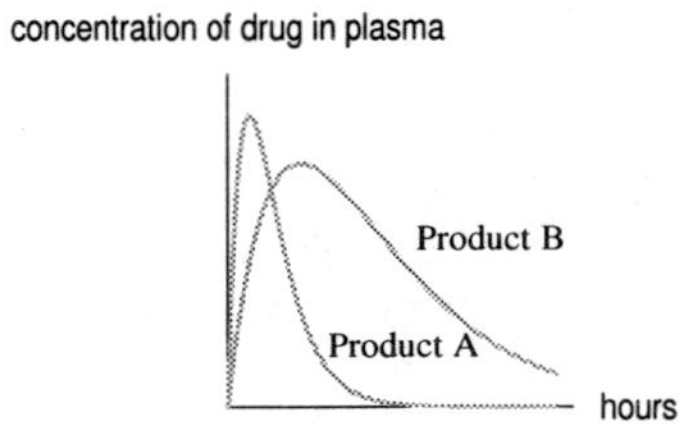

Figure 5.59

37. Draw plasma concentration curves for two drugs A and B if product A has the highest peak concentration, but product B is absorbed more quickly and has greater overall bioavailability.

38. A two-day environmental cleanup started at 9 am on the first day. The number of workers fluctuated as shown in Figure 5.60. If the workers were paid \$10 per hour, how much was the total personnel cost of the cleanup?

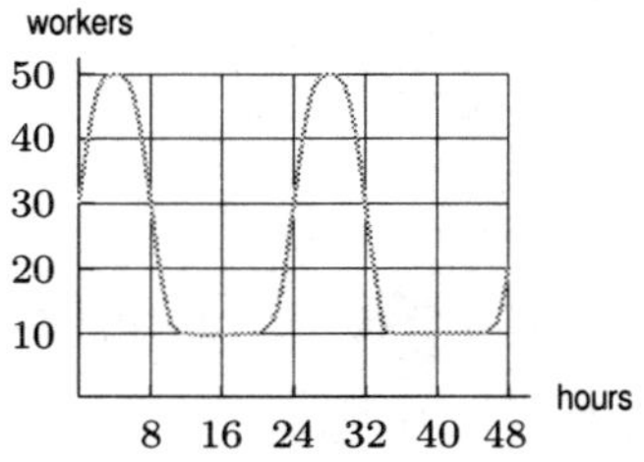

Figure 5.60

39. Suppose in Problem 38 that the workers were paid \$10 per hour for work during the time period 9 am to 5 pm and were paid \$15 per hour for work during the rest of the day. What would the total personnel costs of the clean up have been under these conditions?

40. At the site of a spill of radioactive iodine, radiation levels were four times the maximum acceptable limit, so an evacuation was ordered. If R_0 is the initial radiation level (at $t = 0$) and t is the time in hours, the radiation level $R(t)$, in millirems/hour, is given by

$$R(t) = R_0(0.996)^t.$$

(a) How long does it take for the site to reach the acceptable level of radiation of 0.6 millirems/hour?

(b) How much total radiation (in millirems) has been emitted by that time?

41. If you jump out of an airplane and your parachute fails to open, your downward velocity (in meters per second) t seconds after the jump is approximated by

$$v(t) = 49(1 - (0.8187)^t).$$

(a) Write an expression for the distance you fall in T seconds.

(b) If you jump from 5000 meters above the ground, estimate, using trial and error, how many seconds you fall before hitting the ground.

42. The Montgolfier brothers (Joseph and Etienne) were eighteenth-century pioneers in the field of hot-air ballooning. Had they had the appropriate instruments, they might have left us a record, like that shown in Figure 5.61, of one of their early experiments. The graph shows their vertical velocity, v, with upward as positive.

(a) Over what intervals was the acceleration positive? Negative?

(b) What was the greatest altitude achieved, and at what time?

(c) This particular flight ended on top of a hill. How do you know that it did, and what was the height of the hill above the starting point?

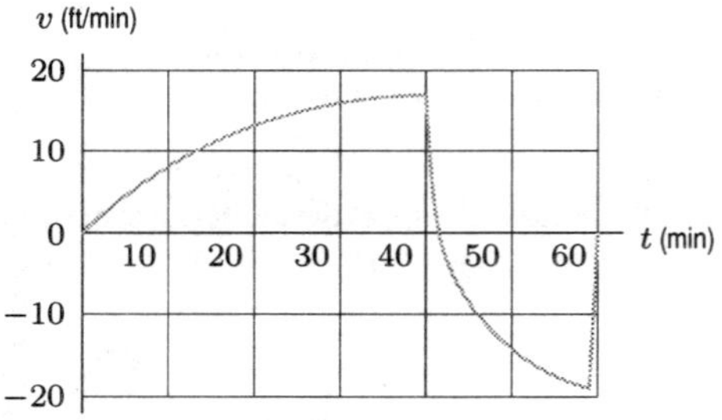

Figure 5.61

5.5 THE FUNDAMENTAL THEOREM OF CALCULUS

In Section 5.4, we saw that the total change of a quantity can be obtained by integrating its rate of change. Since the change in F between a and b is $F(b) - F(a)$ and the rate of change is $F'(t)$ we have the following result:

The Fundamental Theorem of Calculus

If $F'(t)$ is continuous for $a \leq t \leq b$, then

$$\int_a^b F'(t)\,dt = F(b) - F(a).$$

In words:
The definite integral of the derivative of a function gives the total change in the function.

In Section 7.3 we see how to use the Fundamental Theorem to compute a definite integral. The Focus on Theory Section gives another version of the Fundamental Theorem.

Example 1 Figure 5.62 shows $F'(t)$, the rate of change of the value, $F(t)$, of an investment over a 5-month period.

(a) When is the value of the investment increasing in value and when is it decreasing?
(b) Does the investment increase or decrease in value during the 5 months?

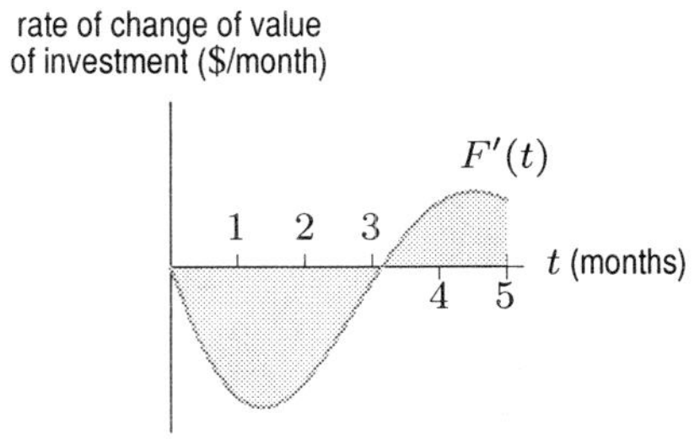

Figure 5.62: Did the investment increase or decrease in value over these 5 months?

Solution

(a) The investment decreased in value during the first 3 months, since the rate of change of value is negative then. The value rose during the last 2 months.
(b) We want to find the total change in the value of the investment between $t = 0$ and $t = 5$. Since the total change is the integral of the rate of change, $F'(t)$, we are looking for

$$\text{Total change in value} = \int_0^5 F'(t)dt.$$

The integral equals the shaded area above the t-axis minus the shaded area below the t-axis. Since in Figure 5.62 the area below the axis is greater than the area above the axis, the integral is negative. The total change in value of the investment during this time is negative, so it decreased in value.

Marginal Cost and Change in Total Cost

Suppose $C(q)$ represents the cost of producing q items. The derivative, $C'(q)$, is the marginal cost. Since marginal cost $C'(q)$ is the rate of change of the cost function with respect to quantity, by the Fundamental Theorem, the integral

$$\int_a^b C'(q)\,dq$$

represents the total change in the cost function between $q = a$ and $q = b$. In other words, the integral gives the amount it costs to increase production from a units to b units.

The cost of producing 0 units is the fixed cost $C(0)$. The area under the marginal cost curve between $q = 0$ and $q = b$ is the total increase in cost between a production of 0 and a production of b. This is called the *total variable cost.* Adding this to the fixed cost gives the total cost to produce b units. In summary,

> If $C'(q)$ is a marginal cost function and $C(0)$ is the fixed cost,
>
> $$\text{Cost to increase production from } a \text{ units to } b \text{ units} = C(b) - C(a) = \int_a^b C'(q)\,dq$$
>
> $$\text{Total variable cost to produce } b \text{ units} = \int_0^b C'(q)\,dq$$
>
> $$\begin{aligned}\text{Total cost of producing } b \text{ units} &= \text{Fixed cost} + \text{Total variable cost}\\ &= C(0) + \int_0^b C'(q)\,dq\end{aligned}$$

Example 2 A marginal cost curve is given in Figure 5.63. If the fixed cost is \$1000, estimate the total cost of producing 250 items.

Solution The total cost of production is Fixed cost + Variable cost. The variable cost of producing 250 items is represented by the area under the marginal cost curve. The area in Figure 5.63 between $q = 0$ and $q = 250$ is about 20 grid squares. Each grid square has area (2 dollars/item)(50 items) = 100 dollars, so

$$\text{Total variable cost} = \int_0^{250} C'(q)\,dq \approx 20(100) = 2000.$$

The total cost to produce 250 items is given by :

$$\begin{aligned}\text{Total cost} &= \text{Fixed cost} + \text{Total variable cost}\\ &\approx \$1000 + \$2000 = \$3000.\end{aligned}$$

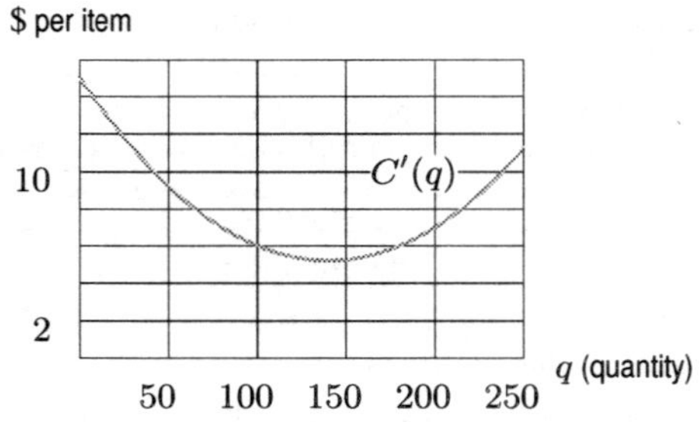

Figure 5.63: A marginal cost curve

Problems for Section 5.5

1. If the marginal cost function $C'(q)$ is measured in dollars per ton, and q gives the quantity in tons, what are the units of measurement for $\int_{800}^{900} C'(q)\,dq$? What does this integral represent?

2. The marginal cost function of a product, in dollars per unit, is $C'(q) = q^2 - 50q + 700$. If fixed costs are \$500, find the total cost to produce 50 items.

3. The total cost in dollars to produce q units of a product is $C(q)$. Fixed costs are \$20,000. The marginal cost is

$$C'(q) = 0.005q^2 - q + 56.$$

(a) On a graph of $C'(q)$, illustrate graphically the total variable cost of producing 150 units.
(b) Estimate $C(150)$, the total cost to produce 150 units.
(c) Find the value of $C'(150)$ and interpret your answer in terms of costs of production.
(d) Use parts (b) and (c) to estimate $C(151)$.

4. A marginal cost function $C'(q)$ is given in Figure 5.64. If the fixed costs are \$10,000, estimate:

(a) The total cost to produce 30 units.
(b) The additional cost if the company increases production from 30 units to 40 units.
(c) The value of $C'(25)$. Interpret your answer in terms of costs of production.

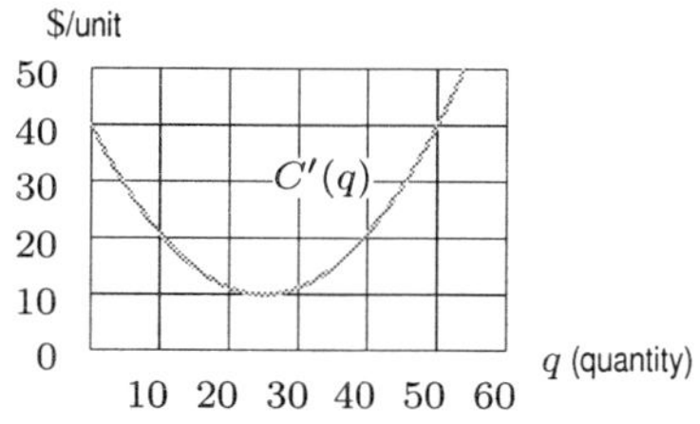

Figure 5.64

5. The population of Tokyo grew at the rate shown in Figure 5.65. Estimate the change in population between 1970 and 1990.

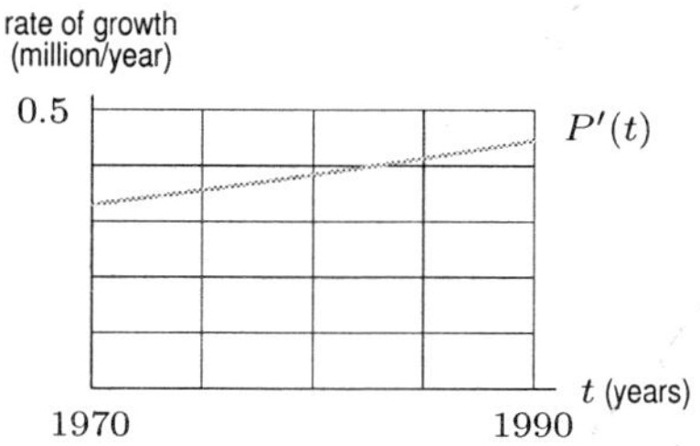

Figure 5.65

6. The marginal cost function for a company is given by

$$C'(q) = q^2 - 16q + 70 \text{ dollars/unit,}$$

where q is the quantity produced. If $C(0) = 500$, find the total cost of producing 20 units. What is the fixed cost and what is the total variable cost for this quantity?

7. The marginal cost $C'(q)$ (in dollars per unit) of producing q units is given in the following table.

(a) If fixed cost is \$10,000, estimate the total cost of producing 400 units.
(b) How much would the total cost increase if production were increased one unit, to 401 units?

q	0	100	200	300	400	500	600
$C'(q)$	25	20	18	22	28	35	45

8. The marginal cost function of producing q mountain bikes is

$$C'(q) = \frac{600}{0.3q + 5}.$$

(a) If the fixed cost in producing the bicycles is \$2000, find the total cost to produce 30 bicycles.
(b) If the bikes are sold for \$200 each, what is the profit (or loss) on the first 30 bicycles?
(c) Find the marginal profit on the 31st bicycle.

9. The marginal revenue function on sales of q units of a product is $R'(q) = 200 - 12\sqrt{q}$ dollars per unit.

(a) Graph $R'(q)$.
(b) Estimate the total revenue if sales are 100 units.
(c) What is the marginal revenue at 100 units? Use this value and your answer to part (b) to estimate the total revenue if sales are 101 units.

10. Figure 5.66 shows $P'(t)$, the rate of change of the price of stock in a certain company at time t.

(a) At what time during this five-week period was the stock at its highest value? At its lowest value?
(b) If $P(t)$ represents the price of the stock, arrange the following quantities in increasing order:

$$P(0),\ P(1),\ P(2),\ P(3),\ P(4),\ P(5).$$

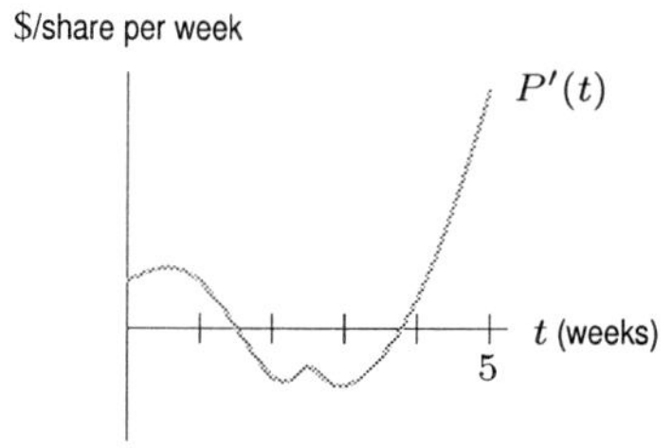

Figure 5.66

11. Ice is forming on a pond at a rate given by

$$\frac{dy}{dt} = \frac{\sqrt{t}}{2} \text{ inches per hour,}$$

where y is the thickness of the ice in inches at time t measured in hours since the ice started forming.

(a) Estimate the thickness of the ice after 8 hours.
(b) At what rate is the thickness of the ice increasing after 8 hours?

12. The net worth, $f(t)$, of a company is growing at a rate of $f'(t) = 2000 - 12t^2$ dollars per year, where t is in years since 2005. How is the net worth of the company expected to change between 2005 and 2015? If the company is worth \$40,000 in 2005, what is it worth in 2015?

13. The graph of a derivative $f'(x)$ is shown in Figure 5.67. Fill in the table of values for $f(x)$ given that $f(0) = 2$.

x	0	1	2	3	4	5	6
$f(x)$	2						

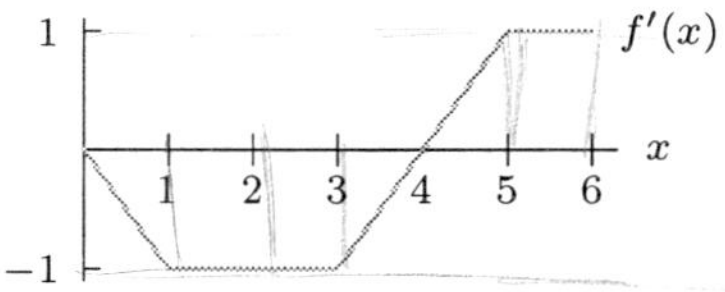

Figure 5.67: Graph of f', not f

CHAPTER SUMMARY

- **Definite integral as limit of right-hand or left-hand sums**
- **Interpretations of the definite integral**
Total change from rate of change, change in position given velocity, area, bioavailability, total variable cost.
- **Working with the definite integral**
Estimate definite integral from graph, table of values, or formula.
- **Fundamental Theorem of Calculus**

REVIEW PROBLEMS FOR CHAPTER FIVE

1. The velocity $v(t)$ in Table 5.9 is decreasing, $2 \le t \le 12$. Using $n = 5$ subdivisions to approximate the total distance traveled, find

(a) An upper estimate **(b)** A lower estimate

Table 5.9

t	2	4	6	8	10	12
$v(t)$	44	42	41	40	37	35

2. The velocity $v(t)$ in Table 5.10 is increasing, $0 \le t \le 12$.

(a) Find an upper estimate for the total distance traveled using

(i) $n = 4$ (ii) $n = 2$

(b) Which of the two answers in part (a) is more accurate? Why?

(c) Find a lower estimate of the total distance traveled using $n = 4$.

Table 5.10

t	0	3	6	9	12
$v(t)$	34	37	38	40	45

3. Use the following table to estimate $\int_{10}^{26} f(x)\,dx$.

x	10	14	18	22	26
$f(x)$	100	88	72	50	28

4. If $f(t)$ is measured in miles per hour and t is measured in hours, what are the units of $\int_a^b f(t)\,dt$?

5. If $f(t)$ is measured in meters/second^2 and t is measured in seconds, what are the units of $\int_a^b f(t)\,dt$?

6. If $f(t)$ is measured in dollars per year and t is measured in years, what are the units of $\int_a^b f(t)\,dt$?

7. If $f(x)$ is measured in pounds and x is measured in feet, what are the units of $\int_a^b f(x)\,dx$?

8. As coal deposits are depleted, it becomes necessary to strip-mine larger areas for each ton of coal. Figure 5.68 shows the number of acres of land per million tons of coal that will be defaced during strip-mining as a function of the number of million tons removed, starting from the present day.

(a) Estimate the total number of acres defaced in extracting the next 4 million tons of coal (measured from the present day). Draw four rectangles under the curve, and compute their area.

(b) Reestimate the number of acres defaced using rectangles above the curve.

(c) Use your answers to parts (a) and (b) to get a better estimate of the actual number of acres defaced.

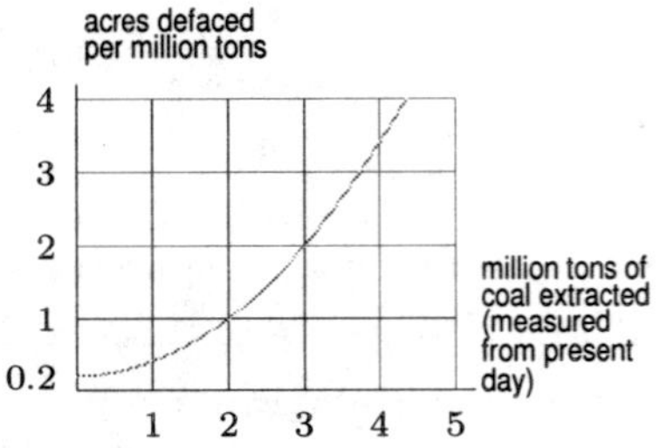

Figure 5.68

For Problems 9–14, use a calculator or computer to evaluate the integral.

9. $\displaystyle\int_0^{10} 2^{-x}\,dx$

10. $\displaystyle\int_1^{5} (x^2+1)\,dx$

11. $\displaystyle\int_0^{1} \sqrt{1+t^2}\,dt$

12. $\displaystyle\int_{-1}^{1} \frac{x^2+1}{x^2-4}\,dx$

13. $\displaystyle\int_2^{3} \frac{-1}{(r+1)^2}\,dr$

14. $\displaystyle\int_1^{3} \frac{z^2+1}{z}\,dz$

15. Find the area under the graph of $f(x) = x^2 + 2$ between $x = 0$ and $x = 6$.

In Problems 16–19, find the given area.

16. Between $y = x^2$ and $y = x^3$ for $0 \le x \le 1$.
17. Between $y = x^{1/2}$ and $y = x^{1/3}$ for $0 \le x \le 1$.
18. Between $y = 3x$ and $y = x^2$.
19. Between $y = x$ and $y = \sqrt{x}$.
20. Coal gas is produced at a gasworks. Pollutants in the gas are removed by scrubbers, which become less and less efficient as time goes on. The following measurements, made at the start of each month, show the rate at which pollutants are escaping (in tons/month) in the gas:

Time (months)	0	1	2	3	4	5	6
Rate pollutants escape	5	7	8	10	13	16	20

(a) Make an overestimate and an underestimate of the total quantity of pollutants that escape during the first month.
(b) Make an overestimate and an underestimate of the total quantity of pollutants that escape during the six months.

21. A student is speeding down Route 11 in his fancy red Porsche when his radar system warns him of an obstacle 400 feet ahead. He immediately applies the brakes, starts to slow down, and spots a skunk in the road directly ahead of him. The "black box" in the Porsche records the car's speed every two seconds, producing the following table. The speed decreases throughout the 10 seconds it takes to stop, although not necessarily at a uniform rate.

Time since brakes applied (sec)	0	2	4	6	8	10
Speed (ft/sec)	100	80	50	25	10	0

(a) What is your best estimate of the total distance the student's car traveled before coming to rest?
(b) Which one of the following statements can you justify from the information given?
 (i) The car stopped before getting to the skunk.
 (ii) The "black box" data is inconclusive. The skunk may or may not have been hit.
 (iii) The skunk was hit by the car.

22. The velocity of a particle moving along the x-axis is given by $f(t) = 6 - 2t$ cm/sec. Use a graph of $f(t)$ to find the exact change in position of the particle from time $t = 0$ to $t = 4$ seconds.
23. A baseball thrown directly upward at 96 ft/sec has velocity $v(t) = 96 - 32t$ ft/sec at time t seconds.
 (a) Graph the velocity from $t = 0$ to $t = 6$.
 (b) When does the baseball reach the peak of its flight? How high does it go?
 (c) How high is the baseball at time $t = 5$?

24. A news broadcast in early 1993 said the typical American's annual income is changing at a rate of $r(t) = 40(1.002)^t$ dollars per month, where t is in months from January 1, 1993. How much did the typical American's income change during 1993?

25. Two species of plants have the same populations at time $t = 0$ and the growth rates shown in Figure 5.69.
 (a) Which species has a larger population at the end of 5 years? At the end of 10 years?
 (b) Which species do you think has the larger population after 20 years? Explain.

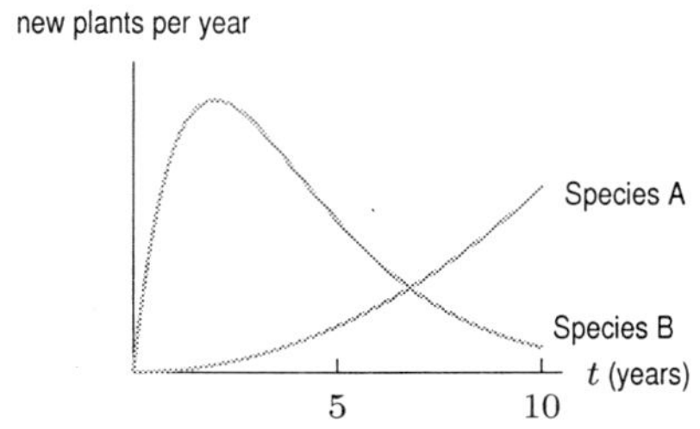

Figure 5.69

26. Figure 5.70 represents your velocity, v, on a bicycle trip along a straight road which starts 10 miles from home. Write a paragraph describing your trip: Do you start out going toward or away from home? How long do you continue in that direction and how far are you from home when you turn around? How many times do you change direction? Do you ever get home? Where are you at the end of the four-hour bike ride?

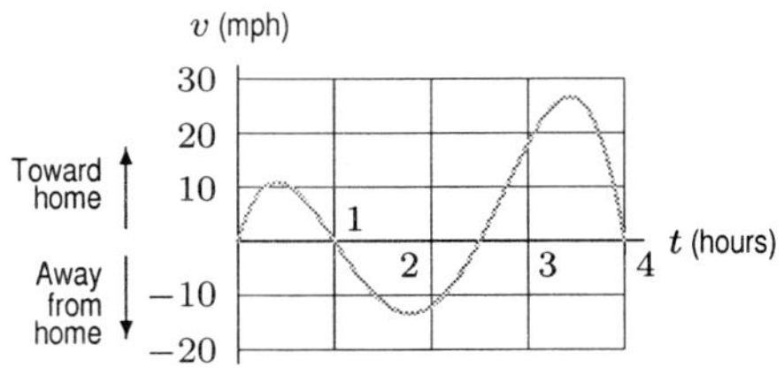

Figure 5.70

27. Figure 5.71 shows the length growth rate of a human fetus.

(a) What feature of a graph of length as a function of age corresponds to the maximum in Figure 5.71?
(b) Estimate the length of a baby born in week 40.

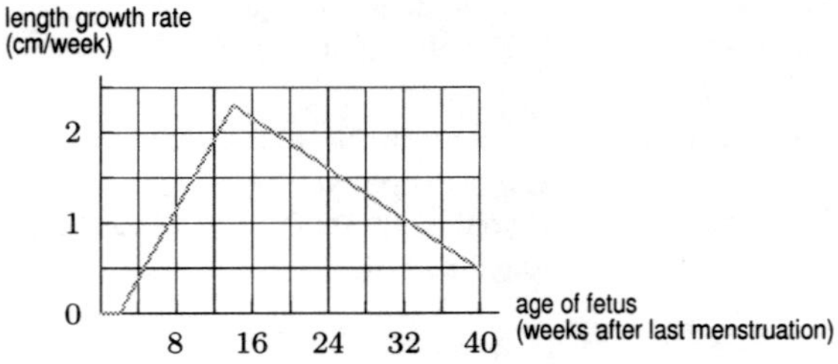

Figure 5.71

28. A bicyclist pedals along a straight road with velocity, v, given in Figure 5.72. She starts 5 miles from a lake; positive velocities take her away from the lake and negative velocities take her toward the lake. When is the cyclist farthest from the lake, and how far away is she then?

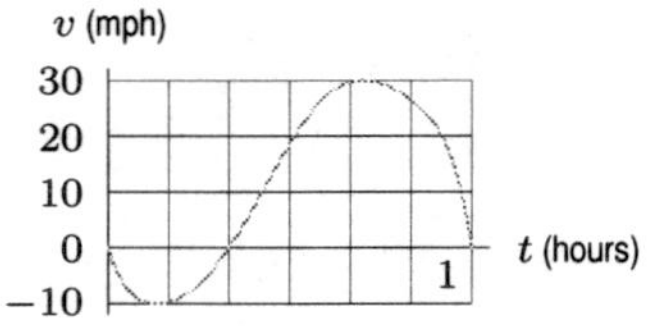

Figure 5.72

29. Using Figure 5.73, decide whether each of the following definite integrals is positive or negative.

(a) $\int_{-5}^{-4} f(x)\,dx$ **(b)** $\int_{-4}^{1} f(x)\,dx$
(c) $\int_{1}^{3} f(x)\,dx$ **(d)** $\int_{-5}^{3} f(x)\,dx$

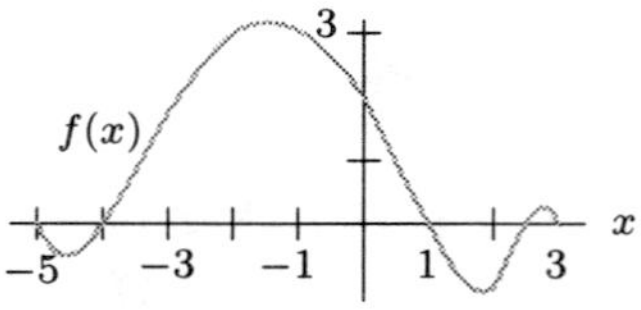

Figure 5.73

30. Using Figure 5.73, arrange the following definite integrals in ascending order:
$\int_{-5}^{-3} f(x)\,dx$, $\int_{-5}^{-1} f(x)\,dx$, $\int_{-5}^{1} f(x)\,dx$, $\int_{-5}^{3} f(x)\,dx$.

31. Use a graph of $y = 2^{-x^2}$ to explain why $\int_{-1}^{1} 2^{-x^2}\,dx$ must be between 0 and 2.

32. Without computation, show that $2 \le \int_0^2 \sqrt{1+x^3}\,dx \le 6$.

33. With t in seconds, the velocity of an object is $v(t) = 10 + 8t - t^2$ m/sec.

(a) Represent the distance traveled during the first 5 seconds as a definite integral and as an area.
(b) Estimate the distance traveled by the object during the first 5 seconds by estimating the area.
(c) Calculate the distance traveled.

34. The world's oil is being consumed at a continuously increasing rate, $f(t)$ (in billions of barrels per year), where t is in years since the start of 1995.

(a) Write a definite integral which represents the total quantity of oil used between the start of 1995 and the start of 2010.
(b) Suppose $r = 32(1.05)^t$. Using a left-hand sum with five subdivisions, find an approximate value for the total quantity of oil used between the start of 1995 and the start of 2010.
(c) Interpret each of the five terms in the sum from part (b) in terms of oil consumption.

35. A car moves along a straight line with velocity, in feet/second, given by

$$v(t) = 6 - 2t \quad \text{for } t \ge 0.$$

(a) Describe the car's motion in words. (When is it moving forward, backward, and so on?)
(b) The car's position is measured from its starting point. When is it farthest forward? Backward?

36. The marginal cost of drilling an oil well depends on the depth at which you are drilling; drilling becomes more expensive, per meter, as you dig deeper into the earth. The fixed costs are 1,000,000 riyals (the riyal is the unit of currency of Saudi Arabia), and, if x is the depth in meters, the marginal costs are

$$C'(x) = 4000 + 10x \quad \text{riyals/meter.}$$

Find the total cost of drilling a 500-meter well.

37. A warehouse charges its customers \$5 per day for every 10 cubic feet of space used for storage. Figure 5.74 records the storage used by one company over a month. How much will the company have to pay?

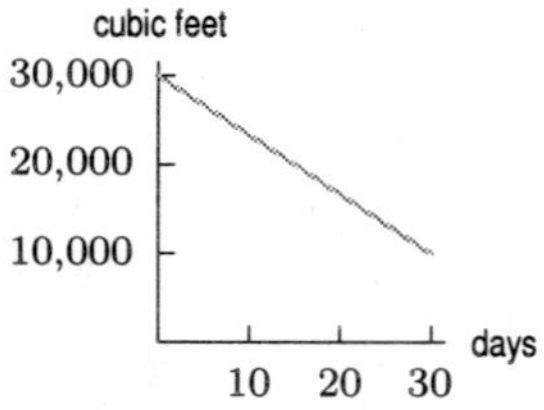

Figure 5.74

38. One of the earliest pollution problems brought to the attention of the Environmental Protection Agency (EPA) was the case of the Sioux Lake in eastern South Dakota. For years a small paper plant located nearby had been discharging waste containing carbon tetrachloride (CCl_4) into the waters of the lake. At the time the EPA learned of the situation, the chemical was entering at a rate of 16 cubic yards/year.

The agency immediately ordered the installation of filters designed to slow (and eventually stop) the flow of CCl_4 from the mill. Implementation of this program took exactly three years, during which the flow of pollutant was steady at 16 cubic yards/year. Once the filters were installed, the flow declined. From the time the filters were installed until the time the flow stopped, the rate of flow was well approximated by

$$\text{Rate (in cubic yards/year)} = t^2 - 14t + 49,$$

where t is time measured in years since the EPA learned of the situation (thus, $t \geq 3$).

(a) Draw a graph showing the rate of CCl_4 flow into the lake as a function of time, beginning at the time the EPA first learned of the situation.
(b) How many years elapsed between the time the EPA learned of the situation and the time the pollution flow stopped entirely?
(c) How much CCl_4 entered the waters during the time shown in the graph in part (a)?

39. (a) Graph $x^3 - 5x^2 + 4x$, marking $x = 1, 2, 3, 4, 5$.
(b) Use your graph and the area interpretation of the definite integral to decide which of the five numbers

$$I_n = \int_0^n (x^3 - 5x^2 + 4x)\,dx \quad \text{for } n = 1, 2, 3, 4, 5$$

is largest. Which is smallest? How many of the numbers are positive? (Don't calculate the integrals.)

40. A mouse moves back and forth in a straight tunnel, attracted to bits of cheddar cheese alternately introduced to and removed from the ends (right and left) of the tunnel. The graph of the mouse's velocity, v, is given in Figure 5.75, with positive velocity corresponding to motion toward the right end. Assuming that the mouse starts ($t = 0$) at the center of the tunnel, use the graph to estimate the time(s) at which:

(a) The mouse changes direction.
(b) The mouse is moving most rapidly to the right; to the left.
(c) The mouse is farthest to the right of center; farthest to the left.
(d) The mouse's speed (i.e., the magnitude of its velocity) is decreasing.
(e) The mouse is at the center of the tunnel.

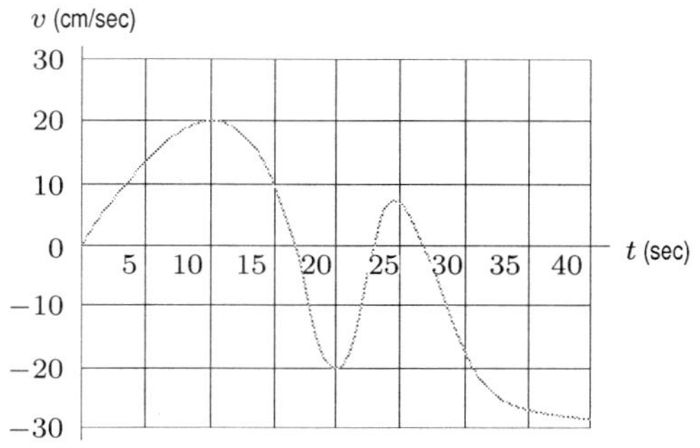

Figure 5.75

41. Pollution is being dumped into a lake at a rate which is increasing at a constant rate from 10 kg/year to 50 kg/year until a total of 270 kg has been dumped. Sketch a graph of the rate at which pollution is being dumped in the lake against time. How long does it take until 270 kg has been dumped?

For Problems 42–44, suppose $F(0) = 0$ and $F'(x) = 4 - x^2$.

42. Calculate $F(b)$ for $b = 0,\ 0.5,\ 1,\ 1.5,\ 2,\ 2.5$.

43. Using a graph of F', decide where F is increasing and where F is decreasing for $0 \leq x \leq 2.5$.

44. Does F have a maximum value for $0 \leq x \leq 2.5$? If so, what is it, and at what value of x does it occur?

CHECK YOUR UNDERSTANDING

In Problems 1–50, indicate whether the statement is true or false.

1. If the velocity is positive, then the total distance traveled is the area under the velocity curve.

2. If velocity is constant and positive, then distance traveled during a time interval is the velocity multiplied by the length of the interval.

3. If velocity is positive and increasing, using the velocity at the beginning of each subinterval gives an overestimate of the distance traveled.

4. A car traveling with increasing velocity (in feet per second) given in the following table travels more than 40 feet between $t = 4$ and $t = 6$ seconds.

t	4	6
$v(t)$	10	20

5. A car traveling with increasing velocity (in feet per second) given in the following table travels less than 40 feet between $t = 4$ and $t = 6$ seconds.

t	4	6
$v(t)$	10	20

6. If water flows out of a tank at an increasing rate starting at 20 gallons per minute at the start of a 10-minute period and ending at 50 gallons per minute at the end of the 10-minute period, then at least 200 gallons have flowed out of the tank.

7. If water flows out of a tank at an increasing rate starting at 20 gallons per minute at the start of a 10-minute period and ending at 50 gallons per minute at the end of the 10-minute period, then at least 500 gallons have flowed out of the tank.

8. At a construction site, money is spent at a continuous, decreasing rate of $r(t)$ dollars per day, where t is in days. If the rates at two particular instants, $t = 3$ and $t = 5$, are given in the table, then at least \$3000 is spent between $t = 3$ and $t = 5$.

t	3	5
$r(t)$	1000	800

9. An object traveling with velocity given by $v(t) = 3t$ feet per second travels exactly 150 feet when $0 \le t \le 10$ seconds.

10. The units of total change and rate of change of a quantity are the same.

11. If $f(t)$ is increasing, then the left-hand sum gives an overestimate of $\int_a^b f(t)\,dt$.

12. If $f(t)$ is concave up, then the left-hand sum gives an overestimate of $\int_a^b f(t)\,dt$.

13. A left-hand sum estimate for $\int_3^5 f(t)\,dt$ is 3000, if $f(t)$ is shown in the following table.

t	3	5
$f(t)$	1000	800

14. A right-hand sum estimate for $\int_3^5 f(t)\,dt$ is 1600, if $f(t)$ is shown in the following table.

t	3	5
$f(t)$	1000	800

15. If the number of terms in a Riemann sum increases, the quantity Δt decreases.

16. A 4-term Riemann sum on the interval $4 \le t \le 6$ has $\Delta t = 2$.

17. The right-hand sum for $\int_a^b f(t)\,dt$ is the total area of rectangles that are above the graph of $f(t)$.

18. If $f(t) \ge 0$ then $\int_a^b f(t)\,dt$ is the area between the t-axis and the graph of $f(t)$, for $a \le t \le b$.

19. If $f(t) = t^2$ then the left-hand sum gives an overestimate for $\int_{-2}^{-1} f(t)\,dt$.

20. The right-hand sum for $\int_5^{25} t^2\,dt$ using $n = 5$ rectangles has $\Delta t = 5$.

21. The integral $\int_0^2 f(x)\,dx$ gives the area between the x-axis and the graph of $f(x)$ between $x = 0$ and $x = 2$.

22. If $\int_0^2 f(x)\,dx = 0$ then $f(x) = 0$ for all $0 \le x \le 2$.

23. If $\int_0^2 f(x)\,dx > 0$ then $f(x) > 0$ for all $0 \le x \le 2$.

24. If $f(x) > 0$ for all $0 \le x \le 2$ then $\int_0^2 f(x)\,dx > 0$.

25. The integral $\int_{-1}^{1}(x^2 - 1)\,dx$ is the negative of the area between the graph of $f(x) = x^2 - 1$ and the x-axis for $-1 \le x \le 1$.

26. If $f(x)$ is both positive and negative when $0 \le x \le 2$ then $\int_0^2 f(x)\,dx$ is the sum of the areas between the graph and the x-axis for $0 \le x \le 2$.

27. The integral $\int_0^3 (1 - x)\,dx$ is the area between the graph of $f(x) = 1 - x$ and the x-axis between $x = 0$ and $x = 3$.

28. The integral $\int_{-1}^{2} e^x\,dx$ is the area between the graph of $f(x) = e^x$ and the x-axis between $x = -1$ and $x = 2$.

29. If $f(x) \ge g(x)$ for all $a \le x \le b$ then $\int_a^b f(x) - g(x)\,dx$ gives the area between the graphs of $f(x)$ and $g(x)$ for $a \le x \le b$.

30. If the graph of $f(x)$ has more area below the x-axis than above the x-axis when $1 \le x \le 10$ then $\int_1^{10} f(x)\,dx$ is negative.

31. If flow rate $r(t)$ has units gallons per hour, and t has units in hours, then $\int_0^{100} r(t)\,dt$ has units of gallons.

32. If expense rate $s(t)$ has units dollars per day, and t has units in days, then $\int_{20}^{35} s(t)\,dt$ has units of days.

33. If acceleration $a(t)$ has units feet/(second)^2, and t has units in seconds, then $\int_{2.1}^{7.2} a(t)\,dt$ has units of feet/second.

34. If $w(t)$ is the number of workers on a building construction site as a function of t, the number of days since the start of construction, then $\int_0^{60} w(t)\,dt$ has units of workers per day.

35. If $C(t)$ is the rate, in dollars per day, for electricity to cool and heat a home on day t, where t is measured in days since January 1, then $\int_0^{365} C(t)\,dt$ has units of dollars.

36. If electricity costs $f(t)$ dollars per day, and t is measured in days, then $\int_0^{30} f(t)\,dt$ gives the total cost of electricity, in dollars, for the 30-day period from $t = 0$ to $t = 30$.

37. If $f(t)$ gives the rate of change of water in a reservoir, in gallons per day, and t is measured in days, then $\int_0^{30} f(t)\,dt$ gives the total volume of water, in gallons, in the reservoir on day $t = 30$.

38. If $f(t)$ gives the rate of change of water in a reservoir, in gallons per day, and t is measured in days, then $\int_0^{30} f(t)\,dt$ gives the total change in the volume of water, in gallons, in the reservoir during the period $t = 0$ to $t = 30$.

39. Bioavailability is the integral of blood concentration as a function of time.

40. If the peak concentration of drug A is greater than the peak concentration of drug B then the bioavailability of drug A is greater than the bioavailability of drug B.

41. The Fundamental Theorem of Calculus says if $F(t)$ is continuous, then $\int_a^b F(t)\,dt = F(b) - F(a)$.

42. The Fundamental Theorem of Calculus says if $g'(t)$ is continuous, then $\int_a^b g'(t)\,dt = g(b) - g(a)$.

43. The Fundamental Theorem of Calculus says if $F'(t)$ is continuous, then $\int_a^b F'(t)\,dt = F'(b) - F'(a)$.

44. If $C(q)$ is the cost to produce q units, then $\int_{100}^{200} C'(q)\,dq$ is the amount it costs to increase production from 100 to 200 units.

45. If $C(q)$ is the cost to produce q units, then $\int_0^{1000} C'(q)\,dq$ is the total amount it costs to produce 1000 units.

46. If the marginal cost function of a product is $C'(q) = 0.03q + 0.1$ dollars, and the fixed costs are \$1500, then the total cost to produce 1000 items is \$15,100.

47. The integral of the marginal cost $C'(q)$ function can be used to find the fixed costs.

48. If $\int_0^5 F'(t)\,dt = 0$, then F is constant on the interval between $t = 0$ and $t = 5$.

49. If $\int_0^5 F'(t)\,dt < 0$, then $F(0) > F(5)$.

50. If $F'(t)$ is constant, then $\int_0^5 F'(t)\,dt = 0$.

PROJECTS FOR CHAPTER FIVE

1. **Carbon Dioxide in Pond Water** Biological activity in a pond is reflected in the rate at which carbon dioxide, CO_2, is added to or withdrawn from the water. Plants take CO_2 out of the water during the day for photosynthesis and put CO_2 into the water at night. Animals put CO_2 into the water all the time as they breathe. Biologists are interested in how the net rate at which CO_2 enters a pond varies during the day. Figure 5.76 shows this rate as a function of time of day.[9] The rate is measured in millimoles (mmol) of CO_2 per liter of water per hour; time is measured in hours past dawn. At dawn, there were 2.600 mmol of CO_2 per liter of water.

   (a) What can be concluded from the fact that the rate is negative during the day and positive at night?

   (b) Some scientists have suggested that plants respire (breathe) at a constant rate at night, and that they photosynthesize at a constant rate during the day. Does Figure 5.76 support this view?

   (c) When was the CO_2 content of the water at its lowest? How low did it go?

   (d) How much CO_2 was released into the water during the 12 hours of darkness? Compare this quantity with the amount of CO_2 withdrawn from the water during the 12 hours of daylight. How can you tell by looking at the graph whether the CO_2 in the pond is in equilibrium?

   (e) Estimate the CO_2 content of the water at three-hour intervals throughout the day. Use your estimates to plot a graph of CO_2 content throughout the day.

[9] Data from R. J. Beyers, *The Pattern of Photosynthesis and Respiration in Laboratory Microsystems* (Mem. 1st Ital. Idrobiol., 1965).

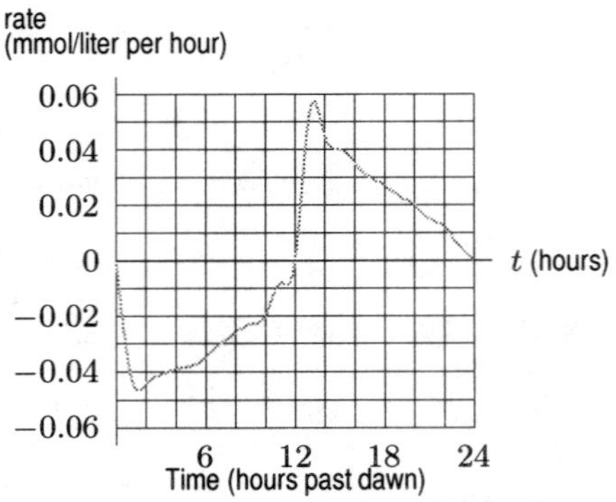

Figure 5.76: Rate at which CO_2 is entering the pond

2. Flooding in the Grand Canyon

The Glen Canyon Dam at the top of the Grand Canyon prevents natural flooding. In 1996, scientists decided an artificial flood was necessary to restore the environmental balance. Water was released through the dam at a controlled rate[10] shown in Figure 5.77. The figure also shows the rate of flow of the last natural flood in 1957.

(a) At what rate was water passing through the dam in 1996 before the artificial flood?

(b) At what rate was water passing down the river in the pre-flood season in 1957?

(c) Estimate the maximum rates of discharge for the 1996 and 1957 floods.

(d) Approximately how long did the 1996 flood last? How long did the 1957 flood last?

(e) Estimate how much additional water passed down the river in 1996 as a result of the artificial flood.

(f) Estimate how much additional water passed down the river in 1957 as a result of the flood.

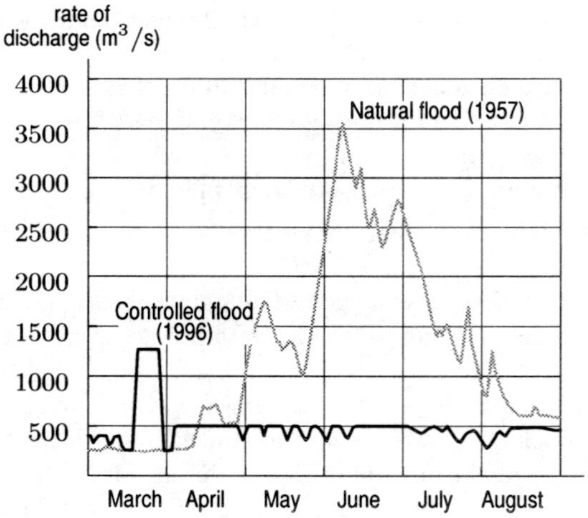

Figure 5.77

[10]Adapted from M. Collier, R. Webb, E. Andrews, "Experimental Flooding in Grand Canyon" in *Scientific American* (January 1997).

THEOREMS ABOUT DEFINITE INTEGRALS

The Second Fundamental Theorem of Calculus

The Fundamental Theorem of Calculus tells us that if we have a function F whose derivative is a continuous function f, then the definite integral of f is given by

$$\int_a^b f(t)\,dt = F(b) - F(a).$$

We now take a different point of view. If a is fixed and the upper limit is x, then the value of the integral is a function of x. We define a new function G on the interval by

$$G(x) = \int_a^x f(t)\,dt.$$

To visualize G, suppose that f is positive and $x > a$. Then $G(x)$ is the area under the graph of f in Figure 5.78. If f is continuous on an interval containing a, then it can be shown that G is defined for all x on that interval.

We now consider the derivative of G. Using the definition of the derivative,

$$G'(x) = \lim_{h\to 0} \frac{G(x+h) - G(x)}{h}.$$

Suppose f and h are positive. Then we can visualize

$$G(x) = \int_a^x f(t)\,dt$$

and

$$G(x+h) = \int_a^{x+h} f(t)\,dt$$

as areas, which leads to representing

$$G(x+h) - G(x) = \int_x^{x+h} f(t)\,dt$$

as a difference of two areas.

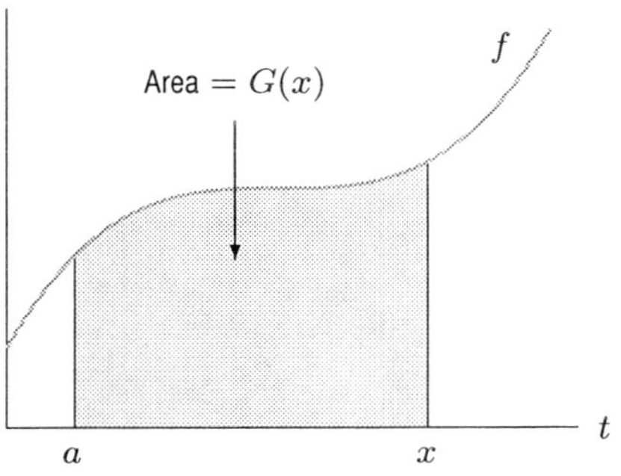

Figure 5.78: Representing $G(x)$ as an area

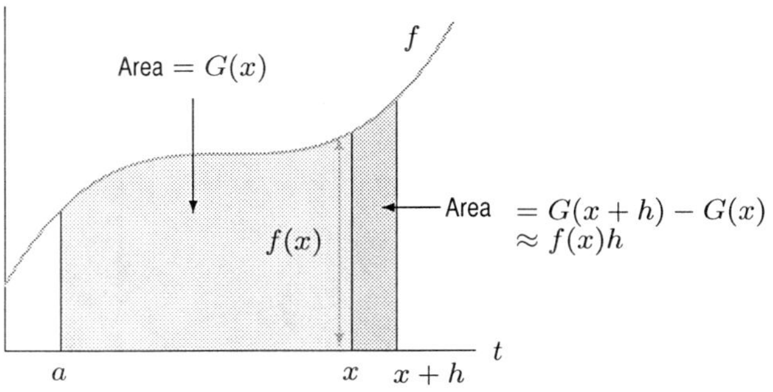

Figure 5.79: $G(x+h) - G(x)$ is the area of a roughly rectangular region

From Figure 5.79, we see that, if h is small, $G(x+h)-G(x)$ is roughly the area of a rectangle of height $f(x)$ and width h (shaded darker in Figure 5.79), so we have

$$G(x+h)-G(x)\approx f(x)h,$$

hence

$$\frac{G(x+h)-G(x)}{h}\approx f(x).$$

The same result holds when h is negative, suggesting that

$$G'(x)=\lim_{h\to 0}\frac{G(x+h)-G(x)}{h}=f(x).$$

This result is another form of the Fundamental Theorem of Calculus. It is usually stated as follows:

Second Fundamental Theorem of Calculus

If f is a continuous function on an interval, and if a is any number in that interval, then the function G defined on the interval by

$$G(x)=\int_a^x f(t)\,dt$$

has derivative f; that is, $G'(x)=f(x)$.

Properties of the Definite Integral

In this chapter, we have used the following properties to break up definite integrals.

Sums and Multiples of Definite Integrals

If a, b, and c are any numbers and f and g are continuous functions, then

1. $\displaystyle\int_a^c f(x)\,dx+\int_c^b f(x)\,dx=\int_a^b f(x)\,dx.$
2. $\displaystyle\int_a^b (f(x)\pm g(x))\,dx=\int_a^b f(x)\,dx\pm\int_a^b g(x)\,dx.$
3. $\displaystyle\int_a^b cf(x)\,dx=c\int_a^b f(x)\,dx.$

In words:

1. The integral from a to c plus the integral from c to b is the integral from a to b.
2. The integral of the sum (or difference) of two functions is the sum (or difference) of their integrals.
3. The integral of a constant times a function is that constant times the integral of the function.

These properties can best be visualized by thinking of the integrals as areas or as the limit of the sum of areas of rectangles.

Problems on the Second Fundamental Theorem of Calculus

For Problems 1–4, find $G'(x)$.

1. $G(x) = \int_a^x t^3\, dt$

2. $G(x) = \int_a^x 3^t\, dt$

3. $G(x) = \int_a^x te^t\, dt$

4. $G(x) = \int_a^x \ln y\, dy$

5. Let $F(b) = \int_0^b 2^x\, dx$.
   - **(a)** What is $F(0)$?
   - **(b)** Does the value of F increase or decrease as b increases? (Assume $b \geq 0$.)
   - **(c)** Estimate $F(1)$, $F(2)$, and $F(3)$.

6. For $x = 0, 0.5, 1.0, 1.5$, and 2.0, make a table of values for $I(x) = \int_0^x \sqrt{t^4+1}\, dt$.

7. Assume that $F'(t) = \sin t \cos t$ and $F(0) = 1$. Find $F(b)$ for $b = 0,\ 0.5,\ 1,\ 1.5,\ 2,\ 2.5$, and 3.

Let $\int_a^b f(x)\, dx = 8$, $\int_a^b (f(x))^2\, dx = 12$, $\int_a^b g(t)\, dt = 2$, and $\int_a^b (g(t))^2\, dt = 3$. Find the integrals in Problems 8–11.

8. $\int_a^b (f(x) + g(x))\, dx$

9. $\int_a^b \left((f(x))^2 - (g(x))^2\right) dx$

10. $\int_a^b (f(x))^2\, dx - \left(\int_a^b f(x)\, dx\right)^2$

11. $\int_a^b cf(z)\, dz$

Chapter Ten

MATHEMATICAL MODELING USING DIFFERENTIAL EQUATIONS

10.1 MATHEMATICAL MODELING: SETTING UP A DIFFERENTIAL EQUATION

Sometimes we do not know a key function, but we do have information about its rate of change, or its derivative. Then we may be able to write a new type of equation, called a *differential equation*, from which we can get information about the original function. For example, we may use what we know about the derivative of a population function (its rate of change) to predict the population in the future.

In this section, we use a verbal description to write a differential equation.

Marine Harvesting

We begin by investigating the effect of fishing on a fish population. Suppose that, left alone, a fish population increases at a continuous rate of 20% per year. Suppose that fish are also being harvested (caught) by fishermen at a constant rate of 10 million fish per year. How does the fish population change over time?

Notice that we have been given information about the rate of change, or derivative, of the fish population. Combined with information about the initial population, we can use this to predict the population in the future. We know that

$$\begin{matrix}\text{Rate of change} \\ \text{of fish population}\end{matrix} = \begin{matrix}\text{Rate of increase} \\ \text{due to breeding}\end{matrix} - \begin{matrix}\text{Rate fish removed} \\ \text{due to harvesting}\end{matrix}.$$

Suppose the fish population, in millions, is P and its derivative is dP/dt, where t is time in years. If left alone, the fish population increases at a continuous rate of 20% per year, so we have

$$\begin{aligned}\text{Rate of increase due to breeding} &= 20\% \cdot \text{Current population} \\ &= 0.20P \text{ million fish/year.}\end{aligned}$$

In addition,

$$\text{Rate fish removed by harvesting} = 10 \text{ million fish/year.}$$

Since the rate of change of the fish population is dP/dt, we have

$$\frac{dP}{dt} = 0.20P - 10.$$

This is a differential equation that models how the fish population changes. The unknown quantity in the equation is the function giving P in terms of t.

Net Worth of a Company

A company earns revenue (income) and also makes payroll payments. Assume that revenue is earned continuously, that payroll payments are made continuously, and that the only factors affecting net worth are revenue and payroll. The company's revenue is earned at a continuous annual rate of 5% times its net worth. At the same time, the company's payroll obligations are paid out at a constant rate of 200 million dollars a year.

We use this information to write a differential equation to model the net worth of the company, W, in millions of dollars, as a function of time, t, in years. We know that

$$\begin{matrix}\text{Rate at which} \\ \text{net worth is changing}\end{matrix} = \begin{matrix}\text{Rate revenue} \\ \text{is earned}\end{matrix} - \begin{matrix}\text{Rate payroll payments} \\ \text{are made}\end{matrix}.$$

Since the company's revenue is earned at a rate of 5% of its net worth, we have

$$\text{Rate revenue is earned} = 5\% \cdot \text{Net worth} = 0.05W \text{ million dollars/year.}$$

Since payroll payments are made at a rate of 200 million dollars a year, we have

$$\text{Rate payroll payments are made} = 200 \text{ million dollars/year.}$$

Putting these two together, since the rate at which net worth is changing is dW/dt, we have

$$\frac{dW}{dt} = 0.05W - 200.$$

This is a differential equation that models how the net worth of the company changes. The unknown quantity in the equation is the function giving net worth W as a function of time t.

Pollution in a Lake

If clean water flows into a polluted lake and a stream takes water out, the level of pollution in the lake will decrease (assuming no new pollutants are added).

Example 1 The quantity of pollutant in the lake decreases at a rate proportional to the quantity present. Write a differential equation to model the quantity of pollutant in the lake. Is the constant of proportionality positive or negative? Use the differential equation to explain why the graph of the quantity of pollutant against time is decreasing and concave up, as in Figure 10.1.

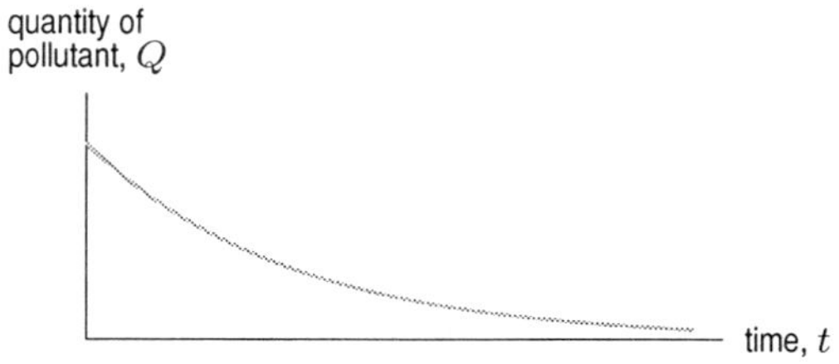

Figure 10.1: Quantity of pollutant in a lake

Solution Let Q denote the quantity of pollutant present in the lake at time t. The rate of change of Q is proportional to Q, so dQ/dt is proportional to Q. Thus, the differential equation is

$$\frac{dQ}{dt} = kQ.$$

Since no new pollutants are being added to the lake, the quantity Q is decreasing over time, so dQ/dt is negative. Thus, the constant of proportionality k is negative.

Why does the differential equation $dQ/dt = kQ$, with k negative, give us the graph shown in Figure 10.1? Since k is negative and Q is positive, we know kQ is negative. Thus, dQ/dt is negative, so the graph of Q against t is decreasing as in Figure 10.1. Why is it concave up? Since Q is getting smaller and k is fixed, as t increases, the product kQ is getting smaller in magnitude, and so the derivative dQ/dt is getting smaller in magnitude. Thus, the graph of Q is more horizontal as t increases. Therefore, the graph is concave up. See Figure 10.1.

The Quantity of a Drug in the Body

In the previous example, the rate at which pollutants leave a lake is proportional to the quantity of pollutants in the lake. This model works for any contaminants flowing in or out of a fluid system with complete mixing. Another example is the quantity of a drug in a person's body.

Example 2 A patient having major surgery is given the antibiotic vancomycin intravenously at a rate of 85 mg per hour. The rate at which the drug is excreted from the body is proportional to the quantity present, with proportionality constant 0.1 if time is in hours. Write a differential equation for the quantity, Q in mg, of vancomycin in the body after t hours.

Solution The quantity of vancomycin, Q, is increasing at a constant rate of 85 mg/hour and is decreasing at a rate of 0.1 times Q. The administration of 85 mg/hour makes a positive contribution to the rate of change dQ/dt. The excretion at a rate of $0.1Q$ makes a negative contribution to dQ/dt. Putting these together, we have

$$\text{Rate of change of a quantity} = \text{Rate in} - \text{Rate out},$$

so

$$\frac{dQ}{dt} = 85 - 0.1Q.$$

The Logistic Model

A population in a confined space grows proportionally to the product of the current population, P, and the difference between the *carrying capacity*, L, and the current population. (The carrying capacity is the maximum population the environment can sustain.) We use this information to write a differential equation for the population P.

The rate of change of P is proportional to the product of P and $L - P$, so

$$\frac{dP}{dt} = kP(L - P), \qquad \text{where } k \text{ is the constant of proportionality.}$$

This is called a *logistic differential equation.* What does it tell us about the graph of P? The derivative dP/dt is the product of k and P and $L - P$, so when P is small, the derivative dP/dt is small and the population grows slowly. As P increases, the derivative dP/dt increases and the population grows more rapidly. However, as P approaches the carrying capacity L, the term $L - P$ is small, and dP/dt is again small and the population grows more slowly. The *logistic growth curve* in Figure 10.2 satisfies these conditions.

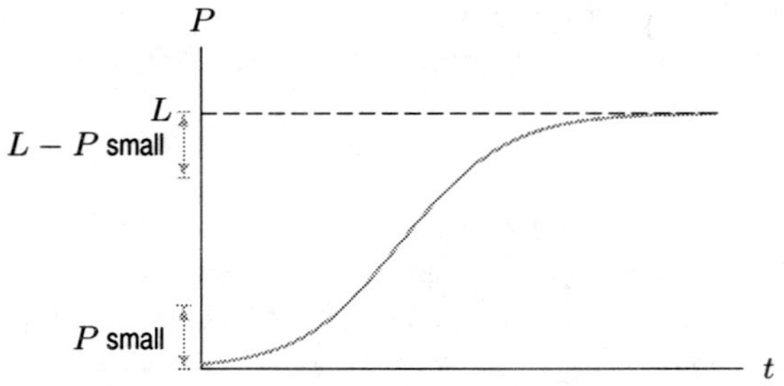

Figure 10.2: The logistic growth curve is a solution to $dP/dt = kP(L - P)$

Problems for Section 10.1

1. Match the graphs in Figure 10.3 with the following descriptions.

(a) The temperature of a glass of ice water left on the kitchen table.

(b) The amount of money in an interest-bearing bank account into which \$50 is deposited.

(c) The speed of a constantly decelerating car.

(d) The temperature of a piece of steel heated in a furnace and left outside to cool.

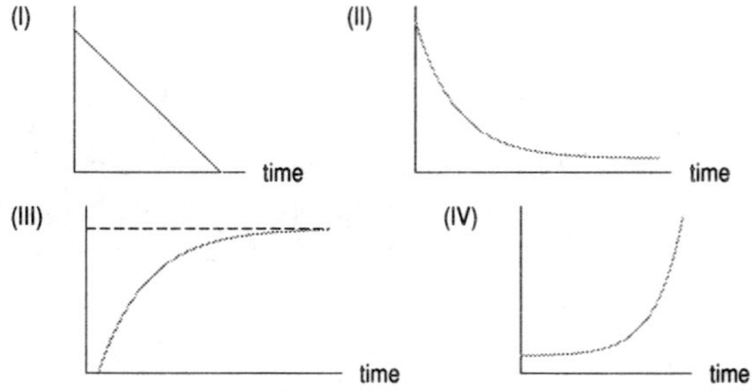

Figure 10.3

2. The graphs in Figure 10.4 represent the temperature, $H(°\text{C})$, of four eggs as a function of time, t, in minutes. Match three of the graphs with the descriptions (a)–(c). Write a similar description for the fourth graph, including an interpretation of any intercepts and asymptotes.

   (a) An egg is taken out of the refrigerator (just above $0°$C) and put into boiling water.
   (b) Twenty minutes after the egg in part (a) is taken out of the fridge and put into boiling water, the same thing is done with another egg.
   (c) An egg is taken out of the refrigerator at the same time as the egg in part (a) and left to sit on the kitchen table.

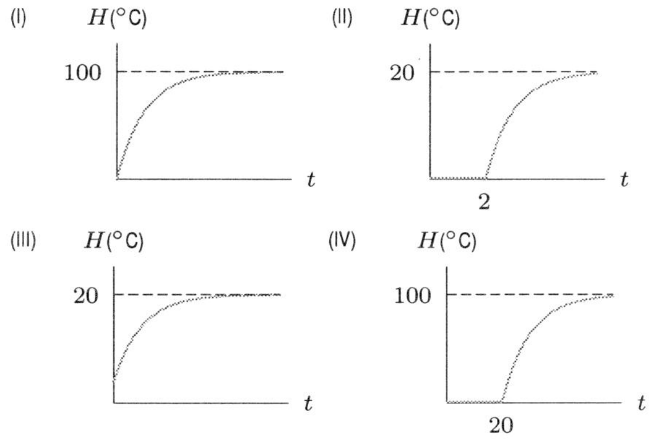

Figure 10.4

3. A population of insects grows at a rate proportional to the size of the population. Write a differential equation for the size of the population, P, as a function of time, t. Is the constant of proportionality positive or negative?

4. Money in a bank account earns interest at a continuous annual rate of 5% times the current balance. Write a differential equation for the balance, B, in the account as a function of time, t, in years.

5. Radioactive substances decay at a rate proportional to the quantity present. Write a differential equation for the quantity, Q, of a radioactive substance present at time t. Is the constant of proportionality positive or negative?

6. A bank account that initially contains \$25,000 earns interest at a continuous rate of 4% per year. Withdrawals are made out of the account at a constant rate of \$2000 per year. Write a differential equation for the balance, B, in the account as a function of the number of years, t.

7. A pollutant spilled on the ground decays at a rate of 8% a day. In addition, clean-up crews remove the pollutant at a rate of 30 gallons a day. Write a differential equation for the amount of pollutant, P, in gallons, left after t days.

8. Morphine is administered to a patient intravenously at a rate of 2.5 mg per hour. About 34.7% of the morphine is metabolized and leaves the body each hour. Write a differential equation for the amount of morphine, M, in milligrams, in the body as a function of time, t, in hours.

9. Alcohol is metabolized and excreted from the body at a rate of about one ounce of alcohol every hour. If some alcohol is consumed, write a differential equation for the amount of alcohol, A (in ounces), remaining in the body as a function of t, the number of hours since the alcohol was consumed.

10. Toxins in pesticides can get into the food chain and accumulate in the body. A person consumes 10 micrograms a day of a toxin, ingested throughout the day. The toxin leaves the body at a continuous rate of 3% every day. Write a differential equation for the amount of toxin, A, in micrograms, in the person's body as a function of the number of days, t.

11. A cup of coffee contains about 100 mg of caffeine. Caffeine is metabolized and leaves the body at a continuous rate of about 17% every hour.

    (a) Write a differential equation for the amount, A, of caffeine in the body as a function of the number of hours, t, since the coffee was consumed.
    (b) Use the differential equation to find dA/dt at the start of the first hour (right after the coffee is consumed). Use your answer to estimate the change in the amount of caffeine during the first hour.

12. A person deposits money into an account at a continuous rate of \$6000 a year, and the account earns interest at a continuous rate of 7% per year.

    (a) Write a differential equation for the balance in the account, B, in dollars, as a function of years, t.
    (b) Use the differential equation to calculate dB/dt if $B = 10{,}000$ and if $B = 100{,}000$. Interpret your answers.

13. A quantity W satisfies the differential equation

$$\frac{dW}{dt} = 5W - 20.$$

    (a) Is W increasing or decreasing at $W = 10$? $W = 2$?
    (b) For what values of W is the rate of change of W equal to zero?

14. A quantity y satisfies the differential equation

$$\frac{dy}{dt} = -0.5y.$$

    Under what conditions is y increasing? Decreasing?

15. An early model of the growth of the Wikipedia assumed that every day a constant number, B, of articles are added by dedicated wikipedians and that other articles are created by the general public at a rate proportional to the number of articles already there. Express this model as a differential equation for $N(t)$, the total number of Wikipedia articles t days after January 1, 2001.

16. A country's infrastructure is its transportation and communication systems, power plants, and other public institutions. The Solow model asserts that the value of national infrastructure K increases due to investment and decreases due to capital depreciation. The rate of increase due to investment is proportional to national income, Y. The rate of decrease due to depreciation is proportional to the value of existing infrastructure. Write a differential equation for K.

10.2 SOLUTIONS OF DIFFERENTIAL EQUATIONS

What does it mean to "solve" a differential equation? A differential equation is an equation involving the derivative of an unknown function. The unknown is not a number but a function. A *solution* to a differential equation is any function that satisfies the differential equation.

In this section, we see how to solve a differential equation numerically and how to check whether or not a function is a solution to a differential equation. In the next section, we see how to visualize a solution.

Another Look at Marine Harvesting

Let's take another look at the fish population discussed in Section 10.1. Left alone, the population increases at a continuous rate of 20% per year. The fish are being harvested at a constant rate of 10 million fish per year. If P is the fish population, in millions, in year t, then we have

$$\frac{dP}{dt} = 0.20P - 10.$$

Solving this differential equation means to find a function giving P in terms of t. Combined with information about the initial population, we can use the equation to predict the population at any time in the future.

Solving the Differential Equation Numerically

Suppose at time $t = 0$, the fish population is 60 million. We can substitute $P = 60$ into the differential equation to compute the derivative, dP/dt:

$$\text{At time } t = 0, \qquad \frac{dP}{dt} = 0.20P - 10 = 0.20(60) - 10 = 12 - 10 = 2.$$

Since at $t = 0$, the fish population is changing at a rate of 2 million fish a year, at the end of the first year, the fish population will have increased by about 2 million fish. So:

$$\text{At } t = 1, \qquad \text{we estimate} \quad P = 60 + 2 = 62.$$

We use this new value of P to estimate dP/dt during the second year:

$$\text{At time } t = 1, \qquad \frac{dP}{dt} = 0.20P - 10 = 12.4 - 10 = 2.4.$$

During the second year, the fish population increased by about 2.4 million fish, so:

$$\text{At } t = 2, \qquad \text{we estimate} \quad P = 62 + 2.4 = 64.4.$$

We use this value of P to estimate the rate of change during the third year, and so on. Continuing in this fashion, we compute the approximate values of P in Table 10.1. This table gives approximate numerical values for P at future times.

Table 10.1 *Approximate values of the fish population as a function of time*

t (years)	0	1	2	3	4	5	...
P (millions)	60	62	64.4	67.28	70.74	74.89	...

A Formula for the Solution to the Differential Equation

A function $P = f(t)$ which satisfies the differential equation

$$\frac{dP}{dt} = 0.20P - 10$$

is called a *solution* of the differential equation. Table 10.1 shows approximate numerical values of a solution. It is sometimes (but not always) possible to find a formula for the solution. In this particular case, there is a formula; it is

$$P = 50 + Ce^{0.20t}, \qquad \text{where } C \text{ is any constant.}$$

We check that this is a solution to the differential equation by substituting it into the left and right sides of the differential equation separately. We find

$$\begin{aligned} \text{Left side} &= \frac{dP}{dt} = 0.20Ce^{0.20t} \\ \text{Right side} = 0.20P - 10 &= 0.20(50 + Ce^{0.20t}) - 10 \\ &= 10 + 0.20Ce^{0.20t} - 10 \\ &= 0.20Ce^{0.20t}. \end{aligned}$$

Since we get the same expression on both sides, we say that $P = 50 + Ce^{0.20t}$ is a solution of this differential equation. Any choice of C works, so the solutions form a family of functions with parameter C. Several members of the family of solutions are graphed in Figure 10.5.

Finding the Arbitrary Constant: Initial Conditions

To find a value for the constant C—in other words, to select a single solution from the family of solutions—we need an additional piece of information, usually the initial population. In this case, we know that $P = 60$ when $t = 0$, so substituting into

$$P = 50 + Ce^{0.20t}$$

gives

$$\begin{aligned} 60 &= 50 + Ce^{0.20(0)} \\ 60 &= 50 + C \cdot 1 \\ C &= 10. \end{aligned}$$

The function $P = 50 + 10e^{0.20t}$ satisfies the differential equation *and* the initial condition that $P = 60$ when $t = 0$.

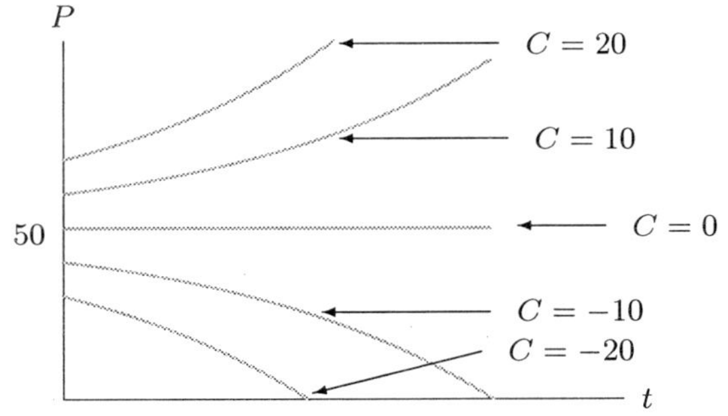

Figure 10.5: Solution curves for $dP/dt = 0.20P - 10$: Members of the family $P = 50 + Ce^{0.20t}$

General Solutions and Particular Solutions

For the differential equation $dP/dt = 0.20P - 10$, it can be shown that every solution is of the form $P = 50 + Ce^{0.20t}$ for some value of C. We say that the *general solution* of the differential equation $dP/dt = 0.20P - 10$ is the family of functions $P = 50 + Ce^{0.20t}$. The solution $P = 50 + 10e^{0.20t}$ that satisfies the differential equation together with the initial condition that $P = 60$ when $t = 0$ is calld a *particular solution*. The differential equation and the initial condition together are called an *initial-value problem*.

Example 1 (a) Check that $P = Ce^{2t}$ is a solution to the differential equation

$$\frac{dP}{dt} = 2P.$$

(b) Find the particular solution satisfying the initial condition $P = 100$ when $t = 0$.

Solution (a) Since $P = Ce^{2t}$ where C is a constant, we find expressions for each side:

$$\text{Left side} = \frac{dP}{dt} = Ce^{2t} \cdot 2 = 2Ce^{2t}$$
$$\text{Right side} = 2P = 2Ce^{2t}.$$

Since the two expressions are equal, $P = Ce^{2t}$ is a solution to the differential equation.

(b) We substitute $P = 100$ and $t = 0$ into the general solution $P = Ce^{2t}$, and solve for C:

$$100 = Ce^{2(0)}$$
$$100 = C \cdot 1$$
$$100 = C.$$

The particular solution for this initial-value problem is $P = 100e^{2t}$.

Example 2 Decide whether or not $y = e^{-2x}$ is a solution of the differential equation $y' - 2y = 0$.

Solution Note that $y' = dy/dx$. Differentiating $y = e^{-2x}$ gives $y' = -2e^{-2x}$. Substituting, we have

$$y' - 2y = -2e^{-2x} - 2e^{-2x} = -4e^{-2x} \neq 0,$$

and so $y = e^{-2x}$ is not a solution to this differential equation.

Example 3 (a) What conditions must be imposed on the constants C and k if $y = Ce^{kt}$ is a solution to the differential equation

$$\frac{dy}{dt} = -0.5y?$$

(b) What additional conditions must be imposed on C and k if $y = Ce^{kt}$ also satisfies the initial condition that $y = 10$ when $t = 0$?

Solution (a) If $y = Ce^{kt}$, then $dy/dt = Cke^{kt}$. Substituting into the equation $dy/dt = -0.5y$ gives

$$Cke^{kt} = -0.5(Ce^{kt}),$$

and therefore, assuming $C \neq 0$, so $Ce^{kt} \neq 0$, we have

$$k = -0.5.$$

So $y = Ce^{-0.5t}$ is a solution to the differential equation. If $C = 0$, then $Ce^{kt} = 0$ is a solution to the differential equation. No conditions are imposed on C.

(b) Since $k = -0.5$, we have $y = Ce^{-0.5t}$. Substituting $y = 10$ when $t = 0$ gives

$$10 = Ce^0$$

so

$$10 = C.$$

So $y = 10e^{-0.5t}$ is a solution to the differential equation together with the initial condition.

Problems for Section 10.2

1. Find the general solution to the differential equation

$$\frac{dy}{dt} = 2t.$$

2. Decide whether or not each of the following is a solution to the differential equation $xy' - 2y = 0$.

(a) $y = x^2$ **(b)** $y = x^3$

3. Check that $y = t^4$ is a solution to the differential equation $t\dfrac{dy}{dt} = 4y$.

In Problems 4–12, use the fact that the derivative gives the slope of a curve to decide which of the graphs (A)–(F) in Figure 10.6 could represent a solution to the differential equation.

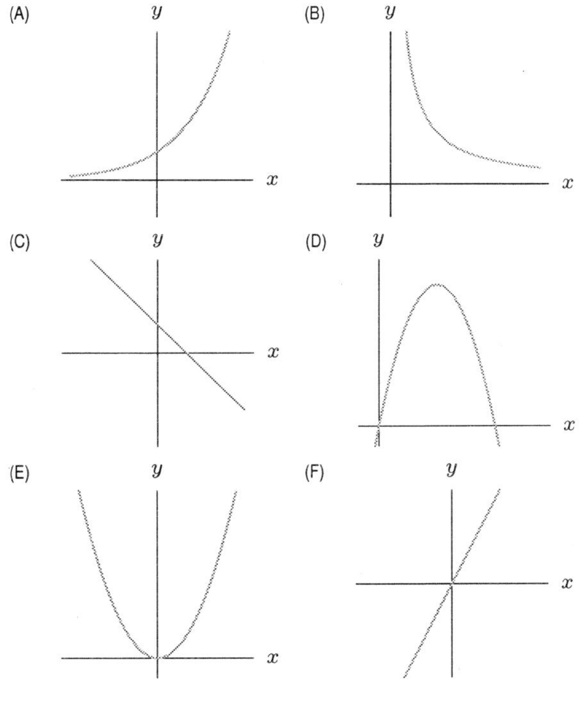

Figure 10.6

4. $\dfrac{dy}{dx} = -1$ **5.** $\dfrac{dy}{dx} = 0.1$ **6.** $\dfrac{dy}{dx} = -y^2$

7. $\dfrac{dy}{dx} = 2x$ **8.** $\dfrac{dy}{dx} = 2$ **9.** $\dfrac{dy}{dx} = y$

10. $\dfrac{dy}{dx} = -\dfrac{1}{x^2}$ **11.** $\dfrac{dy}{dx} = 1 - x$ **12.** $\dfrac{dy}{dx} = 2y$

13. Fill in the missing values in Table 10.2 given that $dy/dt = 0.5t$. Assume the rate of growth, given by dy/dt, is approximately constant over each unit time interval.

Table 10.2

t	0	1	2	3	4
y	8				

14. Fill in the missing values in Table 10.3 given that $dy/dt = 0.5y$. Assume the rate of growth, given by dy/dt, is approximately constant over each unit time interval.

Table 10.3

t	0	1	2	3	4
y	8				

15. Fill in the missing values in Table 10.4 given that $dy/dt = 4 - y$. Assume the rate of growth, given by dy/dt, is approximately constant over each unit time interval.

Table 10.4

t	0	1	2	3	4
y	8				

16. For a certain quantity y, assume that $dy/dt = \sqrt{y}$. Fill in the value of y in Table 10.5. Assume that the rate of growth, dy/dt, is approximately constant over each unit time interval and that the initial value of y is 100.

Table 10.5

t	0	1	2	3	4
y	100				

17. If the initial population of fish is 70 million, use the differential equation $dP/dt = 0.2P - 10$ to estimate the fish population after 1, 2, 3 years.

18. Show that, for any constant P_0, the function $P = P_0 e^t$ satisfies the equation

$$\frac{dP}{dt} = P.$$

19. Suppose $Q = Ce^{kt}$ satisfies the differential equation

$$\frac{dQ}{dt} = -0.03Q.$$

What (if anything) does this tell you about the values of C and k?

20. Is there a value of n which makes $y = x^n$ a solution to the equation $13x(dy/dx) = y$? If so, what value?

21. Find the values of k for which $y = x^2 + k$ is a solution to the differential equation $2y - xy' = 10$.

22. Match solutions and differential equations. (Note: Each equation may have more than one solution, or no solution.)

(a) $\frac{dy}{dx} = \frac{y}{x}$	(I) $y = x^3$
(b) $\frac{dy}{dx} = 3\frac{y}{x}$	(II) $y = 3x$
(c) $\frac{dy}{dx} = 3x$	(III) $y = e^{3x}$
(d) $\frac{dy}{dx} = y$	(IV) $y = 3e^x$
(e) $\frac{dy}{dx} = 3y$	(V) $y = x$

10.4 EXPONENTIAL GROWTH AND DECAY

What is a solution to the differential equation

$$\frac{dy}{dt} = y?$$

A solution is a function that is its own derivative. The function $y = e^t$ has this property, so $y = e^t$ is a solution. In fact, any multiple of e^t also has this property. The family of functions $y = Ce^t$ is the general solution to this differential equation. If k is a constant, the differential equation

$$\frac{dy}{dt} = ky,$$

is similar. This differential equation says that the rate of change of y is proportional to y. The constant k is the constant of proportionality. By substituting $y = Ce^{kt}$ into the differential equation, you can check that $y = Ce^{kt}$ is a solution. For another derivation of the solution, see the Focus on Theory section at the end of the chapter. We have the following result:

> The general solution to the differential equation $\frac{dy}{dt} = ky$ is
>
> $$y = Ce^{kt} \qquad \text{for any constant C.}$$
>
> - This is exponential growth for $k > 0$, and exponential decay for $k < 0$.
> - The constant C is the value of y when t is 0.

Graphs of solution curves for some $k > 0$ are in Figure 10.25. For $k < 0$, the graphs are reflected about the y-axis. See Figure 10.26.

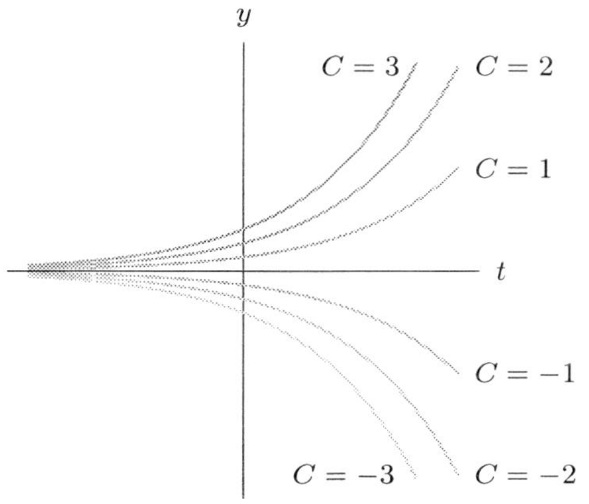

Figure 10.25: Graphs of $y = Ce^{kt}$, which are solutions to $\frac{dy}{dt} = ky$ for some fixed $k > 0$

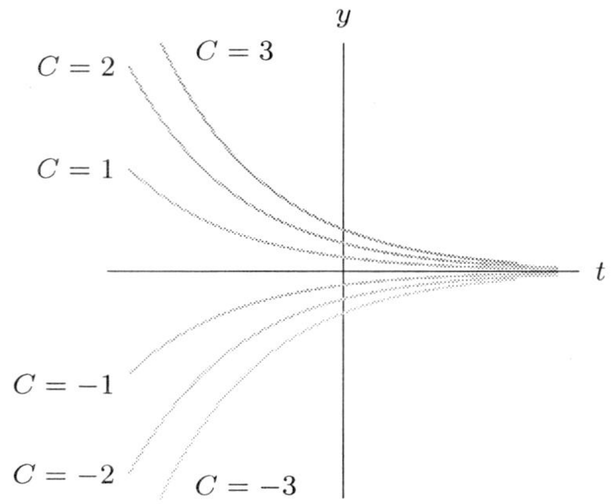

Figure 10.26: Graphs of $y = Ce^{kt}$, which are solutions to $\frac{dy}{dt} = ky$ for some fixed $k < 0$

Example 1 (a) Find the general solution to each of the following differential equations:

(i) $\frac{dy}{dt} = 0.05y$ (ii) $\frac{dP}{dt} = -0.3P$ (iii) $\frac{dw}{dz} = 2z$ (iv) $\frac{dw}{dz} = 2w$

(b) For differential equation (i), find the particular solution satisfying $y = 50$ when $t = 0$.

Solution (a) The differential equations given in (i), (ii), and (iv) are all examples of exponential growth or decay, since each is in the form

$$\text{Derivative} = \text{Constant} \cdot \text{Dependent variable.}$$

Notice that differential equation (iii) is not in this form. Example 1 on page 413 showed that the solution to (iii) is $w = z^2 + C$. The general solutions are

(i) $y = Ce^{0.05t}$ (ii) $P = Ce^{-0.3t}$ (iii) $w = z^2 + C$ (iv) $w = Ce^{2z}$

(b) The general solution to (i) is $y = Ce^{0.05t}$. Substituting $y = 50$ and $t = 0$ gives

$$50 = Ce^{0.05(0)}$$
$$50 = C \cdot 1.$$

So $C = 50$, and the particular solution to this initial-value problem is $y = 50e^{0.05t}$.

Population Growth

Consider the population P of a region where there is no immigration or emigration. The rate at which the population is growing is often proportional to the size of the population. This means larger populations grow faster, as we expect since there are more people to have babies. If the population has a continuous growth rate of 2% per unit time, then we know

$$\text{Rate of growth of population} = 2\% \text{ of Current population,}$$

so

$$\frac{dP}{dt} = 0.02P.$$

This equation is of the form $dP/dt = kP$ for $k = 0.02$ and has the general solution $P = Ce^{0.02t}$. If the initial population at time $t = 0$ is P_0, then $P_0 = Ce^{0.02(0)} = C$. So $C = P_0$ and we have

$$P = P_0 e^{0.02t}.$$

Continuously Compounded Interest

In Chapter 1 we introduced continuous compounding as the limiting case in which interest was added more and more often. Here we approach continuous compounding from a different point of view. We imagine interest being accrued at a rate proportional to the balance at that moment. Thus, the larger the balance, the faster interest is earned and the faster the balance grows.

Example 2 A bank account earns interest continuously at a rate of 5% of the current balance per year. Assume that the initial deposit is \$1000 and that no other deposits or withdrawals are made.

(a) Write a differential equation satisfied by the balance in the account.
(b) Solve the differential equation and graph the solution.

Solution

(a) We are looking for B, the balance in the account in dollars, as a function of t, time in years. Interest is being added continuously to the account at a rate of 5% of the balance at that moment, so

$$\text{Rate at which balance is increasing} = 5\% \text{ of Current balance.}$$

Thus, a differential equation that describes the process is

$$\frac{dB}{dt} = 0.05B.$$

Notice that it does not involve the \$1000, the initial condition, because the initial deposit does not affect the process by which interest is earned.

(b) Since $B_0 = 1000$ is the initial value of B, the solution to this differential equation is

$$B = B_0 e^{0.05t} = 1000e^{0.05t}.$$

This function is graphed in Figure 10.27.

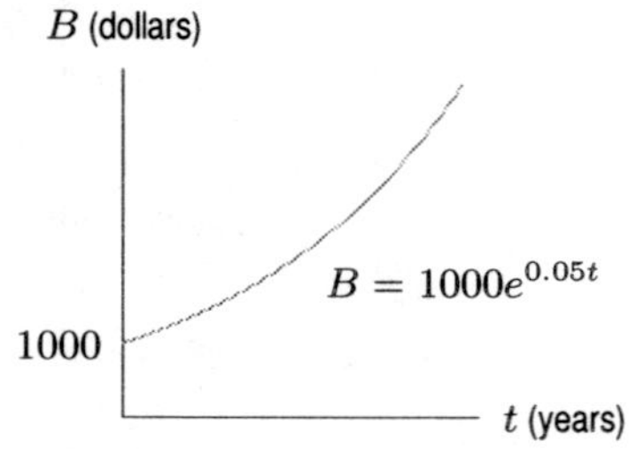

Figure 10.27: Bank balance against time

You may wonder how we can represent an amount of money by a differential equation, since money can only take on discrete values (you can't have fractions of a cent). In fact, the differential equation is only an approximation, but for large amounts of money, it is a pretty good approximation.

Pollution in the Great Lakes

In the 1960s pollution in the Great Lakes became an issue of public concern. We will set up a model for how long it would take the lakes to flush themselves clean, assuming no further pollutants were being dumped in the lake.

Let Q be the total quantity of pollutant in a lake of volume V at time t. Suppose that clean water is flowing into the lake at a constant rate r and that water flows out at the same rate. Assume that the pollutant is evenly spread throughout the lake, and that the clean water coming into the lake immediately mixes with the rest of the water.

How does Q vary with time? First, notice that since pollutants are being taken out of the lake but not added, Q decreases, and the water leaving the lake becomes less polluted, so the rate at which the pollutants leave decreases. This tells us that Q is decreasing and concave up. In addition, the pollutants will never be completely removed from the lake though the quantity remaining will become arbitrarily small. In other words, Q is asymptotic to the t-axis. (See Figure 10.28.)

Setting Up a Differential Equation for the Pollution

To understand how Q changes with time, we write a differential equation for Q. We know that

$$\begin{matrix}\text{Rate } Q \\ \text{changes}\end{matrix} = -\left(\begin{matrix}\text{Rate pollutants} \\ \text{leave in outflow}\end{matrix}\right)$$

where the negative sign represents the fact that Q is decreasing. At time t, the concentration of pollutants is Q/V and water containing this concentration is leaving at rate r. Thus,

$$\begin{matrix}\text{Rate pollutants} \\ \text{leave in outflow}\end{matrix} = \begin{matrix}\text{Rate of} \\ \text{outflow}\end{matrix} \times \text{Concentration} = r \cdot \frac{Q}{V}.$$

So the differential equation is

$$\frac{dQ}{dt} = -\frac{r}{V}Q$$

and its general solution is

$$Q = Q_0 e^{-rt/V}.$$

Table 10.6 contains values of r and V for four of the Great Lakes.[1] We use this data to calculate how long it would take for certain fractions of the pollution to be removed from Lake Erie.

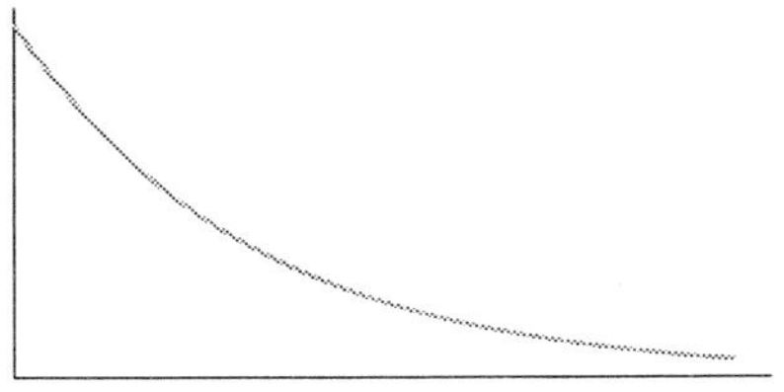

Figure 10.28: Pollutant in lake versus time

Table 10.6 *Volume and outflow in Great Lakes*

	V (km^3)	r (km^3/year)
Superior	12,200	65.2
Michigan	4900	158
Erie	460	175
Ontario	1600	209

Example 3 How long will it take for 90% of the pollution to be removed from Lake Erie? For 99% to be removed?

Solution For Lake Erie, $r/V = 175/460 = 0.38$, so at time t we have

$$Q = Q_0 e^{-0.38t}.$$

When 90% of the pollution has been removed, 10% remains, so $Q = 0.1Q_0$. Substituting gives

$$0.1Q_0 = Q_0 e^{-0.38t}.$$

[1] Data from William E. Boyce and Richard C. DiPrima, *Elementary Differential Equations* (New York: Wiley, 1977).

Canceling Q_0 and solving for t gives

$$t = \frac{-\ln(0.1)}{0.38} \approx 6 \text{ years.}$$

Similarly, when 99% of the pollution has been removed, $Q = 0.01Q_0$, so we solve

$$0.01Q_0 = Q_0 e^{-0.38t},$$

giving

$$t = \frac{-\ln(0.01)}{0.38} \approx 12 \text{ years.}$$

The Quantity of a Drug in the Body

As we saw in Section 10.1, the rate at which a drug leaves a patient's body is proportional to the quantity of the drug left in the body. If we let Q represent the quantity of drug left, then

$$\frac{dQ}{dt} = -kQ.$$

The negative sign indicates the quantity of drug in the body is decreasing. The solution to this differential equation is $Q = Q_0 e^{-kt}$; the quantity decreases exponentially. The constant k depends on the drug and Q_0 is the amount of drug in the body at time zero. Sometimes physicians convey information about the relative decay rate with a *half life*, which is the time it takes for Q to decrease by a factor of $1/2$.

Example 4 Valproic acid is a drug used to control epilepsy; its half-life in the human body is about 15 hours.

(a) Use the half-life to find the constant k in the differential equation $dQ/dt = -kQ$, where Q represents the quantity of drug in the body t hours after the drug is administered.
(b) At what time will 10% of the original dose remain?

Solution (a) Since the half-life is 15 hours, we know that the quantity remaining $Q = 0.5Q_0$ when $t = 15$. We substitute into the solution to the differential equation, $Q = Q_0 e^{-kt}$, and solve for k:

$$\begin{aligned} Q &= Q_0 e^{-kt} \\ 0.5Q_0 &= Q_0 e^{-k(15)} \\ 0.5 &= e^{-15k} && \text{(Divide through by } Q_0\text{)} \\ \ln 0.5 &= -15k && \text{(Take the natural logarithm of both sides)} \\ k &= \frac{-\ln 0.5}{15} = 0.0462. && \text{(Solve for } k\text{)} \end{aligned}$$

(b) To find the time when 10% of the original dose remains in the body, we substitute $0.10Q_0$ for the quantity remaining, Q, and solve for the time, t.

$$\begin{aligned} 0.10Q_0 &= Q_0 e^{-0.0462t} \\ 0.10 &= e^{-0.0462t} \\ \ln 0.10 &= -0.0462t \\ t &= \frac{\ln 0.10}{-0.0462} = 49.84. \end{aligned}$$

There will be 10% of the drug still in the body at $t = 49.84$, or after about 50 hours.

Problems for Section 10.4

Find solutions to the differential equations in Problems 1–6, subject to the given initial condition.

1. $\dfrac{dw}{dr} = 3w, \quad w = 30$ when $r = 0$

2. $\dfrac{dy}{dx} = -0.14y, \quad y = 5.6$ when $x = 0$

3. $\dfrac{dP}{dt} = 0.02P, \quad P(0) = 20$

4. $\dfrac{dp}{dq} = -0.1p, \quad p = 100$ when $q = 5$

5. $\dfrac{dQ}{dt} = \dfrac{Q}{5}, \quad Q = 50$ when $t = 0$

6. $\dfrac{dy}{dx} + \dfrac{y}{3} = 0, \quad y(0) = 10$

7. Money in a bank account grows continuously at an annual rate of r (when the interest rate is 5%, $r = 0.05$, and so on). Suppose \$1000 is put into the account in 2000.

   (a) Write a differential equation satisfied by M, the amount of money in the account at time t, measured in years since 2000.
   (b) Solve the differential equation.
   (c) Sketch the solution until the year 2030 for interest rates of 5% and 10%.

8. A deposit of \$5000 is made to a bank account paying 1.5% annual interest, compounded continuously.

   (a) Write a differential equation for the balance in the account, B, as a function of time, t, in years.
   (b) Solve the differential equation.
   (c) How much money is in the account in 10 years?

9. A bank account that earns 10% interest compounded continuously has an initial balance of zero. Money is deposited into the account at a continuous rate of \$1000 per year.

   (a) Write a differential equation that describes the rate of change of the balance $B = f(t)$.
   (b) Solve the differential equation.

10. The amount of ozone, Q, in the atmosphere is decreasing at a rate proportional to the amount of ozone present. If time t is measured in years, the constant of proportionality is -0.0025. Write a differential equation for Q as a function of t, and give the general solution for the differential equation. If this rate continues, approximately what percent of the ozone in the atmosphere now will decay in the next 20 years?

11. Using the model in the text and the data in Table 10.6 on page 420, find how long it would take for 90% of the pollution to be removed from Lake Michigan and from Lake Ontario, assuming no new pollutants are added. Explain how you can tell which lake will take longer to be purified just by looking at the data in the table.

12. Use the model in the text and the data in Table 10.6 on page 420 to determine which of the Great Lakes would require the longest time and which would require the shortest time for 80% of the pollution to be removed, assuming no new pollutants are being added. Find the ratio of these two times.

13. The rate at which a drug leaves the bloodstream and passes into the urine is proportional to the quantity of the drug in the blood at that time. If an initial dose of Q_0 is injected directly into the blood, 20% is left in the blood after 3 hours.

    (a) Write and solve a differential equation for the quantity, Q, of the drug in the blood after t hours.
    (b) How much of this drug is in a patient's body after 6 hours if the patient is given 100 mg initially?

14. Oil is pumped continuously from a well at a rate proportional to the amount of oil left in the well. Initially there were 1 million barrels of oil in the well; six years later 500,000 barrels remain.

    (a) At what rate was the amount of oil in the well decreasing when there were 600,000 barrels remaining?
    (b) When will there be 50,000 barrels remaining?

15. In some chemical reactions, the rate at which the amount of a substance changes with time is proportional to the amount present. For example, this is the case as δ-glucono-lactone changes into gluconic acid.

    (a) Write a differential equation satisfied by y, the quantity of δ-glucono-lactone present at time t.
    (b) If 100 grams of δ-glucono-lactone is reduced to 54.9 grams in one hour, how many grams will remain after 10 hours?

16. Hydrocodone bitartrate is used as a cough suppressant. After the drug is fully absorbed, the quantity of drug in the body decreases at a rate proportional to the amount left in the body. The half-life of hydrocodone bitartrate in the body is 3.8 hours and the dose is 10 mg.

    (a) Write a differential equation for the quantity, Q, of hydrocodone bitartrate in the body at time t, in hours since the drug was fully absorbed.
    (b) Solve the differential equation given in part (a).
    (c) Use the half-life to find the constant of proportionality, k.
    (d) How much of the 10-mg dose is still in the body after 12 hours?

17. The amount of land in use for growing crops increases as the world's population increases. Suppose $A(t)$ represents the total number of hectares of land in use in year t. (A hectare is about $2\frac{1}{2}$ acres.)

    (a) Explain why it is plausible that $A(t)$ satisfies the equation $A'(t) = kA(t)$. What assumptions are you making about the world's population and its relation to the amount of land used?

(b) In 1950 about $1 \cdot 10^9$ hectares of land were in use; in 1980 the figure was $2 \cdot 10^9$. If the total amount of land available for growing crops is thought to be $3.2 \cdot 10^9$ hectares, when does this model predict it is exhausted? (Let $t = 0$ in 1950.)

Chapter Six

USING THE DEFINITE INTEGRAL

6.1 AVERAGE VALUE

In this section we show how to interpret the definite integral as the average value of a function.

The Definite Integral as an Average

We know how to find the average of n numbers: Add them and divide by n. But how do we find the average value of a continuously varying function? Let us consider an example. Suppose $f(t)$ is the temperature at time t, measured in hours since midnight, and that we want to calculate the average temperature over a 24-hour period. One way to start would be to average the temperatures at n equally spaced times, $t_1, t_2, \ldots, t_n$, during the day.

$$\text{Average temperature} \approx \frac{f(t_1) + f(t_2) + \cdots + f(t_n)}{n}.$$

The larger we make n, the better the approximation. We can rewrite this expression as a Riemann sum over the interval $0 \le t \le 24$ if we use the fact that $\Delta t = 24/n$, so $n = 24/\Delta t$:

$$\begin{aligned}\text{Average temperature} &\approx \frac{f(t_1) + f(t_2) + \cdots + f(t_n)}{24/\Delta t} \\ &= \frac{f(t_1)\Delta t + f(t_2)\Delta t + \cdots + f(t_n)\Delta t}{24} \\ &= \frac{1}{24}\sum_{i=1}^{n} f(t_i)\Delta t.\end{aligned}$$

As $n \to \infty$, the Riemann sum tends toward an integral, and the approximation gets better. We expect that

$$\begin{aligned}\text{Average temperature} &= \lim_{n\to\infty} \frac{1}{24}\sum_{i=1}^{n} f(t_i)\Delta t \\ &= \frac{1}{24}\int_0^{24} f(t)\,dt.\end{aligned}$$

Generalizing for any function f, if $a < b$, we have

$$\begin{array}{c}\text{Average value of } f \\ \text{on the interval from } a \text{ to } b\end{array} = \frac{1}{b-a}\int_a^b f(x)\,dx.$$

The units of $f(x)$ are the same as the units of the average value of $f(x)$.

How to Visualize the Average on a Graph

The definition of average value tells us that

$$(\text{Average value of } f)\cdot(b-a) = \int_a^b f(x)\,dx.$$

Let's interpret the integral as the area under the graph of f. If $f(x)$ is positive, then the average value of f is the height of a rectangle whose base is $(b-a)$ and whose area is the same as the area between the graph of f and the x-axis. (See Figure 6.1.)

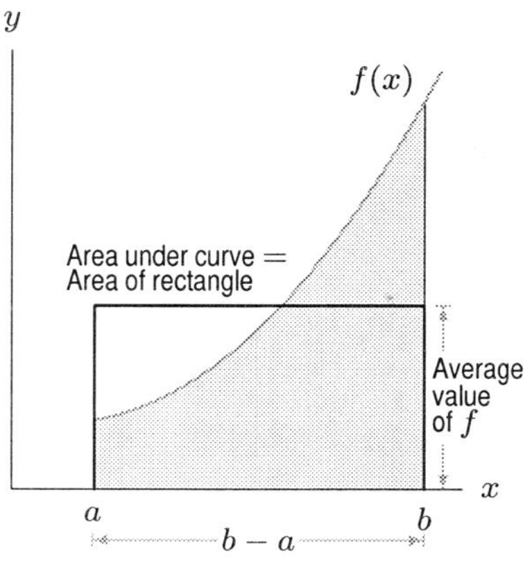

Figure 6.1: Area and average value

Example 1 Suppose that $C(t)$ represents the daily cost of heating your house, in dollars per day, where t is time in days and $t = 0$ corresponds to January 1, 2010. Interpret $\dfrac{1}{90-0}\displaystyle\int_0^{90} C(t)\,dt$.

Solution The units for the integral $\int_0^{90} C(t)\,dt$ are (dollars/day)×(days) = dollars. The integral represents the total cost in dollars to heat your house for the first 90 days of 2010, namely the months of January, February, and March. The expression $\frac{1}{90-0}\int_0^{90} C(t)\,dt$ represents the average cost per day to heat your house during the first 90 days of 2010. It is measured in (1/days) × (dollars) = dollars/day, the same units as $C(t)$.

Example 2 The population of McAllen, Texas can be modeled by the function

$$P = f(t) = 570(1.037)^t,$$

where P is in thousands of people and t is in years since 2000. Use this function to predict the average population of McAllen between the years 2020 and 2040.

Solution We want the average value of $f(t)$ between $t = 20$ and $t = 40$. Using a calculator to evaluate the integral, we get

$$\text{Average population} = \frac{1}{40-20}\int_{20}^{40} f(t)\,dt = \frac{1}{20}(34{,}656.2) = 1732.81.$$

The average population of McAllen between 2020 and 2040 is predicted to be about 1733 thousand people.

Example 3 (a) For the function $f(x)$ graphed in Figure 6.2, evaluate $\displaystyle\int_0^5 f(x)\,dx$.

(b) Find the average value of $f(x)$ on the interval $x = 0$ to $x = 5$. Check your answer graphically.

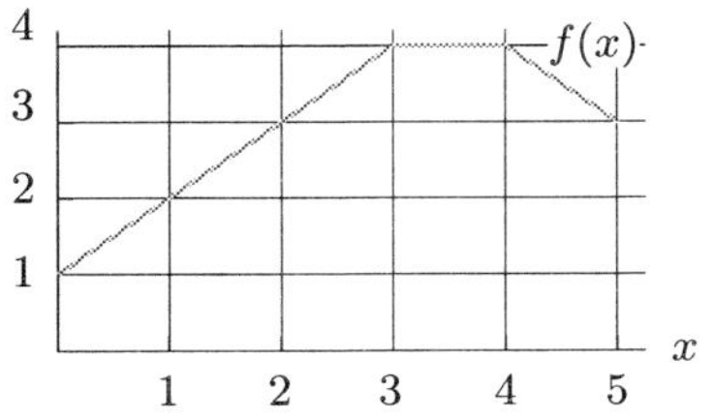

Figure 6.2: Estimate $\int_0^5 f(x)dx$

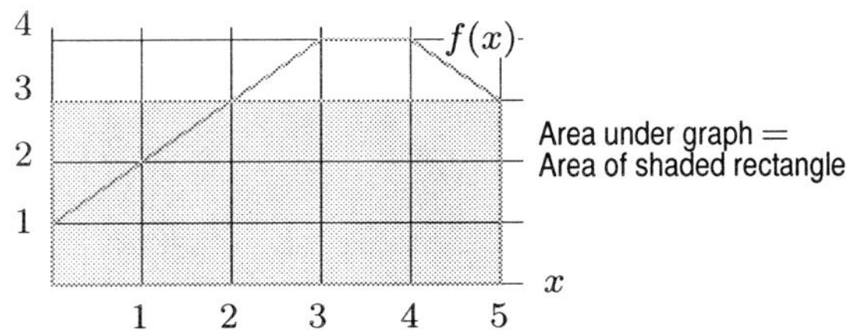

Figure 6.3: Average value of $f(x)$ is 3

Solution (a) Since $f(x) \geq 0$, the definite integral is the area of the region under the graph of $f(x)$ between $x = 0$ and $x = 5$. Figure 6.2 shows that this region consists of 13 full grid squares and 4 half grid squares, each grid square of area 1, for a total area of 15, so

$$\int_0^5 f(x)\,dx = 15.$$

(b) The average value of $f(x)$ on the interval from 0 to 5 is given by

$$\text{Average value} = \frac{1}{5-0}\int_0^5 f(x)\,dx = \frac{1}{5}(15) = 3.$$

To check the answer graphically, draw a horizontal line at $y = 3$ on the graph of $f(x)$. (See Figure 6.3.) Then observe that, between $x = 0$ and $x = 5$, the area under the graph of $f(x)$ is equal to the area of the rectangle with height 3.

Problems for Section 6.1

1. What is the average value of the function f in Figure 6.4 over the interval $1 \leq x \leq 6$?

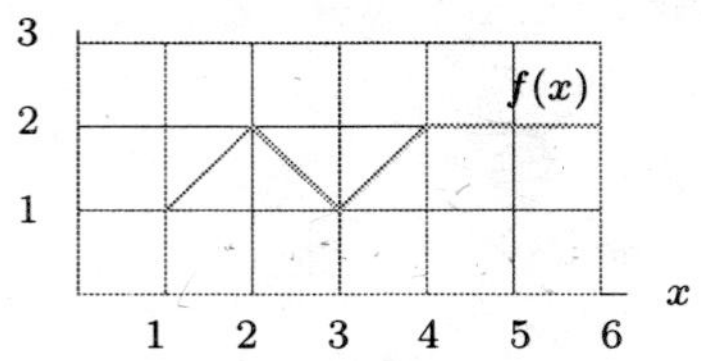

Figure 6.4

2. Use Figure 6.5 to estimate the following:

(a) The integral $\int_0^5 f(x)\,dx$.
(b) The average value of f between $x = 0$ and $x = 5$ by estimating visually the average height.
(c) The average value of f between $x = 0$ and $x = 5$ by using your answer to part (a).

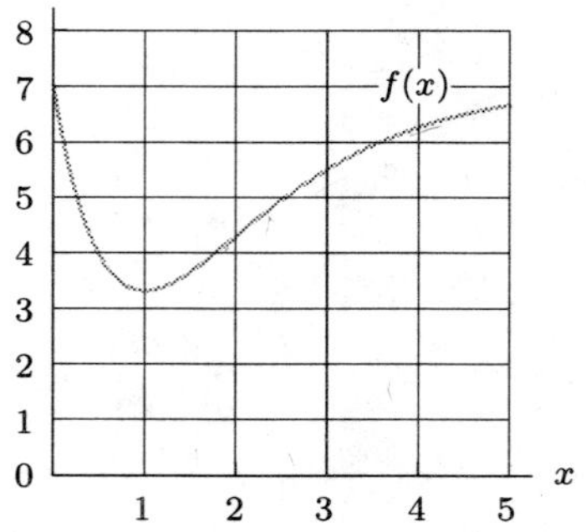

Figure 6.5

3. Find the average value of $g(t) = 1 + t$ over the interval $[0, 2]$

4. Find the average value of the function $f(x) = 5+4x-x^2$ between $x = 0$ and $x = 3$.

In Problems 5–6, estimate the average value of the function between $x = 0$ and $x = 7$.

5.

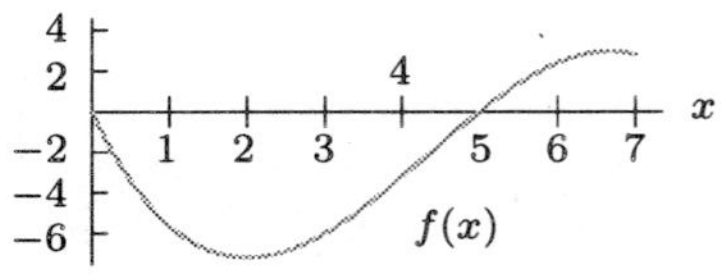

6.

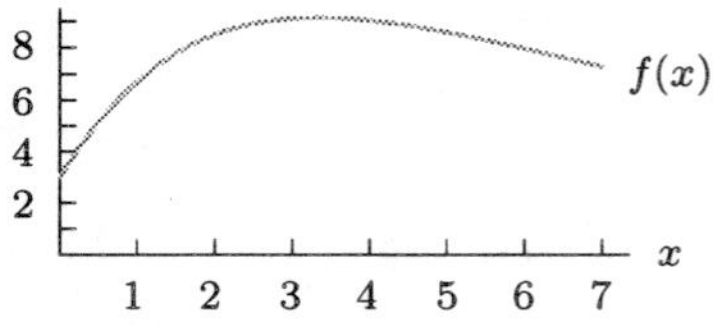

In Problems 7–8, estimate the average value of the function $f(x)$ on the interval from $x = a$ to $x = b$.

7.

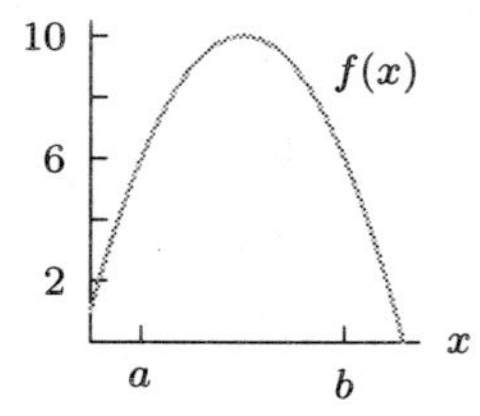

8.

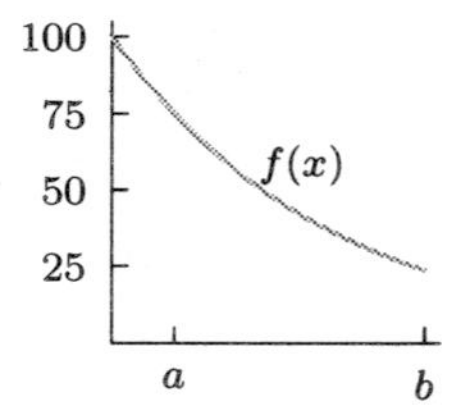

In Problems 9–10 annual income for ages 25 to 85 is given graphically. People sometimes spend less than their income (to save for retirement) or more than their income (taking out a loan). The process of spreading out spending over a lifetime is called consumption smoothing.

(a) Find the average annual income for these years.
(b) Assuming that the person spends at a constant rate equal to their average income, when are they spending less than they earn, and when are they spending more?

9.

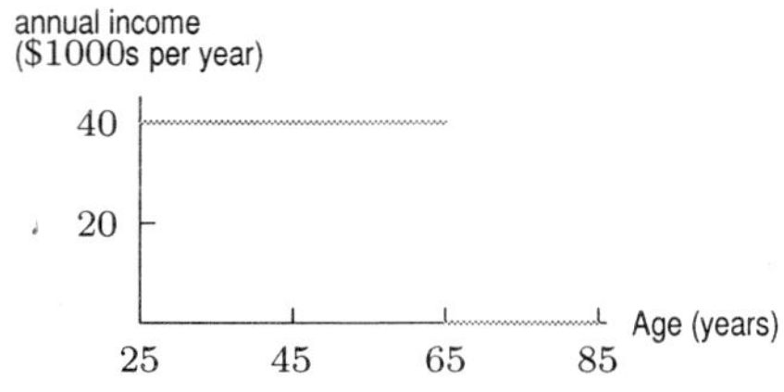

10.

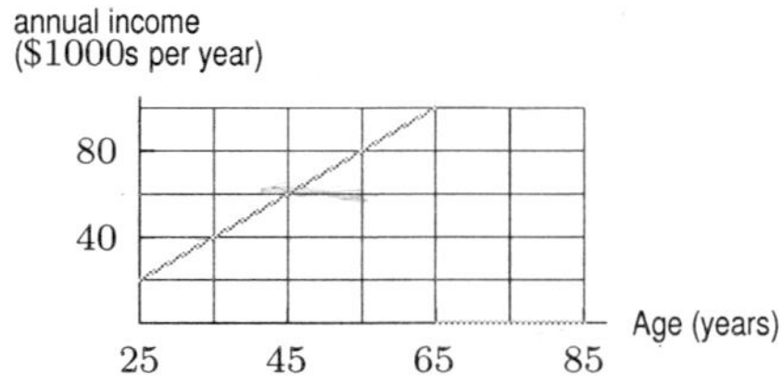

11. The value, V, of a Tiffany lamp, worth \$225 in 1975, increases at 15% per year. Its value in dollars t years after 1975 is given by

$$V = 225(1.15)^t.$$

Find the average value of the lamp over the period 1975–2010.

12. If t is measured in days since June 1, the inventory $I(t)$ for an item in a warehouse is given by

$$I(t) = 5000(0.9)^t.$$

(a) Find the average inventory in the warehouse during the 90 days after June 1.
(b) Graph $I(t)$ and illustrate the average graphically.

Problems 13–14 refer to Figure 6.6, which shows human arterial blood pressure during the course of one heartbeat.

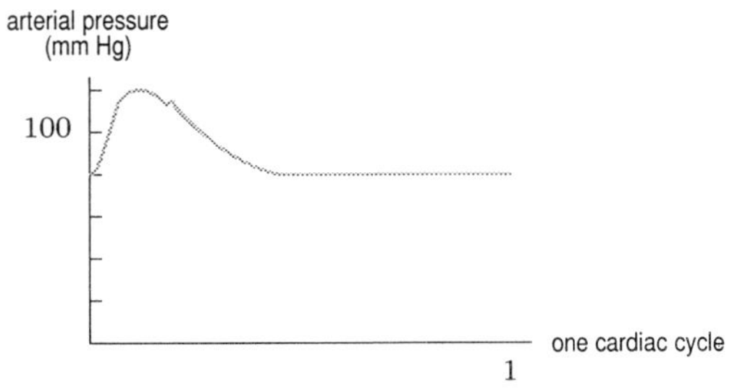

Figure 6.6

13. (a) Estimate the maximum blood pressure, called the systolic pressure.
(b) Estimate the minimum blood pressure, called the diastolic pressure.
(c) Calculate the average of the systolic and diastolic pressures.
(d) Is the average arterial pressure over the entire cycle greater than, less than, or equal to the answer for part (c)?

14. Estimate the average arterial blood pressure over one cardiac cycle.

15. Figure 6.7 shows the rate, $f(x)$, in thousands of algae per hour, at which a population of algae is growing, where x is in hours.

(a) Estimate the average value of the rate over the interval $x = -1$ to $x = 3$.
(b) Estimate the total change in the population over the interval $x = -3$ to $x = 3$.

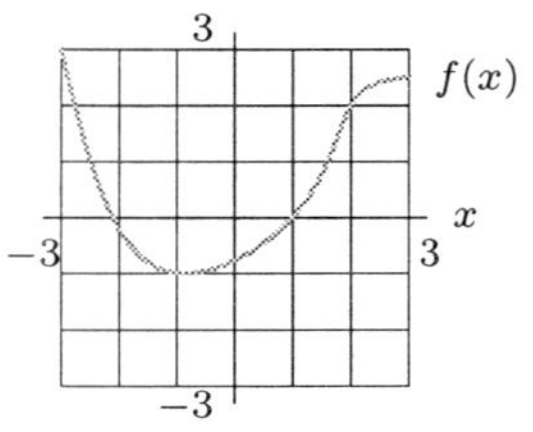

Figure 6.7

16. The population of the world t years after 2000 is predicted to be $P = 6.1e^{0.0125t}$ billion.

(a) What population is predicted in 2010?
(b) What is the predicted average population between 2000 and 2010?

17. The number of hours, H, of daylight in Madrid as a function of date is approximated by the formula

$$H = 12 + 2.4\sin[0.0172(t - 80)],$$

where t is the number of days since the start of the year. Find the average number of hours of daylight in Madrid:

(a) in January (b) in June (c) over a year

(d) Explain why the relative magnitudes of your answers to parts (a), (b), and (c) are reasonable.

18. A bar of metal is cooling from 1000°C to room temperature, 20°C. The temperature, H, of the bar t minutes after it starts cooling is given, in °C, by

$$H = 20 + 980e^{-0.1t}.$$

(a) Find the temperature of the bar at the end of one hour.

(b) Find the average value of the temperature over the first hour.

(c) Is your answer to part (b) greater or smaller than the average of the temperatures at the beginning and the end of the hour? Explain this in terms of the concavity of the graph of H.

19. The rate of sales (in sales per month) of a company is given, for t in months since January 1, by

$$r(t) = t^4 - 20t^3 + 118t^2 - 180t + 200.$$

(a) Graph the rate of sales per month during the first year ($t = 0$ to $t = 12$). Does it appear that more sales were made during the first half of the year, or during the second half?

(b) Estimate the total sales during the first 6 months of the year and during the last 6 months of the year.

(c) What are the total sales for the entire year?

(d) Find the average sales per month during the year.

20. Throughout much of the 20th century, the yearly consumption of electricity in the US increased exponentially at a continuous rate of 7% per year. Assume this trend continues and that the electrical energy consumed in 1900 was 1.4 million megawatt-hours.

(a) Write an expression for yearly electricity consumption as a function of time, t, in years since 1900.

(b) Find the average yearly electrical consumption throughout the 20th century.

(c) During what year was electrical consumption closest to the average for the century?

(d) Without doing the calculation for part (c), how could you have predicted which half of the century the answer would be in?

21. Using Figure 6.8, list the following numbers from least to greatest:

(a) $f'(1)$

(b) The average value of f on $0 \le x \le 4$

(c) $\int_0^1 f(x)dx$

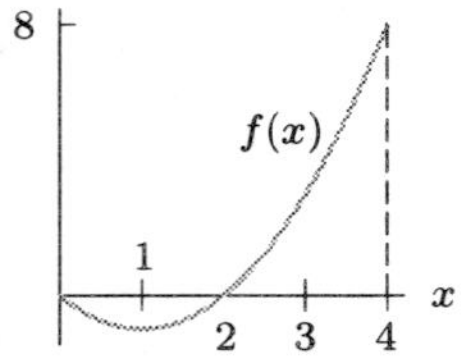

Figure 6.8

22. Using Figure 6.9, list from least to greatest,

(a) $f'(1)$.

(b) The average value of $f(x)$ on $0 \le x \le a$.

(c) The average value of the rate of change of $f(x)$, for $0 \le x \le a$.

(d) $\int_0^a f(x)\, dx$.

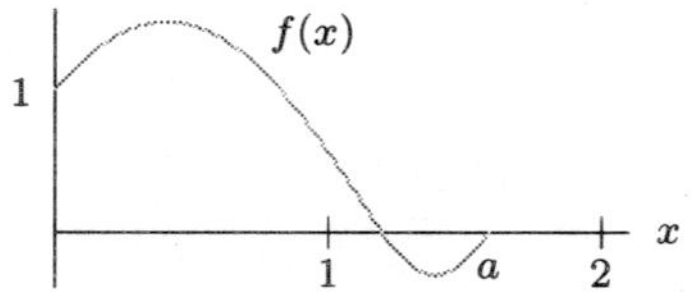

Figure 6.9

6.4 INTEGRATING RELATIVE GROWTH RATES

Population Growth Rates

In Chapter 5, we saw how to calculate change in a population, P, from its derivative, dP/dt, using the Fundamental Theorem of Calculus. However, population growth rates are often given by the relative rate of change $(1/P)(dP/dt)$, rather than by the derivative. For example, we saw that the population of Nevada is growing at 3.6% per year.

We can use the relative growth rate to find the percentage change in a population. Recall that

$$\text{Relative growth rate} = \frac{1}{P}\frac{dP}{dt} = \frac{d}{dt}(\ln P).$$

By the Fundamental Theorem of Calculus, the integral of the relative growth rate gives the total change in $\ln(P)$:

$$\int_a^b \frac{P'(t)}{P(t)}\, dt = \int_a^b \frac{d}{dt}(\ln P(t))\, dt = \ln(P(b)) - \ln(P(a)) = \ln\left(\frac{P(b)}{P(a)}\right).$$

[2]www.intel.com, accessed May 28, 2005.

[3]2007 Harley-Davidson Annual Report, accessed from www.harley-davidson.com.

[4]McDonald's Annual Report 2007 accessed at www.mcdonalds.com.

Example 1 The relative rate of growth $P'(t)/P(t)$ of a population $P(t)$ over a 50-year period is given in Figure 6.28. By what factor did the population increase during the period?

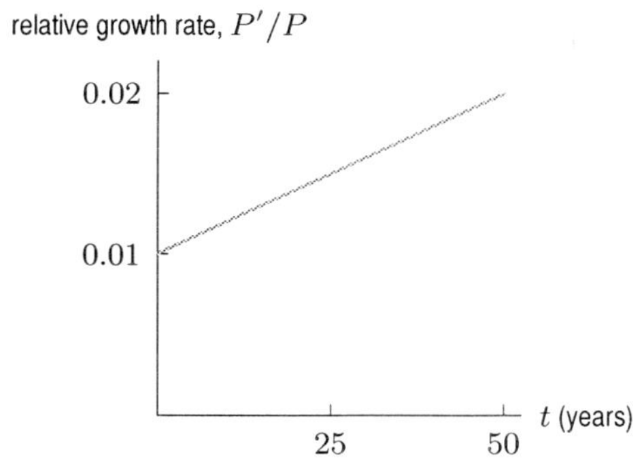

Figure 6.28: The relative growth rate of population

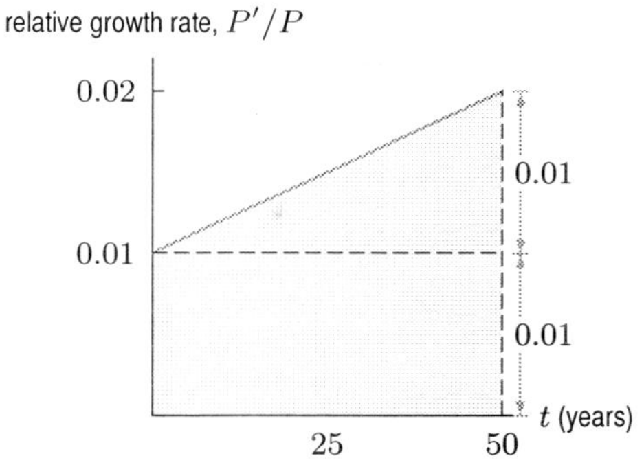

Figure 6.29: Calculating the percentage change in a population from the relative growth rate

Solution We have

$$\ln\left(\frac{P(50)}{P(0)}\right) = \int_0^{50} \frac{P'(t)}{P(t)}\,dt.$$

This integral equals the area under the graph of $P'(t)/P(t)$ between $t = 0$ and $t = 50$. See Figure 6.29. The area of a rectangle and triangle gives

$$\text{Area} = 50(0.01) + \frac{1}{2} \cdot 50(0.01) = 0.75.$$

Thus,

$$\ln\left(\frac{P(50)}{P(0)}\right) = 0.75$$
$$\frac{P(50)}{P(0)} = e^{0.75} = 2.1$$
$$P(50) = 2.1P(0).$$

The population more than doubled during the 50 years, increasing by a factor of about 2.1. We cannot determine the amount by which the population increased unless we know how large the population was initially.

If everything else remains constant, the relative growth rate increases if either the relative birth rate increases or the relative death rate decreases. Even if the relative birth rate decreases, we can still see an increase in the relative growth rate if the relative death rate decreases faster. This is the case with the population of the world today. The difference between the birth rate and the death rate is an important variable.

Example 2 Figure 6.30 shows relative birth rates and relative death rates for developed and developing countries.[5]

(a) Which is changing faster, the birth rate or the death rate? What does this tell you about how the populations are changing?

[5] *Food and Population: A Global Concern*, Elaine Murphy, Washington, DC: Population Reference Bureau, Inc., 1984, p. 2.

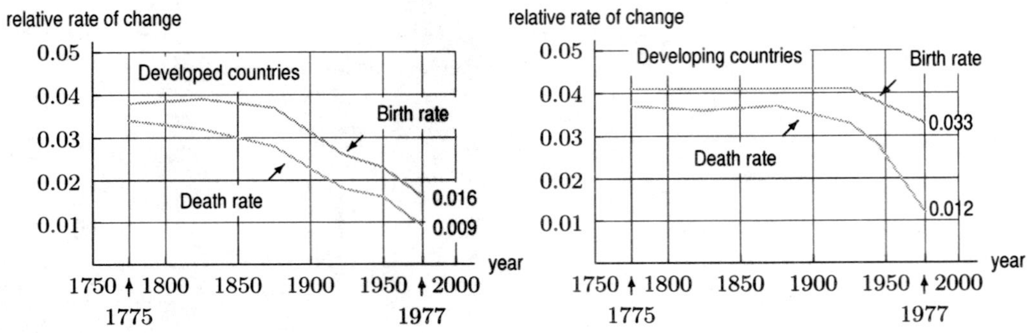

Figure 6.30: Birth and death rates in developed and developing countries, 1775-1977

(b) By what percentage did the population of developing countries increase between 1800 and 1900?

(c) By what percentage did the population of developing countries increase between 1950 and 1977?

Solution

(a) In developed countries, the birth rate is higher than the death rate, so the population is increasing. The birth rate and the death rate are both decreasing, and at approximately the same rate, so the population of developed countries is increasing at a constant relative rate.

In developing countries, the birth rate is higher than the death rate and both have been decreasing since about 1925. In recent years, the death rate has been decreasing faster than the birth rate, so the relative rate of population growth is increasing. The decline in the death rate has contributed significantly to the recent population growth in developing countries.

(b) The relative rate of growth of the population is the difference between the relative birth rate and the relative death rate, so it is represented in Figure 6.30 by the vertical distance between the birth-rate curve and the death-rate curve. The area of the region between the two curves from 1800 to 1900, shown in Figure 6.31, gives the change in $\ln P(t)$. The region has area approximately equal to a rectangle of height 0.005 and width 100, so has area 0.5. We have

$$\ln\left(\frac{P(1900)}{P(1800)}\right) = \int_{1800}^{1900} \frac{P'(t)}{P(t)}\,dt \approx 0.5$$
$$\frac{P(1900)}{P(1800)} \approx e^{0.5} = 1.65$$

During the nineteenth century the population of developing countries increased by about 65%, a factor of 1.65.

(c) The shaded region between the birth rate and death rate curves from 1950 to 1977 in Figure 6.31 consists of approximately 1.5 rectangles, each of which has area $(0.01)(27) = 0.27$. The area of this region is approximately (1.5) (0.27) = 0.405. We have

$$\ln\left(\frac{P(1977)}{P(1950)}\right) = \int_{1950}^{1977} \frac{P'(t)}{P(t)}\,dt \approx 0.405,$$

so

$$\frac{P(1977)}{P(1950)} \approx e^{0.405} = 1.50$$

Between 1950 and 1977, the population of developing countries increased by about 50%, a factor of 1.5.

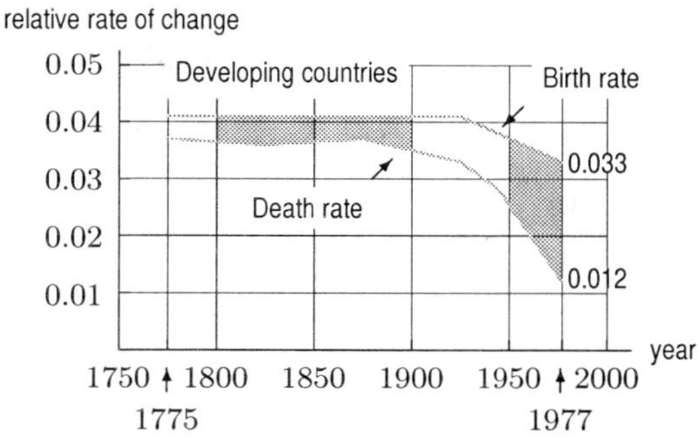

Figure 6.31: Relative growth rate of population = Relative birth rate − Relative death rate

Problems for Section 6.4

1. A town has a population of 1000. Fill in the table assuming that the town's population grows by

(a) 50 people per year **(b)** 5% per year

Year	0	1	2	3	4	...	10
Population	1000					...	

2. Birth and death rates are often reported as births or deaths per thousand members of the population. What is the relative rate of growth of a population with a birth rate of 30 births per 1000 and a death rate of 20 deaths per 1000?

3. The relative growth rate of a population is given in Figure 6.32. By what percentage does the population change over the 10-year period?

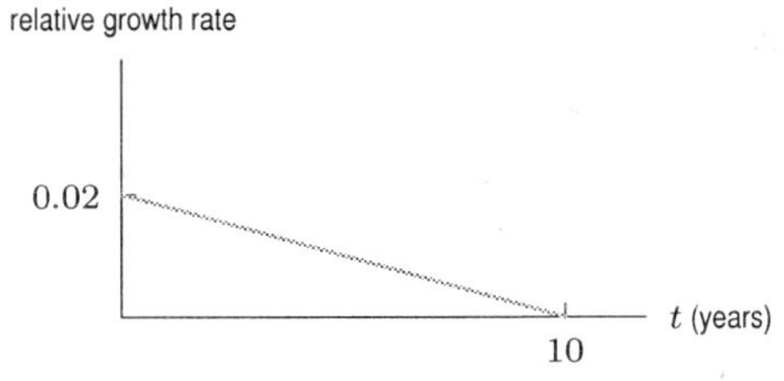

Figure 6.32

4. The article "Job Scene: What's Hot, What's Not in the Next 10 Years," appearing in the *Chicago Tribune* in July 1996, contained the following quote about predicted change in jobs between 1995 and 2005:

- Employment is projected to increase to 144.7 million from 127 million – only **14** percent. The growth rate was **24** percent from 1983 to 1994.
- Jobs in services and retail trades are expected to increase by **16.2** million. Business services, health and education will account for **9.1** million new jobs.
- Manufacturing will lose **1.3** million jobs, a continuation of its decline.
- Jobs for those with master's degrees will grow the most, **28** percent.

(a) The first number in bold is 14. Is this an absolute change, a relative change, an absolute rate of change, or a relative rate of change? Find each of the other three measures of change, corresponding to 14.

(b) Identify each of the numbers in bold as an absolute change, a relative change, an absolute rate of change, or a relative rate of change.

5. Table 6.3 shows the cumulative number of AIDS deaths worldwide.[6] Find the absolute increase in AIDS deaths between 2003 and 2004 and between 2006 and 2007. Find the relative increase between 2003 and 2004 and between 2006 and 2007.

Table 6.3 *Cumulative AIDS deaths worldwide, in millions*

Year	2003	2004	2005	2006	2007
Cases	30.2	33.3	35.5	37.6	39.6

6. A population is 100 at time $t = 0$, with t in years.

(a) If the population has a constant absolute growth rate of 10 people per year, find a formula for the size of the population at time t.

(b) If the population has a constant relative growth rate of 10% per year, find a formula for the size of the population at time t.

(c) Graph both functions on the same axes.

[6] www.unaids.org, May 2009.

7. The size of a bacteria population is 4000. Find a formula for the size, P, of the population t hours later if the population is decreasing by

(a) 100 bacteria per hour **(b)** 5% per hour

In which case does the bacteria population reach 0 first?

8. The relative growth rate of a population $P(t)$ is given in Figure 6.33. By what percentage does the population change over the 8-year period? If the population was 10,000 at time $t = 0$, what is the population 8 years later?

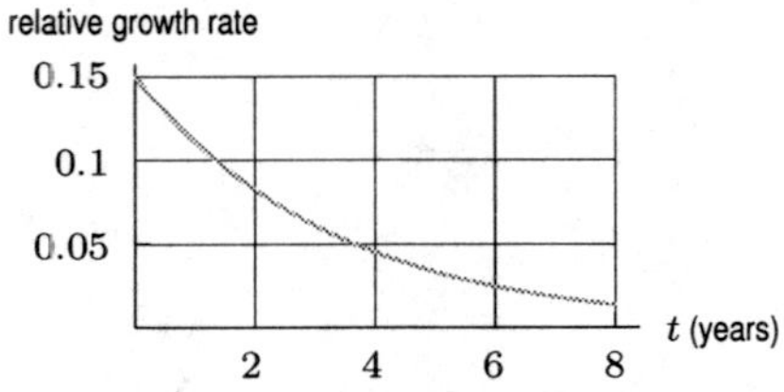

Figure 6.33

In Problems 9–12, a graph of the relative rate of change of a population is given with t in years. By approximately what percentage does the population change over the 10-year period?

9.
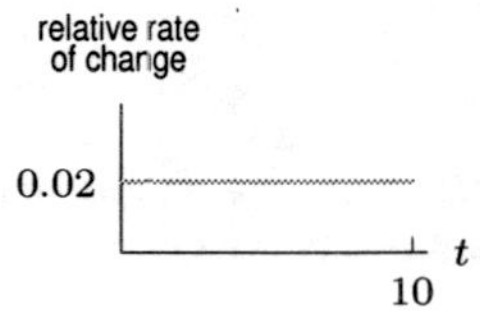

10.
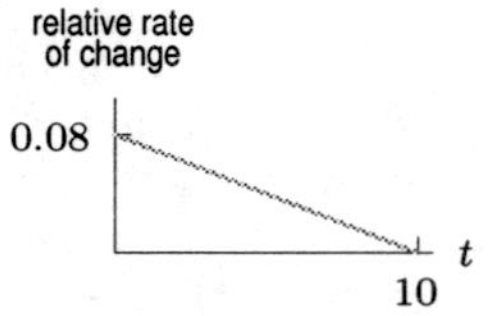

11.
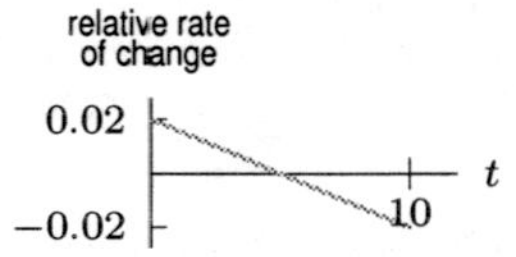

12.
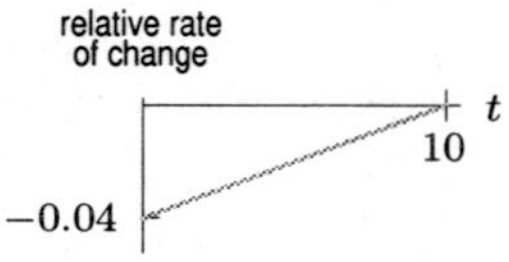

Problems 13–16 show the relative rate of change of f for $0 \le t \le 10$. Give the intervals on which f is increasing and on which f is decreasing.

13.
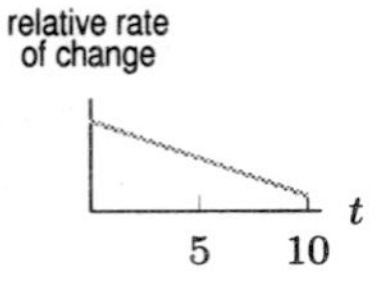

14.
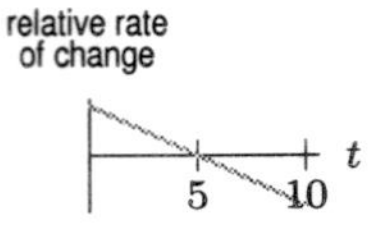

15.
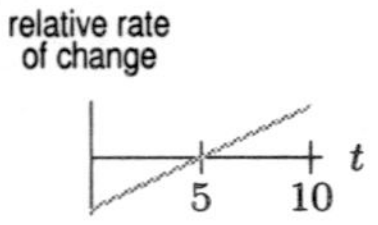

16.
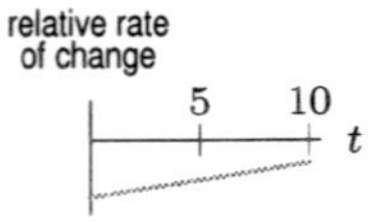

17. The population, P, in millions, of Nicaragua was 5.78 million in 2008 and growing at an annual rate of 1.8%.

(a) Write a formula for P as a function of t, where t is years since 2008.
(b) Find the projected average rate of change (or absolute growth rate) in Nicaragua between 2008 and 2009, and between 2009 and 2010. Explain why your answers are different.
(c) Use your answers to part (b) to confirm that the relative rate of change (or relative growth rate) over both time intervals was 1.8%.

In Problems 18–19, the graph shows relative birth and death rates for a population $P(t)$. Determine whether the population is increasing or decreasing, and find the percentage by which the population changes during the 10-year period.

18.
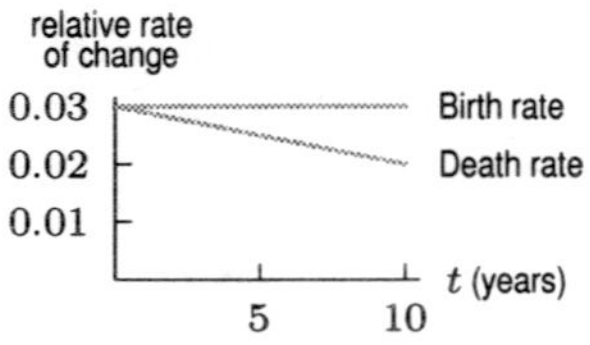

19.
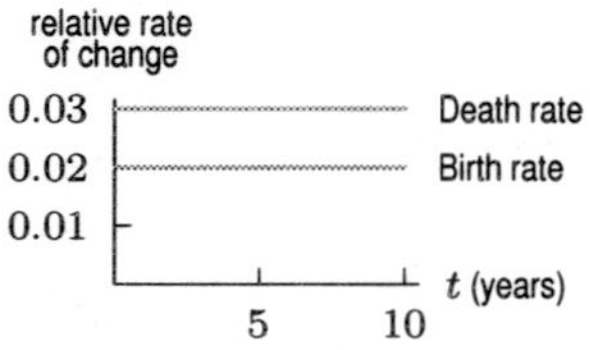

Chapter Ten

MATHEMATICAL MODELING USING DIFFERENTIAL EQUATIONS

10.5 APPLICATIONS AND MODELING

In the last section, we considered several situations modeled by the differential equation

$$\frac{dy}{dt} = ky.$$

In this section, we consider situations where the rate of change of y is a linear function of y of the form

$$\frac{dy}{dt} = k(y - A), \qquad \text{where } k \text{ and } A \text{ are constants.}$$

The Quantity of a Drug in the Body

A patient is given the drug warfarin, an anticoagulant, intravenously at the rate of 0.5 mg/hour. Warfarin is metabolized and leaves the body at the rate of about 2% per hour. A differential equation for the quantity, Q (in mg), of warfarin in the body after t hours is given by

$$\begin{aligned}\text{Rate of change} &= \text{Rate in} - \text{Rate out}\\ \frac{dQ}{dt} &= 0.5 - 0.02Q.\end{aligned}$$

What does this tell us about the quantity of warfarin in the body for different initial values of Q?

If Q is small, then $0.02Q$ is also small and the rate the drug is excreted is less than the rate at which the drug is entering the body. Since the rate in is greater than the rate out, the rate of change is positive and the quantity of drug in the body is increasing. If Q is large enough that $0.02Q$ is greater than 0.5, then $0.5 - 0.02Q$ is negative, so dQ/dt is negative and the quantity is decreasing.

For small Q, the quantity will increase until the rate in equals the rate out. For large Q, the quantity will decrease until the rate in equals the rate out. What is the value of Q at which the rate in exactly matches the rate out? We have

$$\begin{aligned}\text{Rate in} &= \text{Rate out}\\ 0.5 &= 0.02Q\\ Q &= 25.\end{aligned}$$

If the amount of warfarin in the body is initially 25 mg, then the amount being excreted exactly matches the amount being added. The quantity of drug Q will stay constant at 25 mg. Notice also that when $Q = 25$, the derivative dQ/dt is zero, since

$$\frac{dQ}{dt} = 0.5 - 0.02(25) = 0.5 - 0.5 = 0.$$

If the initial quantity is 25, then the solution is the horizontal line $Q = 25$. This solution is called an *equilibrium solution*.

The slope field for this differential equation is shown in Figure 10.29, with solution curves drawn for $Q_0 = 20$, $Q_0 = 25$, and $Q_0 = 30$. In each case, we see that the quantity of drug in the body is approaching the equilibrium solution of 25 mg. The solution curve with $Q_0 = 30$ should remind you of an exponential decay function. It is, in fact, an exponential decay function that has been shifted up 25 units.

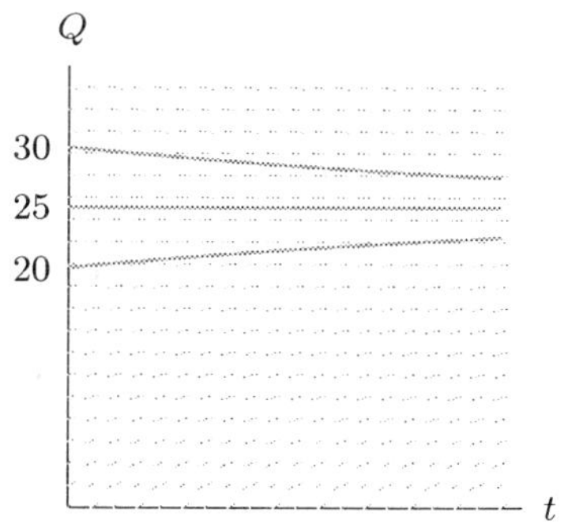

Figure 10.29: Slope field for $dQ/dt = 0.5 - 0.02Q$

Solving the Differential Equation $dy/dt = k(y - A)$

The drug concentration in the previous example satisfies a differential equation of the form

$$\frac{dy}{dt} = k(y - A).$$

Let us find the general solution to this equation. Since A is a constant, $dA/dt = 0$ so that we have

$$\frac{d}{dt}(y - A) = \frac{dy}{dt} - \frac{dA}{dt} = \frac{dy}{dt} - 0 = k(y - A).$$

Thus $y - A$ satisfies an exponential differential equation, so $y - A$ must be of the form

$$y - A = Ce^{kt}.$$

For an alternative derivation of the solution, see the Focus on Theory section at the end of the chapte.

> The general solution to the differential equation
>
> $$\frac{dy}{dt} = k(y - A)$$
>
> is
>
> $$y = A + Ce^{kt}, \qquad \text{for any constant } C.$$

Warning: Notice that, for differential equations of this form, the arbitrary constant C is *not* the initial value of the variable, but rather the initial value of $y - A$.

Example 1 Give the solution to each of the following differential equations:

(a) $\dfrac{dy}{dt} = 0.02(y - 50)$

(b) $\dfrac{dP}{dt} = 5(P - 10), \quad P = 8$ when $t = 0$

(c) $\dfrac{dy}{dt} = 3y - 300$

(d) $\dfrac{dW}{dt} = 500 - 0.1W$

Solution

(a) The general solution is $y = 50 + Ce^{0.02t}$.

(b) The general solution is $P = 10 + Ce^{5t}$. Use the initial condition to solve for C:

$$8 = 10 + C(e^0)$$
$$8 = 10 + C$$

So $C = -2$, and the particular solution is $P = 10 - 2e^{5t}$.

(c) First rewrite the right-hand side of the equation in the form $k(y - A)$ by factoring out a 3:

$$\frac{dy}{dt} = 3(y - 100).$$

The general solution to this differential equation is $y = 100 + Ce^{3t}$.

(d) We begin by factoring out the coefficient of W:

$$\frac{dW}{dt} = 500 - 0.1W = -0.1\left(W - \frac{500}{0.1}\right) = -0.1(W - 5000).$$

The general solution to this differential equation is $W = 5000 + Ce^{-0.1t}$.

Example 2 At the start of this section, we gave the following differential equation for the quantity of warfarin in the body:

$$\frac{dQ}{dt} = 0.5 - 0.02Q.$$

Write the general solution to this differential equation. Find particular solutions for $Q_0 = 20$, $Q_0 = 25$, and $Q_0 = 30$.

Solution We first rewrite the differential equation in the form $dQ/dt = k(Q - A)$ by factoring out -0.02:

$$\frac{dQ}{dt} = 0.5 - 0.02Q = -0.02(Q - 25).$$

The general solution to this differential equation is

$$Q = 25 + Ce^{-0.02t}.$$

To find the particular solution when $Q_0 = 20$, we use the initial condition to solve for C:

$$20 = 25 + C(e^0)$$
$$C = -5.$$

The particular solution when $Q_0 = 20$ is $Q = 25 - 5e^{-0.02t}$.

When $Q_0 = 25$, we have $C = 0$ and the particular solution is the horizontal line $Q = 25$. When $Q_0 = 30$, we have $C = 5$ and the particular solution is $Q = 25 + 5e^{-0.02t}$. These three solutions are the three we saw earlier in Figure 10.29.

Example 3 A company's revenue is earned at a continuous annual rate of 5% of its net worth. At the same time, the company's payroll obligations are paid out at a constant rate of 200 million dollars a year. We saw in Section 10.1 that the differential equation governing the net worth, W (in millions of dollars), of this company in year t is given by

$$\text{Rate of change of } W = \text{Rate in} - \text{Rate out}$$
$$\frac{dW}{dt} = 0.05W - 200.$$

(a) Solve the differential equation, assuming an initial net worth of W_0 million dollars.

(b) Sketch the solution for $W_0 = 3000$, 4000 and 5000. For which of these values of W_0 does the company go bankrupt? In which year?

Solution (a) Factor out 0.05 to get

$$\frac{dW}{dt} = 0.05(W - 4000).$$

The general solution is

$$W = 4000 + Ce^{0.05t}.$$

To find C we use the initial condition that $W = W_0$ when $t = 0$.

$$W_0 = 4000 + Ce^0$$
$$W_0 - 4000 = C$$

Substituting this value for C into $W = 4000 + Ce^{0.05t}$ gives

$$W = 4000 + (W_0 - 4000)e^{0.05t}.$$

(b) If $W_0 = 4000$, then $W = 4000$, the equilibrium solution.
If $W_0 = 5000$, then $W = 4000 + 1000e^{0.05t}$.
If $W_0 = 3000$, then $W = 4000 - 1000e^{0.05t}$. The graphs of these functions are in Figure 10.30. Notice that if the net worth starts with W_0 near, but not equal to, \$4000 million, then W moves further away. We see that if $W_0 = 3000$, the value of W goes to 0, and the company goes bankrupt. Solving $W = 0$ gives $t \approx 27.7$, so the company goes bankrupt in its twenty-eighth year.

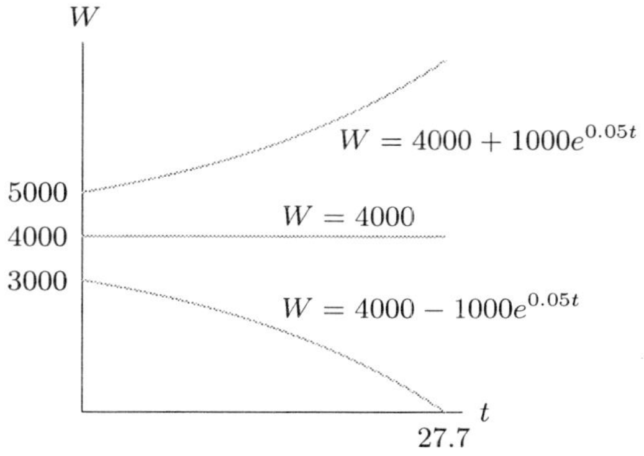

Figure 10.30: Solutions to $\frac{dW}{dt} = 0.05W - 200$

Equilibrium Solutions

Figure 10.29 shows the quantity of warfarin in the body for several different initial quantities. All these curves are solutions to the differential equation

$$\frac{dQ}{dt} = 0.5 - 0.02Q = -0.02(Q - 25),$$

and all the solutions have the form

$$Q = 25 + Ce^{-0.02t}$$

for some C. Notice that $Q \to 25$ as $t \to \infty$ for all solutions because $e^{-0.02t} \to 0$ as $t \to \infty$. In other words, in the long run, the quantity approaches the *equilibrium solution* of $Q = 25$ no matter what the initial quantity.

Notice that the equilibrium solution can be found directly from the differential equation by solving $dQ/dt = 0$:

$$\frac{dQ}{dt} = -0.02(Q - 25) = 0,$$

giving $Q = 25$. Because Q always gets closer and closer to the equilibrium value of 25 as $t \to \infty$, we call $Q = 25$ a *stable* equilibrium for Q.

A different situation is shown in Figure 10.30 with the solutions to the differential equation $dW/dt = 0.05W - 200$. We find the equilibrium by looking at the solution curves or by setting $dW/dt = 0$:

$$\frac{dW}{dt} = 0.05W - 200 = 0.05(W - 4000) = 0,$$

giving $W = 4000$ as the equilibrium solution. This equilibrium solution is called *unstable* because if W starts near, but not equal to, 4000, the net worth W moves further away from 4000 as $t \to \infty$.

- An **equilibrium solution** is constant for all values of the independent variable. The graph is a horizontal line. Equilibrium solutions can be identified by setting the derivative of the function to zero.
- An equilibrium solution is **stable** if a small change in the initial conditions gives a solution which tends toward the equilibrium as the independent variable tends to positive infinity.
- An equilibrium solution is **unstable** if a small change in the initial conditions gives a solution curve which veers away from the equilibrium as the independent variable tends to positive infinity.

In general, a differential equation may have more than one equilibrium solution or no equilibrium solution.

Example 4 Find the equilibrium solution for each of the following differential equations. Determine whether the equilibrium solution is stable or unstable.

(a) $\dfrac{dH}{dt} = -2(H - 20)$ (b) $\dfrac{dB}{dt} = 2(B - 10)$

Solution (a) To find equilibrium solutions, we set $dH/dt = 0$:

$$\frac{dH}{dt} = -2(H - 20) = 0,$$

giving $H = 20$ as the equilibrium solution. The general solution to this differential equation is $H = 20 + Ce^{-2t}$. The solution curves for $H_0 = 10$, $H_0 = 20$, and $H_0 = 30$ are shown in Figure 10.31. We see that the equilibrium solution is stable.

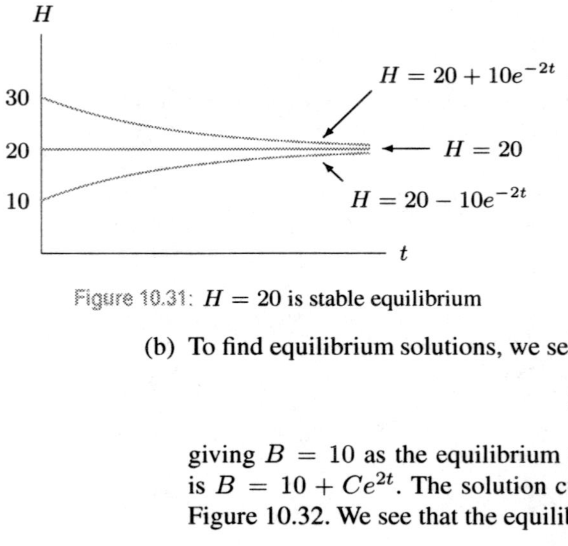

Figure 10.31: $H = 20$ is stable equilibrium

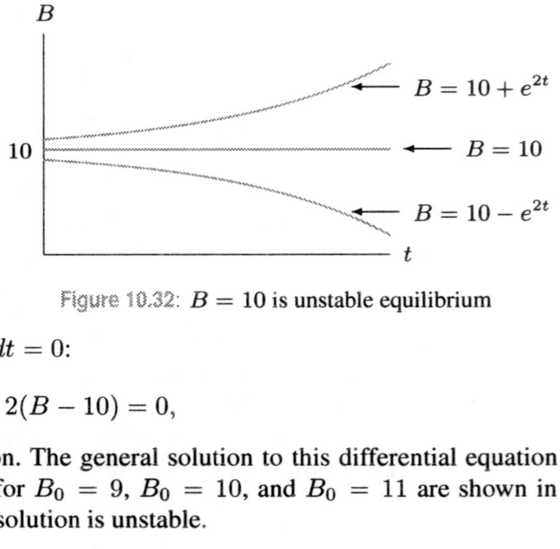

Figure 10.32: $B = 10$ is unstable equilibrium

(b) To find equilibrium solutions, we set $dB/dt = 0$:

$$\frac{dB}{dt} = 2(B - 10) = 0,$$

giving $B = 10$ as the equilibrium solution. The general solution to this differential equation is $B = 10 + Ce^{2t}$. The solution curves for $B_0 = 9$, $B_0 = 10$, and $B_0 = 11$ are shown in Figure 10.32. We see that the equilibrium solution is unstable.

Newton's Law of Heating and Cooling

Newton proposed that the temperature of a hot object decreases at a rate proportional to the difference between its temperature and that of its surroundings. Similarly, a cold object heats up at a rate proportional to the temperature difference between the object and its surroundings.

For example, a hot cup of coffee standing on a table cools at a rate proportional to the temperature difference between the coffee and the surrounding air. As the coffee cools, the rate at which it cools decreases because the temperature difference between the coffee and the air decreases. In the long run, the rate of cooling tends to zero and the temperature of the coffee approaches room temperature. Figure 10.33 shows the temperature of two cups of coffee against time, one starting at a higher temperature than the other, but both tending toward room temperature in the long run.

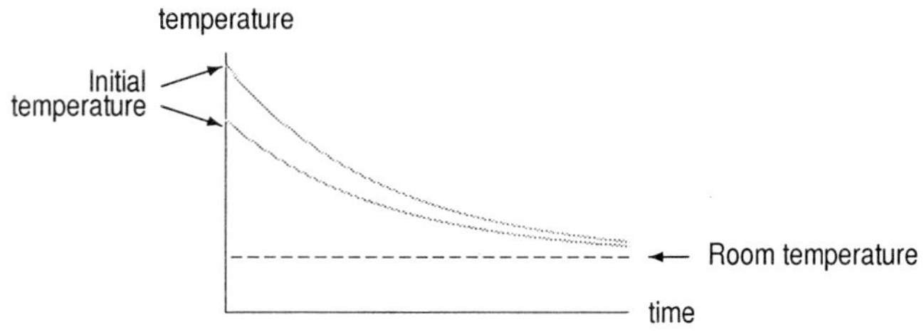

Figure 10.33: Temperature of coffee versus time

Let H be the temperature at time t of a cup of coffee in a 70°F room. Newton's Law says that the rate of change of H is proportional to the temperature difference between the coffee and the room:

$$\text{Rate of change of temperature} = \text{Constant} \cdot \text{Temperature difference.}$$

The rate of change of temperature is dH/dt. The temperature difference between the coffee and the room is $(H - 70)$, so

$$\frac{dH}{dt} = \text{Constant} \cdot (H - 70).$$

What about the sign of the constant? If the coffee starts out hotter than 70° (that is, $H - 70 > 0$), then the temperature of the coffee decreases (i.e., $dH/dt < 0$) and so the constant must be negative:

$$\frac{dH}{dt} = -k(H - 70) \qquad k > 0.$$

What can we learn from this differential equation? Suppose we take $k = 1$. The slope field for this differential equation in Figure 10.34 shows several solution curves. Notice that, as we expect, the temperature of the coffee is approaching the temperature of the room. The general solution to this differential equation is

$$H = 70 + Ce^{-t},$$

where C is an arbitrary constant.

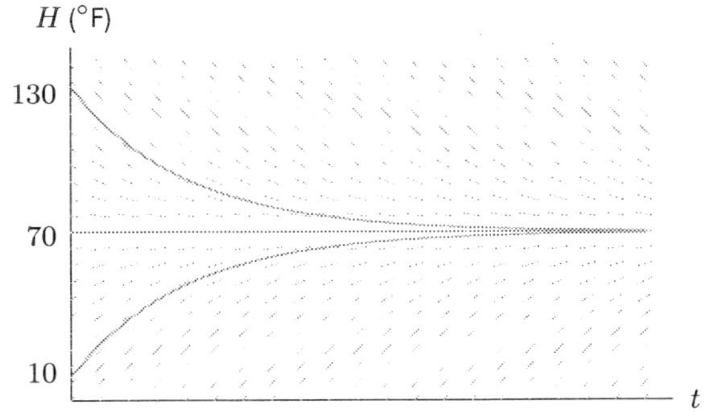

Figure 10.34: Slope field for $\frac{dH}{dt} = -(H - 70)$

Example 5 The body of a murder victim is found at noon in a room with a constant temperature of 20°C. At noon the temperature of the body is 35°C; two hours later the temperature of the body is 33°C.

(a) Find the temperature, H, of the body as a function of t, the time in hours since it was found.
(b) Graph H against t. What happens to the temperature in the long run?
(c) At the time of the murder, the victim's body had the normal body temperature, 37°C. When did the murder occur?

Solution

(a) Newton's Law of Cooling says that

$$\text{Rate of change of temperature} = \text{Constant} \cdot \text{Temperature difference}.$$

Since the temperature difference is $H - 20$, we have for some constant k

$$\frac{dH}{dt} = -k(H - 20).$$

The general solution is

$$H = 20 + Ce^{-kt}.$$

To determine C, we use the fact that $H = 35$ at $t = 0$:

$$35 = 20 + C \cdot e^0$$
$$35 = 20 + C$$

So $C = 15$ and we have

$$H = 20 + 15e^{-kt}.$$

To find k, we use the fact that $H = 33$ when $t = 2$:

$$33 = 20 + 15e^{-k(2)}.$$

We isolate the exponential and solve for k:

$$13 = 15e^{-2k}$$
$$\frac{13}{15} = e^{-2k}$$
$$\ln\left(\frac{13}{15}\right) = -2k$$
$$k = -\frac{\ln(13/15)}{2} = 0.072.$$

Therefore, the temperature, H, of the body as a function of time, t, is given by

$$H = 20 + 15e^{-0.072t}.$$

(b) The graph of $H = 20 + 15e^{-0.072t}$ has a vertical intercept of $H = 35$, the initial temperature. The temperature decays exponentially with a horizontal asymptote of $H = 20$. (See Figure 10.35.) "In the long run" means as $t \to \infty$. The graph shows that $H \to 20$ as $t \to \infty$.

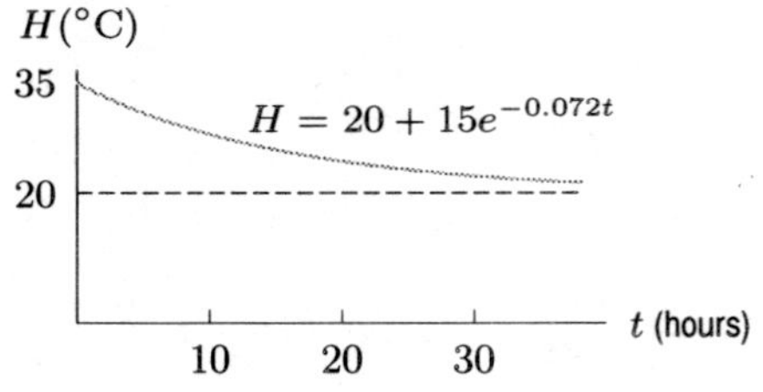

Figure 10.35: Temperature of dead body

(c) We want to know when the temperature was 37°C. We substitute $H = 37$ and solve for t:

$$37 = 20 + 15e^{-0.072t}$$
$$\frac{17}{15} = e^{-0.072t}.$$

Taking natural logs on both sides gives

$$\ln\left(\frac{17}{15}\right) = -0.072t$$

so

$$t = -\frac{\ln(17/15)}{0.072} = -1.74 \text{ hours}.$$

The murder occurred about 1.74 hours before noon, that is, about 10:15 am.

Problems for Section 10.5

Find particular solutions in Problems 1–8.

1. $\frac{dH}{dt} = 3(H - 75), \quad H = 0$ when $t = 0$

2. $\frac{dy}{dt} = 0.5(y - 200), \quad y = 50$ when $t = 0$

3. $\frac{dP}{dt} = P + 4, \quad P = 100$ when $t = 0$

4. $\frac{dB}{dt} = 4B - 100, \quad B = 20$ when $t = 0$

5. $\frac{dQ}{dt} = 0.3Q - 120, \quad Q = 50$ when $t = 0$

6. $\frac{dm}{dt} = 0.1m + 200, \quad m(0) = 1000$

7. $\frac{dB}{dt} + 2B = 50, \quad B(1) = 100$

8. $\frac{dB}{dt} + 0.1B - 10 = 0 \qquad B(2) = 3$

9. Check that $y = A + Ce^{kt}$ is a solution to the differential equation
$$\frac{dy}{dt} = k(y - A).$$

10. A bank account earns 7% annual interest compounded continuously. You deposit \$10,000 in the account, and withdraw money continuously from the account at a rate of \$1000 per year.
    - **(a)** Write a differential equation for the balance, B, in the account after t years.
    - **(b)** What is the equilibrium solution to the differential equation? (This is the amount that must be deposited now for the balance to stay the same over the years.)
    - **(c)** Find the solution to the differential equation.
    - **(d)** How much is in the account after 5 years?
    - **(e)** Graph the solution. What happens to the balance in the long run?

11. A company earns 2% per month on its assets, paid continuously, and its expenses are paid out continuously at a rate of \$80,000 per month.
    - **(a)** Write a differential equation for the value, V, of the company as a function of time, t, in months.
    - **(b)** What is the equilibrium solution for the differential equation? What is the significance of this value for the company?
    - **(c)** Solve the differential equation found in part (a).
    - **(d)** If the company has assets worth \$3 million at time $t = 0$, what are its assets worth one year later?

12. Money in an account earns interest at a continuous rate of 8% per year, and payments are made continuously out of the account at the rate of \$5000 a year. The account initially contains \$50,000. Write a differential equation for the amount of money in the account, B, in t years. Solve the differential equation. Does the account ever run out of money? If so, when?

13. A bank account earns 5% annual interest, compounded continuously. Money is deposited in a continuous cash flow at a rate of \$1200 per year into the account.
    - **(a)** Write a differential equation that describes the rate at which the balance $B = f(t)$ is changing.
    - **(b)** Solve the differential equation given an initial balance $B_0 = 0$.
    - **(c)** Find the balance after 5 years.

14. A patient is given the drug theophylline intravenously at a rate of 43.2 mg/hour to relieve acute asthma. The rate at which the drug leaves the patient's body is proportional to the quantity there, with proportionality constant 0.082 if time, t, is in hours. The patient's body contains none of the drug initially.
    - **(a)** Describe in words how you expect the quantity of theophylline in the patient to vary with time.
    - **(b)** Write a differential equation satisfied by the quantity of theophylline in the body, $Q(t)$.
    - **(c)** Solve the differential equation and graph the solution. What happens to the quantity in the long run?

15. One theory on the speed an employee learns a new task claims that the more the employee already knows, the more slowly he or she learns. Suppose that the rate at which a person learns is equal to the percentage of the task not yet learned. If y is the percentage learned by time t, the percentage not yet learned by that time is $100 - y$, so we can model this situation with the differential equation
$$\frac{dy}{dt} = 100 - y.$$
    - **(a)** Find the general solution to this differential equation.
    - **(b)** Sketch several solutions.
    - **(c)** Find the particular solution if the employee starts learning at time $t = 0$ (so $y = 0$ when $t = 0$).

16. A chain smoker smokes five cigarettes every hour. From each cigarette, 0.4 mg of nicotine is absorbed into the person's bloodstream. Nicotine leaves the body at a rate proportional to the amount present, with constant of proportionality -0.346 if t is in hours.
    - **(a)** Write a differential equation for the level of nicotine in the body, N, in mg, as a function of time, t, in hours.
    - **(b)** Solve the differential equation from part (a). Initially there is no nicotine in the blood.
    - **(c)** The person wakes up at 7 am and begins smoking. How much nicotine is in the blood when the person goes to sleep at 11 pm (16 hours later)?

17. As you know, when a course ends, students start to forget the material they have learned. One model (called the Ebbinghaus model) assumes that the rate at which a student forgets material is proportional to the difference between the material currently remembered and some positive constant, a.

 (a) Let $y = f(t)$ be the fraction of the original material remembered t weeks after the course has ended. Set up a differential equation for y. Your equation will contain two constants; the constant a is less than y for all t.
 (b) Solve the differential equation.
 (c) Describe the practical meaning (in terms of the amount remembered) of the constants in the solution $y = f(t)$.

18. **(a)** What are the equilibrium solutions for the differential equation

$$\frac{dy}{dt} = 0.2(y-3)(y+2)?$$

 (b) Use a graphing calculator or computer to sketch a slope field for this differential equation. Use the slope field to determine whether each equilibrium solution is stable or unstable.

19. **(a)** Find the equilibrium solution of the equation

$$\frac{dy}{dt} = 0.5y - 250.$$

 (b) Find the general solution of this equation.
 (c) Graph several solutions with different initial values.
 (d) Is the equilibrium solution stable or unstable?

20. Figure 10.36 gives the slope field for a differential equation. Estimate all equilibrium solutions and indicate whether each is stable or unstable.

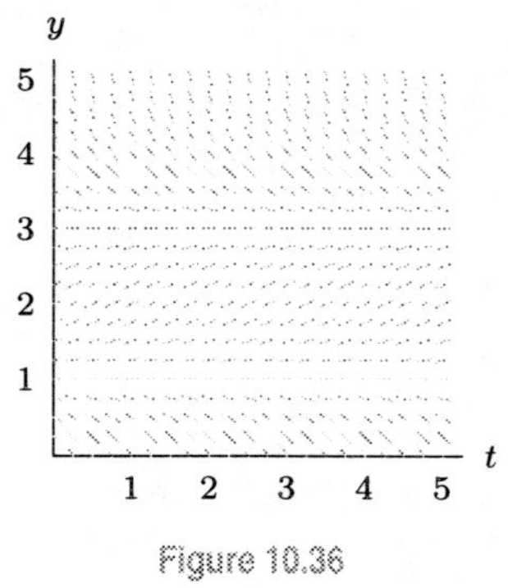

Figure 10.36

21. A yam is put in a 200°C oven and heats up according to the differential equation

$$\frac{dH}{dt} = -k(H-200), \quad \text{for } k \text{ a positive constant.}$$

 (a) If the yam is at 20°C when it is put in the oven, solve the differential equation.
 (b) Find k using the fact that after 30 minutes the temperature of the yam is 120°C.

22. At 1:00 pm one winter afternoon, there is a power failure at your house in Wisconsin, and your heat does not work without electricity. When the power goes out, it is 68°F in your house. At 10:00 pm, it is 57°F in the house, and you notice that it is 10°F outside.

 (a) Assuming that the temperature, T, in your home obeys Newton's Law of Cooling, write the differential equation satisfied by T.
 (b) Solve the differential equation to estimate the temperature in the house when you get up at 7:00 am the next morning. Should you worry about your water pipes freezing?
 (c) What assumption did you make in part (a) about the temperature outside? Given this (probably incorrect) assumption, would you revise your estimate up or down? Why?

23. A detective finds a murder victim at 9 am. The temperature of the body is measured at 90.3°F. One hour later, the temperature of the body is 89.0°F. The temperature of the room has been maintained at a constant 68°F.

 (a) Assuming the temperature, T, of the body obeys Newton's Law of Cooling, write a differential equation for T.
 (b) Solve the differential equation to estimate the time the murder occurred.

24. A drug is administered intravenously at a constant rate of r mg/hour and is excreted at a rate proportional to the quantity present, with constant of proportionality $\alpha > 0$.

 (a) Solve a differential equation for the quantity, Q, in milligrams, of the drug in the body at time t hours. Assume there is no drug in the body initially. Your answer will contain r and α. Graph Q against t. What is Q_∞, the limiting long-run value of Q?
 (b) What effect does doubling r have on Q_∞? What effect does doubling r have on the time to reach half the limiting value, $\frac{1}{2}Q_\infty$?
 (c) What effect does doubling α have on Q_∞? On the time to reach $\frac{1}{2}Q_\infty$?

25. Some people write the solution of the initial value problem

$$\frac{dy}{dt} = k(y-A) \qquad y = y_0 \text{ at } t = 0$$

in the form

$$\frac{y-A}{y_0 - A} = e^{kt}.$$

Show that this formula gives the correct solution for y, assuming $y_0 \neq A$.

Chapter Seven

ANTIDERIVATIVES

Contents

7.1 CONSTRUCTING ANTIDERIVATIVES ANALYTICALLY

What is an Antiderivative?

If the derivative of $F(x)$ is $f(x)$, that is, if $F'(x) = f(x)$, then we call $F(x)$ an *antiderivative* of $f(x)$. For example, the derivative of x^2 is $2x$, so we say that

$$x^2 \text{ is an antiderivative of } 2x.$$

Can you think of another function whose derivative is $2x$? How about $x^2 + 1$? Or $x^2 + 17$? Since, for any constant C,

$$\frac{d}{dx}(x^2 + C) = 2x + 0 = 2x,$$

any function of the form $x^2 + C$ is an antiderivative of $2x$. It can be shown that all antiderivatives of $2x$ are of this form, so we say that

$$x^2 + C \text{ is the family of antiderivatives of } 2x.$$

Once we know one antiderivative $F(x)$ for a function $f(x)$ on an interval, then all other antiderivatives of $f(x)$ are of the form $F(x) + C$.

The Indefinite Integral

We introduce a notation for the family of antiderivatives that looks like a definite integral without the limits. We call $\int f(x)\,dx$ the *indefinite integral* of $f(x)$. If all antiderivatives of $f(x)$ are of the form $F(x) + C$, we write

$$\int f(x)\,dx = F(x) + C.$$

It is important to understand the difference between

$$\int_a^b f(x)\,dx \qquad \text{and} \qquad \int f(x)\,dx.$$

The first is a number and the second is a family of functions. Because the notation is similar, the word "integration" is frequently used for the process of finding antiderivatives as well as of finding definite integrals. The context usually makes clear which is intended.

Finding Formulas for Antiderivatives

Finding antiderivatives of functions is like taking square roots of numbers: if we pick a number at random, such as 7 or 493, we may have trouble figuring out its square root without a calculator. But if we happen to pick a number such as 25 or 64, which we know is a perfect square, then we can easily guess its square root. Similarly, if we pick a function which we recognize as a derivative, then we can find its antiderivative easily.

For example, noticing that $2x$ is the derivative of x^2 tells us that x^2 is an antiderivative of $2x$. If we divide by 2, then we guess that

$$\text{An antiderivative of } x \text{ is } \frac{x^2}{2}.$$

To check this statement, take the derivative of $x^2/2$:

$$\frac{d}{dx}\left(\frac{x^2}{2}\right) = \frac{1}{2}\cdot\frac{d}{dx}x^2 = \frac{1}{2}\cdot 2x = x.$$

What about an antiderivative of x^2? The derivative of x^3 is $3x^2$, so the derivative of $x^3/3$ is $3x^2/3 = x^2$. Thus,

$$\text{An antiderivative of } x^2 \text{ is } \frac{x^3}{3}.$$

Can you see the pattern? It looks like

$$\text{An antiderivative of } x^n \text{ is } \frac{x^{n+1}}{n+1}.$$

(We assume $n \neq -1$, or we would have $x^0/0$, which does not make sense.) It is easy to check this formula by differentiation:

$$\frac{d}{dx}\left(\frac{x^{n+1}}{n+1}\right) = \frac{(n+1)x^n}{n+1} = x^n.$$

Thus, in indefinite integral notation, we see that

$$\int x^n\,dx = \frac{x^{n+1}}{n+1} + C, \quad n \neq -1.$$

Can you think of an antiderivative of the function $f(x) = 5$? We know that the derivative of $5x$ is 5, so $F(x) = 5x$ is an antiderivative of $f(x) = 5$. In general, if k is a constant, the derivative of kx is k, so we have the result:

If k is constant,

$$\int k\,dx = kx + C.$$

Example 1 Find $\int (3x + x^2)\,dx$.

Solution We know that $x^2/2$ is an antiderivative of x and that $x^3/3$ is an antiderivative of x^2, so we expect

$$\int (3x + x^2)\,dx = 3\left(\frac{x^2}{2}\right) + \frac{x^3}{3} + C.$$

Again, check your antiderivatives by differentiation—it's easy to do. Here

$$\frac{d}{dx}\left(\frac{3}{2}x^2 + \frac{x^3}{3} + C\right) = \frac{3}{2}\cdot 2x + \frac{3x^2}{3} = 3x + x^2.$$

The preceding example illustrates that the sum and constant multiplication rules of differentiation work in reverse:

Properties of Antiderivatives: Sums and Constant Multiples

In indefinite integral notation,

1. $\int [f(x) \pm g(x)]\,dx = \int f(x)\,dx \pm \int g(x)\,dx$
2. $\int cf(x)\,dx = c\int f(x)\,dx.$

In words,

1. An antiderivative of the sum (or difference) of two functions is the sum (or difference) of their antiderivatives.
2. An antiderivative of a constant times a function is the constant times an antiderivative of the function.

Example 2 Find antiderivatives of each of the following: (a) x^5 (b) t^8 (c) $12x^3$ (d) $q^3 - 6q^2$

Solution
(a) $\int x^5\,dx = \frac{x^6}{6} + C.$
(b) $\int t^8\,dt = \frac{t^9}{9} + C.$
(c) $\int 12x^3\,dx = 12\left(\frac{x^4}{4}\right) + C = 3x^4 + C.$
(d) $\int q^3 - 6q^2\,dq = \frac{q^4}{4} - 6\left(\frac{q^3}{3}\right) + C = \frac{q^4}{4} - 2q^3 + C.$
To check, differentiate the antiderivative; you should get the original function.

What is the antiderivative of x^n when $n = -1$? In other words, what is an antiderivative of $1/x$? Fortunately, we know a function whose derivative is $1/x$, namely, the natural logarithm. Thus, since

$$\frac{d}{dx}(\ln x) = \frac{1}{x},$$

we know that

$$\int \frac{1}{x}\,dx = \ln x + C, \quad \text{for } x > 0.$$

If $x < 0$, then $\ln x$ is not defined, so it can't be an antiderivative of $1/x$. In this case, we can try $\ln(-x)$:

$$\frac{d}{dx}\ln(-x) = (-1)\frac{1}{-x} = \frac{1}{x}$$

so

$$\int \frac{1}{x}\,dx = \ln(-x) + C, \quad \text{for } x < 0.$$

This means $\ln x$ is an antiderivative of $1/x$ if $x > 0$, and $\ln(-x)$ is an antiderivative of $1/x$ if $x < 0$. Since $|x| = x$ when $x > 0$ and $|x| = -x$ when $x < 0$, we can collapse these two formulas into:

$$\text{An antiderivative of } \frac{1}{x} \text{ is } \ln|x|$$

on any interval that does not contain 0. Therefore

$$\int \frac{1}{x}\,dx = \ln|x| + C.$$

Since the exponential function e^x is its own derivative, it is also its own antiderivative; thus

$$\int e^x\,dx = e^x + C.$$

What about e^{kx}? We know that the derivative of e^{kx} is ke^{kx}, so, for $k \neq 0$, we have

$$\int e^{kx}\,dx = \frac{1}{k}e^{kx} + C.$$

Example 3 Find antiderivatives of each of the following: (a) $8x^3 + \frac{1}{x}$ (b) $12e^{0.2t}$

Solution (a) $\int 8x^3 + \frac{1}{x}\, dx = 8\left(\frac{x^4}{4}\right) + \ln|x| + C = 2x^4 + \ln|x| + C.$

(b) $\int 12e^{0.2t}\, dt = 12\left(\frac{1}{0.2}e^{0.2t}\right) + C = 60e^{0.2t} + C.$

Differentiate your answers to check them.

Antiderivatives of Periodic Functions

The antiderivatives of the sine and cosine are easy to guess. Since

$$\frac{d}{dx}\sin x = \cos x \quad \text{and} \quad \frac{d}{dx}\cos x = -\sin x,$$

we get

$$\int \cos x\, dx = \sin x + C \quad \text{and} \quad \int \sin x\, dx = -\cos x + C.$$

Since the derivative of $\sin(kx)$ is $k\cos(kx)$ and the derivative of $\cos(kx)$ is $-k\sin(kx)$, we have, for $k \neq 0$,

$$\int \cos(kx)\, dx = \frac{1}{k}\sin(kx) + C \quad \text{and} \quad \int \sin(kx)\, dx = -\frac{1}{k}\cos(kx) + C.$$

Example 4 Find $\int (\sin x + 3\cos(5x))\, dx.$

Solution We break the antiderivative into two terms:

$$\int (\sin x + 3\cos(5x))\, dx = \int \sin x\, dx + 3\int \cos(5x)\, dx = -\cos x + \frac{3}{5}\sin(5x) + C.$$

Check by differentiating:

$$\frac{d}{dx}(-\cos x + \frac{3}{5}\sin(5x) + C) = \sin x + 3\cos(5x).$$

Problems for Section 7.1

In Problems 1–26, find an antiderivative.

1. $f(x) = 5$

2. $f(t) = 5t$

3. $g(t) = t^2 + t$

4. $f(x) = x^2$

5. $f(x) = x^4$

6. $g(t) = t^7 + t^3$

7. $g(x) = 6x^3 + 4$

8. $f(q) = 5q^2$

9. $h(y) = 3y^2 - y^3$

10. $k(x) = 10 + 8x^3$

11. $f(x) = 3x^2 + 5$

12. $f(x) = x + x^5 + x^{-5}$

13. $p(x) = x^2 - 6x + 17$

14. $g(z) = \sqrt{z}$

15. $f(x) = 5x - \sqrt{x}$

16. $p(z) = (\sqrt{z})^3$

17. $h(z) = \frac{1}{z}$

18. $p(t) = t^3 - \frac{t^2}{2} - t$

19. $g(z) = \frac{1}{z^3}$

20. $q(y) = y^4 + \frac{1}{y}$

21. $f(x) = x^6 - \dfrac{1}{7x^6}$

22. $g(x) = \dfrac{1}{x} + \dfrac{1}{x^2} + \dfrac{1}{x^3}$

23. $g(t) = e^{-3t}$

24. $h(t) = \cos t$

25. $g(t) = 5 + \cos t$

26. $g(\theta) = \sin\theta - 2\cos\theta$

In Problems 27–32, find an antiderivative $F(x)$ with $F'(x) = f(x)$ and $F(0) = 0$. Is there only one possible solution?

27. $f(x) = 3$

28. $f(x) = -7x$

29. $f(x) = 2 + 4x + 5x^2$

30. $f(x) = x^2$

31. $f(x) = \sqrt{x}$

32. $f(x) = e^x$

Find the indefinite integrals in Problems 33–62.

33. $\int (5x + 7)\, dx$

34. $\int 9x^2\, dx$

35. $\int 6x^2\, dx$

36. $\int t^{12}\, dt$

37. $\int (x + 1)^2\, dx$

38. $\int (x^2 + \frac{1}{x^2})\, dx$

39. $\int (t^2 + 5t + 1)\, dt$

40. $\int 5e^z\, dz$

41. $\int (t^3 + 6t^2)\, dt$

42. $\int (x^5 - 12x^3) dx$

43. $\int 3\sqrt{w}\, dw$

44. $\int (x^2 + 4x - 5) dx$

45. $\int \left(\dfrac{3}{t} - \dfrac{2}{t^2}\right) dt$

46. $\int e^{2t}\, dt$

47. $\int \left(x + \dfrac{1}{\sqrt{x}}\right) dx$

48. $\int (x^3 + 5x^2 + 6) dx$

49. $\int (e^x + 5)\, dx$

50. $\int \left(x^2 + \dfrac{1}{x}\right) dx$

51. $\int e^{3r}\, dr$

52. $\int \cos\theta\, d\theta$

53. $\int \sin t\, dt$

54. $\int 25e^{-0.04q}\, dq$

55. $\int 100e^{4x}\, dx$

56. $\int (2e^x - 8\cos x)\, dx$

57. $\int (3\cos x - 7\sin x)\, dx$

58. $\int \sin(3x) dx$

59. $\int x\cos(x^2 + 4) dx$

60. $\int 6\cos(3x) dx$

61. $\int (10 + 8\sin(2x)) dx$

62. $\int (12\sin(2x) + 15\cos(5x)) dx$

For Problems 63–66, find an antiderivative $F(x)$ with $F'(x) = f(x)$ and $F(0) = 5$.

63. $f(x) = 6x - 5$

64. $f(x) = x^2 + 1$

65. $f(x) = 8\sin(2x)$

66. $f(x) = 6e^{3x}$

67. A firm's marginal cost function is $MC = 3q^2 + 4q + 6$. Find the total cost function if the fixed costs are 200.

68. The marginal revenue function of a monopolistic producer is $MR = 20 - 4q$.

(a) Find the total revenue function.
(b) Find the corresponding demand curve.

7.2 INTEGRATION BY SUBSTITUTION

In Chapter 3, we learned rules to differentiate any function obtained by combining constants, powers of x, $\sin x$, $\cos x$, e^x, and $\ln x$, using addition, multiplication, division, or composition of functions. Such functions are called *elementary*.

In this section, we introduce integration by substitution. However, there is a great difference between looking for derivatives and looking for antiderivatives. Every elementary function has elementary derivatives, but many elementary functions do not have elementary antiderivatives. Some examples are $\sqrt{x^3 + 1}$, $(\sin x)/x$, and e^{-x^2}. These are ordinary functions that arise naturally, not exotic functions, yet they do not have elementary antiderivatives.

Integration by substitution reverses the chain rule. According to the chain rule,

$$\frac{d}{dx}(f(g(x))) = \underbrace{f'}_{\text{Derivative of outside}}(\overbrace{g(x)}^{\text{Inside}}) \cdot \underbrace{g'(x)}_{\text{Derivative of inside}}.$$

Thus, any function which is the result of differentiating with the chain rule is the product of two factors: the "derivative of the outside" and the "derivative of the inside." If a function has this form, its antiderivative is $f(g(x))$.

Example 1 Use the chain rule to find $f'(x)$ and then write the corresponding antidifferentiation formula.

(a) $f(x) = e^{x^2}$ (b) $f(x) = \frac{1}{6}(x^2+1)^6$ (c) $f(x) = \ln(x^2+4)$

Solution (a) Using the chain rule, we see

$$\frac{d}{dx}\left(e^{x^2}\right) = e^{x^2} \cdot 2x \quad \text{so} \quad \int e^{x^2} \cdot 2x \, dx = e^{x^2} + C.$$

(b) Using the chain rule, we see

$$\frac{d}{dx}\left(\frac{1}{6}(x^2+1)^6\right) = (x^2+1)^5 \cdot 2x \quad \text{so} \quad \int (x^2+1)^5 \cdot 2x \, dx = \frac{1}{6}(x^2+1)^6 + C.$$

(c) Using the chain rule, we see

$$\frac{d}{dx}(\ln(x^2+4)) = \frac{1}{x^2+4} \cdot 2x \quad \text{so} \quad \int \frac{1}{x^2+4} \cdot 2x \, dx = \ln(x^2+4) + C.$$

In Example 1, the derivative of each inside function is $2x$. Notice that the derivative of the inside function is a factor in the integrand in each antidifferentiation formula.

Finding an inside function whose derivative appears as a factor is key to the method of substitution. We formalize this method as follows:

To Make a Substitution in an Integral

Let w be the "inside function" and $dw = w'(x)\,dx = \frac{dw}{dx}dx$. Then express the integrand in terms of w.

Example 2 Make a substitution to find each of the following integrals:

(a) $\int e^{x^2} \cdot 2x \, dx$ (b) $\int (x^2+1)^5 \cdot 2x \, dx$ (c) $\int \frac{1}{x^2+4} \cdot 2x \, dx$

Solution (a) We look for an inside function whose derivative appears as a factor. In this case, the inside function is x^2, with derivative $2x$. We let $w = x^2$. Then $dw = w'(x)\,dx = 2x\,dx$. The original integrand can now be rewritten in terms of w:

$$\int e^{x^2} \cdot 2x \, dx = \int e^w \, dw = e^w + C = e^{x^2} + C.$$

By changing the variable to w, we simplified the integrand. The final step, after antidifferentiating, is to convert back to the original variable, x.

(b) Here, the inside function is $x^2 + 1$, with derivative $2x$. We let $w = x^2 + 1$. Then $dw = w'(x)\,dx = 2x\,dx$. Rewriting the original integral in terms of w, we have

$$\int (x^2+1)^5 \cdot 2x \, dx = \int w^5 \, dw = \frac{1}{6}w^6 + C = \frac{1}{6}(x^2+1)^6 + C.$$

Again, by changing the variable to w, we simplified the integrand.

(c) The inside function is x^2+4, so we let $w = x^2+4$. Then $dw = w'(x)\,dx = 2x\,dx$. Substituting, we have

$$\int \frac{1}{x^2+4} \cdot 2x \, dx = \int \frac{1}{w} \, dw = \ln(w) + C = \ln(x^2+4) + C.$$

Notice that the derivative of the inside function must be present in the integral for this method to work. The method works, however, even when the derivative is missing a constant factor, as in the next two examples.

Example 3 Find $\int te^{(t^2+1)}\,dt$.

Solution Here the inside function is t^2+1, with derivative $2t$. Since there is a factor of t in the integrand, we try $w = t^2+1$. Then $dw = w'(t)\,dt = 2t\,dt$. Notice, however, that the original integrand has only $t\,dt$, not $2t\,dt$. We therefore write

$$\frac{1}{2}\,dw = t\,dt$$

and then substitute:

$$\int te^{(t^2+1)}\,dt = \int e^{\overbrace{(t^2+1)}^{w}} \cdot \underbrace{t\,dt}_{\frac{1}{2}dw} = \int e^w\,\frac{1}{2}\,dw = \frac{1}{2}\int e^w\,dw = \frac{1}{2}e^w + C = \frac{1}{2}e^{(t^2+1)} + C.$$

Why didn't we put $\frac{1}{2}\int e^w\,dw = \frac{1}{2}e^w + \frac{1}{2}C$ in the preceding example? Since the constant C is arbitrary, it does not really matter whether we add C or $\frac{1}{2}C$. The convention is always to add C to whatever antiderivative we have calculated.

Example 4 Find $\int x^3\sqrt{x^4+5}\,dx$.

Solution The inside function is x^4+5, with derivative $4x^3$. The integrand has a factor of x^3, and since the only thing missing is a constant factor, we try

$$w = x^4 + 5.$$

Then

$$dw = w'(x)\,dx = 4x^3\,dx,$$

giving

$$\frac{1}{4}dw = x^3\,dx.$$

Thus,

$$\int x^3\sqrt{x^4+5}\,dx = \int \sqrt{w}\,\frac{1}{4}dw = \frac{1}{4}\int w^{1/2}\,dw = \frac{1}{4}\cdot\frac{w^{3/2}}{3/2} + C = \frac{1}{6}(x^4+5)^{3/2} + C.$$

Warning

We saw in the preceding examples that we can apply the substitution method when a *constant* factor is missing from the derivative of the inside function. However, we may not be able to use substitution if anything other than a constant factor is missing. For example, setting $w = x^4+5$ to find

$$\int x^2\sqrt{x^4+5}\,dx$$

does us no good because $x^2\,dx$ is not a constant multiple of $dw = 4x^3\,dx$. In order to use substitution, it helps if the integrand contains the derivative of the inside function, *to within a constant factor*.

Example 5 Find $\displaystyle\int \frac{t^2}{1+t^3}\,dt.$

Solution Observing that the derivative of $1+t^3$ is $3t^2$, we take $w = 1+t^3$, $dw = 3t^2\,dt$, so $\frac{1}{3}\,dw = t^2\,dt$. Thus,

$$\int \frac{t^2}{1+t^3}\,dt = \int \frac{\frac{1}{3}\,dw}{w} = \frac{1}{3}\ln|w| + C = \frac{1}{3}\ln|1+t^3| + C.$$

Since the numerator is $t^2\,dt$, we might have tried $w = t^3$. This substitution leads to the integral $\frac{1}{3}\int 1/(1+w)\,dw$. To evaluate this integral we would have to make a second substitution $u = 1+w$. There is often more than one way to do an integral by substitution.

Using Substitution with Periodic Functions

The method of substitution can be used for integrals involving periodic functions.

Example 6 Find $\displaystyle\int 3x^2\cos(x^3)\,dx.$

Solution We look for an inside function whose derivative appears—in this case x^3. We let $w = x^3$. Then $dw = w'(x)\,dx = 3x^2\,dx$. The original integrand can now be completely rewritten in terms of the new variable w:

$$\int 3x^2\cos(x^3)\,dx = \int \cos\underbrace{(x^3)}_{w}\cdot\underbrace{3x^2\,dx}_{dw} = \int \cos w\,dw = \sin w + C = \sin(x^3) + C.$$

By changing the variable to w, we have simplified the integrand to $\cos w$, which can be antidifferentiated more easily. The final step, after antidifferentiating, is to convert back to the original variable, x.

Example 7 Find $\displaystyle\int e^{\cos\theta}\sin\theta\,d\theta.$

Solution We let $w = \cos\theta$ since its derivative is $-\sin\theta$ and there is a factor of $\sin\theta$ in the integrand. This gives

$$dw = w'(\theta)\,d\theta = -\sin\theta\,d\theta,$$

so

$$-dw = \sin\theta\,d\theta.$$

Thus,

$$\int e^{\cos\theta}\sin\theta\,d\theta = \int e^w\,(-dw) = (-1)\int e^w\,dw = -e^w + C = -e^{\cos\theta} + C.$$

Problems for Section 7.2

Find the integrals in Problems 1–40. Check your answers by differentiation.

1. $\displaystyle\int 3x^2(x^3+1)^4 dx$
2. $\displaystyle\int \frac{2x}{x^2+1}dx$
3. $\displaystyle\int (x+10)^3 dx$
4. $\displaystyle\int x(x^2+9)^6 dx$
5. $\displaystyle\int 2qe^{q^2+1}dq$
6. $\displaystyle\int 5e^{5t+2}dt$
7. $\displaystyle\int te^{t^2}\,dt$
8. $\displaystyle\int e^{-x}\,dx$
9. $\displaystyle\int t^2(t^3-3)^{10}\,dt$
10. $\displaystyle\int x^2(1+2x^3)^2\,dx$
11. $\displaystyle\int x(x^2-4)^{7/2}\,dx$
12. $\displaystyle\int x(x^2+3)^2\,dx$
13. $\displaystyle\int \frac{1}{\sqrt{4-x}}\,dx$
14. $\displaystyle\int \frac{dy}{y+5}$

15. $\int 12x^2\cos(x^3)dx$

16. $\int (2t-7)^{73}\,dt$

17. $\int (x^2+3)^2\,dx$

18. $\int y^2(1+y)^2\,dy$

19. $\int \sin(3-t)\,dt$

20. $\int \sin\theta(\cos\theta+5)^7\,d\theta$

21. $\int \sqrt{\cos 3t}\sin 3t\,dt$

22. $\int \frac{t}{1+3t^2}\,dt$

23. $\int \sin^6\theta\cos\theta\,d\theta$

24. $\int x^2e^{x^3+1}\,dx$

25. $\int \sin^2 x\cos x dx$

26. $\int \sin^3\alpha\cos\alpha\,d\alpha$

27. $\int x\sin(4x^2)dx$

28. $\int e^{3x-4}dx$

29. $\int xe^{3x^2}\,dx$

30. $\int x\sqrt{3x^2+4}dx$

31. $\int \frac{q}{5q^2+8}dq$

32. $\int \frac{(\ln z)^2}{z}\,dz$

33. $\int \frac{y}{y^2+4}\,dy$

34. $\int \frac{e^t+1}{e^t+t}\,dt$

35. $\int \frac{e^{\sqrt{y}}}{\sqrt{y}}\,dy$

36. $\int \frac{\cos\sqrt{x}}{\sqrt{x}}\,dx$

37. $\int \frac{1+e^x}{\sqrt{x+e^x}}\,dx$

38. $\int \frac{e^t}{e^t+1}dt$

39. $\int \frac{x+1}{x^2+2x+19}dx$

40. $\int \frac{e^x-e^{-x}}{e^x+e^{-x}}\,dx$

41. If appropriate, evaluate the following integrals by substitution. If substitution is not appropriate, say so, and do not evaluate.

(a) $\int x\sin(x^2)\,dx$ **(b)** $\int x^2\sin x\,dx$

(c) $\int \frac{x^2}{1+x^2}\,dx$ **(d)** $\int \frac{x}{(1+x^2)^2}\,dx$

(e) $\int x^3e^{x^2}\,dx$ **(f)** $\int \frac{\sin x}{2+\cos x}\,dx$

42. **(a)** Find $\int (x+5)^2\,dx$ in two ways:

(i) By multiplying out

(ii) By substituting $w=x+5$

(b) Are the results the same? Explain.

43. Find $\int 4x(x^2+1)\,dx$ using two methods:

(a) Do the multiplication first, and then antidifferentiate.
(b) Use the substitution $w=x^2+1$.
(c) Explain how the expressions from parts (a) and (b) are different. Are they both correct?

7.3 USING THE FUNDAMENTAL THEOREM TO FIND DEFINITE INTEGRALS

In the previous section we calculated antiderivatives. In this section we see how antiderivatives are used to calculate definite integrals exactly. This calculation is based on the Fundamental Theorem:

> **Fundamental Theorem of Calculus** If F', the derivative of F, is continuous, then
>
> $$\int_a^b F'(x)\,dx = F(b)-F(a).$$

So far we have approximated definite integrals using a graph or left- and right-hand sums. The Fundamental Theorem gives us another method of calculating definite integrals. To find $\int_a^b F'(x)\,dx$, we first try to find F, and then calculate $F(b)-F(a)$. This method of computing definite integrals has an important advantage: it gives an exact answer. However, the method works only when we can find a formula for an antiderivative $F(x)$.

Example 1 Compute $\int_1^3 2x\,dx$ numerically and using the Fundamental Theorem.

Solution Using a calculator, we obtain

$$\int_1^3 2x\,dx = 8.000.$$

The Fundamental Theorem allows us to compute the integral exactly. We take $F'(x) = 2x$, so $F(x) = x^2$ and we obtain

$$\int_1^3 2x\,dx = F(3) - F(1) = 3^2 - 1^2 = 8.$$

In Example 1 we used the antiderivative $F(x) = x^2$, but $F(x) = x^2 + C$ works just as well for any constant C, because the constant cancels out when we subtract $F(a)$ from $F(b)$:

$$\int_1^3 2x\,dx = F(3) - F(1) = (3^2 + C) - (1^2 + C) = 8.$$

It is helpful to introduce a shorthand notation for $F(b) - F(a)$: we write

$$F(x)\Big|_a^b = F(b) - F(a).$$

For example:

$$\int_1^3 2x\,dx = x^2\Big|_1^3 = 3^2 - 1^2 = 8.$$

Example 2 Use the Fundamental Theorem to compute the following definite integrals:

(a) $\int_0^2 6x^2\,dx$ (b) $\int_0^2 t^3\,dt$ (c) $\int_1^2 (8x+5)\,dx$ (d) $\int_0^1 8e^{2t}\,dt$

Solution (a) Since $F'(x) = 6x^2$, we take $F(x) = 6(x^3/3) = 2x^3$. So

$$\int_0^2 6x^2 dx = 6\left(\frac{x^3}{3}\right)\Big|_0^2 = 2x^3\Big|_0^2 = 2\cdot 2^3 - 2\cdot 0^3 = 16.$$

(b) Since $F'(t) = t^3$, we take $F(t) = t^4/4$, so

$$\int_0^2 t^3\,dt = F(t)\Big|_0^2 = F(2) - F(0) = \frac{2^4}{4} - \frac{0^4}{4} = \frac{16}{4} - 0 = 4.$$

(c) Since $F'(x) = 8x + 5$, we take $F(x) = 4x^2 + 5x$, giving

$$\begin{aligned}\int_1^2 (8x+5)dx = (4x^2+5x)\Big|_1^2 &= (4\cdot 2^2 + 5\cdot 2) - (4\cdot 1^2 + 5\cdot 1)\\ &= 26 - 9 = 17.\end{aligned}$$

(d) Since $F'(t) = 8e^{2t}$, we take $F(t) = (8e^{2t})/2 = 4e^{2t}$, so

$$\int_0^1 8e^{2t}dt = 8\left(\frac{1}{2}e^{2t}\right)\Big|_0^1 = 4e^{2t}\Big|_0^1 = 4e^2 - 4e^0 = 25.556.$$

Example 3 Write a definite integral to represent the area under the graph of $f(t) = e^{0.5t}$ between $t = 0$ and $t = 4$. Use the Fundamental Theorem to calculate the area.

Solution The function is graphed in Figure 7.1. We have

$$\text{Area} = \int_0^4 e^{0.5t}dt = 2e^{0.5t}\Big|_0^4 = 2e^{0.5(4)} - 2e^{0.5(0)} = 2e^2 - 2 = 12.778.$$

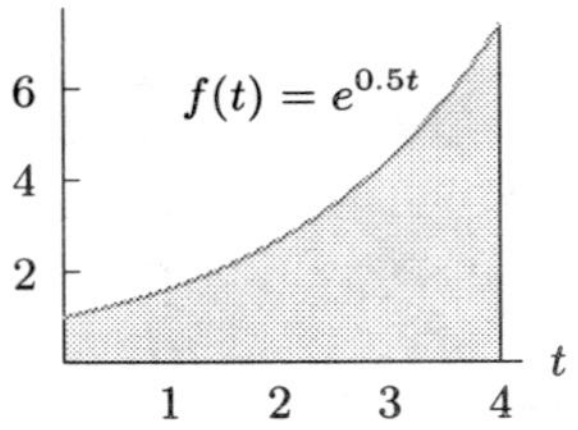

Figure 7.1: Shaded area $= \int_0^4 e^{0.5t}dt$

Definite Integrals by Substitution

The following example shows two ways of computing a definite integral by substitution.

Example 4 Compute $\displaystyle\int_0^2 xe^{x^2}\,dx.$

Solution To evaluate this definite integral using the Fundamental Theorem of Calculus, we first need to find an antiderivative of $f(x) = xe^{x^2}$. The inside function is x^2, so we let $w = x^2$. Then $dw = 2x\,dx$, so $\frac{1}{2}\,dw = x\,dx$. Thus,

$$\int xe^{x^2}\,dx = \int e^w \frac{1}{2}\,dw = \frac{1}{2}e^w + C = \frac{1}{2}e^{x^2} + C.$$

Now we find the definite integral:

$$\int_0^2 xe^{x^2}\,dx = \frac{1}{2}e^{x^2}\Big|_0^2 = \frac{1}{2}(e^4 - e^0) = \frac{1}{2}(e^4 - 1).$$

There is another way to look at the same problem. After we established that

$$\int xe^{x^2}\,dx = \frac{1}{2}e^w + C,$$

our next two steps were to replace w by x^2, and then x by 2 and 0. We could have directly replaced the original limits of integration, $x = 0$ and $x = 2$, by the corresponding w limits. Since $w = x^2$, the w limits are $w = 0^2 = 0$ (when $x = 0$) and $w = 2^2 = 4$ (when $x = 2$), so we get

$$\int_{x=0}^{x=2} xe^{x^2}\,dx = \frac{1}{2}\int_{w=0}^{w=4} e^w\,dw = \frac{1}{2}e^w\Big|_0^4 = \frac{1}{2}\left(e^4 - e^0\right) = \frac{1}{2}(e^4 - 1).$$

As we would expect, both methods give the same answer.

To Use Substitution to Find Definite Integrals

Either

- Compute the indefinite integral, expressing an antiderivative in terms of the original variable, and then evaluate the result at the original limits,

or

- Convert the original limits to new limits in terms of the new variable and do not convert the antiderivative back to the original variable.

Improper Integrals

So far, in our discussion of the definite integral $\int_a^b f(x)\,dx$, we have assumed that the interval $a \le x \le b$ is of finite length and the integrand f is continuous. An *improper integral* is a definite integral in which one (or both) of the limits of integration is infinite or the integrand is unbounded. An example of an improper integral is

$$\int_1^\infty \frac{1}{x^2}\,dx$$

This integral represents the area under the graph of $\frac{1}{x^2}$ from $x = 1$ infinitely far to the right. (See Figure 7.2.)

We estimate this area by letting the upper limit of integration get larger and larger. We see that

$$\int_1^{10} \frac{1}{x^2}\,dx = 0.9, \qquad \int_1^{100} \frac{1}{x^2}\,dx = 0.99, \qquad \int_1^{1000} \frac{1}{x^2}\,dx = 0.999,$$

and so on. These calculations suggest that as the upper limit of integration tends to infinity, the area tends to 1. We say that the improper integral $\int_1^\infty \frac{1}{x^2}\,dx$ *converges* to 1. To show that the integral converges to exactly 1 (and not to 1.0001, say), we need to use the Fundamental Theorem of Calculus. (See Problem 38.) It may seem strange that the region shaded in Figure 7.2 (which has infinite length) can have *finite* area. The area is finite because the values of the function $1/x^2$ shrink to zero so fast as $x \to \infty$. In other examples (where the integrand does not shrink to zero so fast), the area represented by an improper integral may not be finite. In that case, we say the improper integral *diverges*. (See Problem 52.)

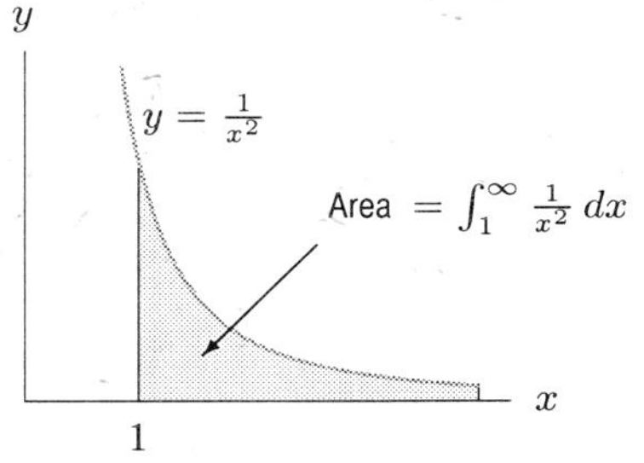

Figure 7.2: Area representation of improper integral

Problems for Section 7.3

Using the Fundamental Theorem, evaluate the definite integrals in Problems 1–20 exactly.

1. $\int_1^3 5\,dx$

2. $\int_0^4 6x\,dx$

3. $\int_1^2 (2x+3)\,dx$

4. $\int_0^2 (3t^2+4t+3)\,dt$

5. $\int_1^2 \frac{1}{t^2}\,dt$

6. $\int_1^4 \frac{1}{\sqrt{x}}\,dx$

7. $\int_0^5 3x^2\,dx$

8. $\int_0^3 t^3\,dt$

9. $\int_1^3 6x^2\,dx$

10. $\int_1^2 5t^3\,dt$

11. $\int_4^9 \sqrt{x}\,dx$

12. $\int_0^1 (y^2+y^4)\,dy$

13. $\int_1^2 \frac{1}{2t}\,dt$

14. $\int_2^5 (x^3-\pi x^2)\,dx$

15. $\int_0^1 e^{-0.2t}\,dt$

16. $\int_0^1 2e^x\,dx$

17. $\int_{-1}^1 \cos t\,dt$

18. $\int_0^{\pi/4} (\sin t + \cos t)\,dt$

19. $\int_0^3 e^{0.05t}\,dt$

20. $\int_0^1 (6q^2+4)\,dq$

21. Use substitution to express each of the following integrals as a multiple of $\int_a^b (1/w)\,dw$ for some a and b. Then evaluate the integrals.

(a) $\displaystyle\int_0^1 \frac{x}{1+x^2}\,dx$ **(b)** $\displaystyle\int_0^{\pi/4} \frac{\sin x}{\cos x}\,dx$

Use integration by substitution and the Fundamental Theorem to evaluate the definite integrals in Problems 22–25.

22. $\displaystyle\int_0^2 x(x^2+1)^2\,dx$ **23.** $\displaystyle\int_0^3 \frac{2x}{x^2+1}\,dx$

24. $\displaystyle\int_0^1 2te^{-t^2}\,dt$ **25.** $\displaystyle\int_0^3 \frac{1}{\sqrt{t+1}}\,dt$

26. Use the Fundamental Theorem to find the area under $f(x) = x^2$ between $x = 1$ and $x = 4$.

27. Write the definite integral for the area under the graph of $f(x) = 6x^2 + 1$ between $x = 0$ and $x = 2$. Use the Fundamental Theorem of Calculus to evaluate it.

28. Use the Fundamental Theorem to find the average value of $f(x) = x^2 + 1$ on the interval $x = 0$ to $x = 10$. Illustrate your answer on a graph of $f(x)$.

29. Use the Fundamental Theorem of Calculus to find the average value of $f(x) = e^{0.5x}$ between $x = 0$ and $x = 3$. Show the average value on a graph of $f(x)$.

30. Find the exact area of the region bounded by the x-axis and the graph of $y = x^3 - x$.

31. Use the Fundamental Theorem to determine the value of b if the area under the graph of $f(x) = 4x$ between $x = 1$ and $x = b$ is equal to 240. Assume $b > 1$.

32. Use the Fundamental Theorem to determine the value of b if the area under the graph of $f(x) = 8x$ between $x = 1$ and $x = b$ is equal to 192. Assume $b > 1$.

33. Use the Fundamental Theorem to determine the value of b if the area under the graph of $f(x) = x^2$ between $x = 0$ and $x = b$ is equal to 100. Assume $b > 0$.

34. Oil is leaking out of a ruptured tanker at the rate of $r(t) = 50e^{-0.02t}$ thousand liters per minute.

(a) At what rate, in liters per minute, is oil leaking out at $t = 0$? At $t = 60$?

(b) How many liters leak out during the first hour?

35. With t in years since 2000, the population, P, of the world in billions can be modeled by $P = 6.1e^{0.012t}$.

(a) What does this model predict for the world population in 2010? In 2020?

(b) Use the Fundamental Theorem to predict the average population of the world between 2000 and 2010.

36. **(a)** Graph $f(x) = e^{-x^2}$ and shade the area represented by the improper integral $\int_{-\infty}^{\infty} e^{-x^2}\,dx$.

(b) Find $\int_{-a}^{a} e^{-x^2}\,dx$ for $a = 1$, $a = 2$, $a = 3$, $a = 5$.

(c) The improper integral $\int_{-\infty}^{\infty} e^{-x^2}\,dx$ converges to a finite value. Use your answers from part (b) to estimate that value.

37. Graph $y = 1/x^2$ and $y = 1/x^3$ on the same axes. Which do you think is larger: $\int_1^{\infty} 1/x^2\,dx$ or $\int_1^{\infty} 1/x^3\,dx$? Why?

38. In this problem, you will show that the following improper integral converges to 1.

$$\int_1^{\infty} \frac{1}{x^2}\,dx.$$

(a) Use the Fundamental Theorem to find $\int_1^b 1/x^2\,dx$. Your answer will contain b.

(b) Now take the limit as $b \to \infty$. What does this tell you about the improper integral?

39. Decide if the improper integral $\int_0^{\infty} e^{-2t}\,dt$ converges, and if so, to what value, by the following method.

(a) Evaluate $\int_0^b e^{-2t}\,dt$ for $b = 3, 5, 7, 10$. What do you observe? Make a guess about the convergence of the improper integral.

(b) Find $\int_0^b e^{-2t}\,dt$ using the Fundamental Theorem. Your answer will contain b.

(c) Take a limit as $b \to \infty$. Does your answer confirm your guess?

40. **(a)** Evaluate $\int_0^b xe^{-x/10}\,dx$ for $b = 10, 50, 100, 200$.

(b) Assuming that it converges, estimate the value of $\int_0^{\infty} xe^{-x/10}\,dx$.

41. The rate, r, at which people get sick during an epidemic of the flu can be approximated by

$$r = 1000te^{-0.5t},$$

where r is measured in people/day and t is measured in days since the start of the epidemic.

(a) Write an improper integral representing the total number of people that get sick.

(b) Use a graph of r to represent the improper integral from part (a) as an area.

42. Find the exact area below the curve $y = x^3(1-x)$ and above the x-axis.

43. Find the exact area enclosed by the curve $y = x^2(1-x)^2$ and the x-axis.

44. A car moves with velocity, v, at time t in hours by

$$v(t) = \frac{60}{50^t} \text{ miles/hour.}$$

(a) Does the car ever stop?

(b) Write an integral representing the total distance traveled for $t \geq 0$.

(c) Do you think the car goes a finite distance for $t \geq 0$? If so, estimate that distance.

45. An island has a carrying capacity of 1 million rabbits. (That is, no more than 1 million rabbits can be supported by the island.) The rabbit population is two at time $t = 1$ day and grows at a rate of $r(t)$ thousand rabbits/day until the carrying capacity is reached. For each of the following formulas for $r(t)$, is the carrying capacity ever reached? Explain your answer.

(a) $r(t) = 1/t^2$ **(b)** $r(t) = t$
(c) $r(t) = 1/\sqrt{t}$

46. (a) Between 1995 and 2005, ACME Widgets sold widgets at a continuous rate of $R = R_0 e^{0.15t}$ widgets per year, where t is time in years since January 1, 1995. Suppose they were selling widgets at a rate of 1000 per year on January 1, 1995. How many widgets did they sell between 1995 and 2005? How many did they sell if the rate on January 1, 1995 was 150,000,000 widgets per year?

(b) In the first case (1000 widgets per year on January 1, 1995), how long did it take for half the widgets in the ten-year period to be sold? In the second case (150,000,000 widgets per year on January 1, 1995), when had half the widgets in the ten-year period been sold?

(c) In 2005, ACME advertised that half the widgets it had sold in the previous ten years were still in use. Based on your answer to part (b), how long must a widget last in order to justify this claim?

7.4 INTEGRATION BY PARTS

Now we introduce *integration by parts*, which is a technique for finding integrals based on the product rule. To evaluate

$$\int xe^x\,dx,$$

we look for a function whose derivative is xe^x. The product rule might lead us to guess xe^x, because we know that its derivative has two terms, one of which is xe^x:

$$\frac{d}{dx}(xe^x) = \frac{d}{dx}(x)e^x + x\frac{d}{dx}(e^x) = e^x + xe^x.$$

Of course, our guess is wrong because of the extra e^x. But we can adjust our guess by subtracting e^x; this leads us to try $xe^x - e^x$. Let's check it:

$$\frac{d}{dx}(xe^x - e^x) = \frac{d}{dx}(xe^x) - \frac{d}{dx}(e^x) = e^x + xe^x - e^x = xe^x.$$

It works, so $\int xe^x\,dx = xe^x - e^x + C$.

The General Formula for Integration by Parts

We can formalize this process in the following way. We begin with the product rule:

$$\frac{d}{dx}(uv) = u'v + uv'$$

where u and v are functions of x with derivatives u' and v', respectively. We rewrite this as:

$$uv' = \frac{d}{dx}(uv) - u'v$$

and then integrate both sides:

$$\int uv'\,dx = \int \frac{d}{dx}(uv)\,dx - \int u'v\,dx.$$

Since an antiderivative of $\frac{d}{dx}(uv)$ is just uv, we get the following formula:

Integration by Parts

$$\int uv'\,dx = uv - \int u'v\,dx.$$

This technique is useful when the integrand can be viewed as a product and when the integral on the right-hand side is simpler than that on the left.

Example 1 Use integration by parts to find $\int xe^x\,dx$.

Solution We let $xe^x = (x)\cdot(e^x) = uv'$, and choose $u = x$ and $v' = e^x$. Thus, $u' = 1$ and $v = e^x$, so

$$\int \underbrace{(x)}_{u}\underbrace{(e^x)}_{v'}\,dx = \underbrace{(x)}_{u}\underbrace{(e^x)}_{v} - \int \underbrace{(1)}_{u'}\underbrace{(e^x)}_{v}\,dx = xe^x - e^x + C.$$

Let's look again at Example 1. Notice what would have happened if we took $v = e^x + C_1$. Then

$$\begin{aligned}\int xe^x\,dx &= x(e^x + C_1) - \int (e^x + C_1)\,dx \\ &= xe^x + C_1x - e^x - C_1x + C \\ &= xe^x - e^x + C,\end{aligned}$$

as before. Thus, it is not necessary to include an arbitrary constant in the antiderivative for v; any antiderivative will do.

What would have happened if we had picked u and v' the other way around in Example 1? If $u = e^x$ and $v' = x$, then $u' = e^x$ and $v = x^2/2$. The formula for integration by parts then gives

$$\int xe^x\,dx = \frac{x^2}{2}e^x - \int \frac{x^2}{2}\cdot e^x\,dx,$$

which is true but not helpful, since the new integral on the right seems harder than the original one on the left. To use this method, we must choose u and v' to make the integral on the right no harder to find than the integral on the left.

How to Choose u and v'

- Whatever you let v' be, you need to be able to find v.
- It helps if $u'v$ is simpler (or at least no more complicated) than uv'.

Example 2 Find $\int xe^{3x}\,dx$.

Solution We let $xe^{3x} = (x)\cdot(e^{3x}) = uv'$, and choose $u = x$ and $v' = e^{3x}$. Thus, $u' = 1$ and $v = \frac{1}{3}e^{3x}$. We have

$$\int \underbrace{(x)}_{u}\underbrace{(e^{3x})}_{v'}\,dx = \underbrace{(x)}_{u}\underbrace{\left(\frac{1}{3}e^{3x}\right)}_{v} - \int \underbrace{(1)}_{u'}\underbrace{\left(\frac{1}{3}e^{3x}\right)}_{v}\,dx = \frac{1}{3}xe^{3x} - \frac{1}{9}e^{3x} + C.$$

There are some examples which don't look like good candidates for integration by parts because they don't appear to involve products, but for which the method works well. Such examples often involve $\ln x$ or the inverse trigonometric functions. Here is one:

Example 3 Find $\int_2^3 \ln x\,dx$.

Solution This does not look like a product unless we write $\ln x = (1)(\ln x)$. Then we might say $u = 1$ so $u' = 0$, which certainly makes things simpler. But if $v' = \ln x$, what is v? If we knew, we would not need integration by parts. Let's try the other way: if $u = \ln x$, $u' = 1/x$ and if $v' = 1$, $v = x$, so

$$\int_2^3 \underbrace{(\ln x)}_{u}\underbrace{(1)}_{v'}\,dx = \underbrace{(\ln x)}_{u}\underbrace{(x)}_{v}\Big|_2^3 - \int_2^3 \underbrace{\left(\frac{1}{x}\right)}_{u'}\cdot\underbrace{(x)}_{v}\,dx$$

$$= x\ln x\Big|_2^3 - \int_2^3 1\,dx = (x\ln x - x)\Big|_2^3$$

$$= 3\ln 3 - 3 - 2\ln 2 + 2 = 3\ln 3 - 2\ln 2 - 1.$$

Notice that when doing a definite integral by parts, we must remember to put the limits of integration (here 2 and 3) on the uv term (in this case $x\ln x$) as well as on the integral $\int u'v\,dx$.

Example 4 Find $\int x^6 \ln x\,dx$.

Solution We can write $x^6\ln x$ as uv' where $u = \ln x$ and $v' = x^6$. Then $v = \frac{1}{7}x^7$ and $u' = 1/x$, so integration by parts gives us:

$$\int x^6\ln x\,dx = \int(\ln x)x^6\,dx = (\ln x)\left(\frac{1}{7}x^7\right) - \int \frac{1}{7}x^7\cdot\frac{1}{x}\,dx$$

$$= \frac{1}{7}x^7\ln x - \frac{1}{7}\int x^6\,dx$$

$$= \frac{1}{7}x^7\ln x - \frac{1}{49}x^7 + C.$$

In Example 4 we did not choose $v' = \ln x$, because it is not immediately clear what v would be. In fact, we used integration by parts in Example 3 to find the antiderivative of $\ln x$. Also, using $u = \ln x$, as we have done, gives $u'v = x^6/7$, which is simpler to integrate than $uv' = x^6\ln x$. This example shows that u does not have to be the first factor in the integrand (here x^6).

Problems for Section 7.4

Find the integrals in Exercises 1–14.

1. $\int te^{5t}\,dt$

2. $\int pe^{-0.1p}\,dp$

3. $\int (z+1)e^{2z}\,dz$

4. $\int y\ln y\,dy$

5. $\int x^3\ln x\,dx$

6. $\int q^5\ln 5q\,dq$

7. $\int y\sqrt{y+3}\,dy$

8. $\int (t+2)\sqrt{2+3t}\,dt$

9. $\int \frac{z}{e^z}\,dz$

10. $\int \frac{\ln x}{x^2}\,dx$

11. $\int \frac{y}{\sqrt{5-y}}\,dy$

12. $\int \frac{t+7}{\sqrt{5-t}}\,dt$

13. $\int t\sin t\,dt$

14. $\int_3^5 x\cos x\,dx$

In Problems 15–16, use integration by parts twice to evaluate the integral.

15. $\int t^2e^{5t}\,dt$

16. $\int (\ln t)^2\,dt$

Evaluate the integrals in Problems 17–20 both exactly [e.g. $\ln(3\pi)$] and numerically [e.g. $\ln(3\pi)\approx 2.243$].

17. $\int_1^5 \ln t\,dt$

18. $\int_0^{10} ze^{-z}\,dz$

19. $\int_1^3 t\ln t\,dt$

20. $\int_0^5 \ln(1+t)\,dt$

21. For each of the following integrals, indicate whether integration by substitution or integration by parts is more appropriate. Do not evaluate the integrals.

(a) $\displaystyle\int \frac{x^2}{1+x^3}\,dx$ **(b)** $\displaystyle\int xe^{x^2}\,dx$

(c) $\displaystyle\int x^2\ln(x^3+1)\,dx$ **(d)** $\displaystyle\int \frac{1}{\sqrt{3x+1}}\,dx$

(e) $\displaystyle\int x^2\ln x\,dx$ **(f)** $\displaystyle\int \ln x\,dx$

22. Find $\int_1^2 \ln x\,dx$ numerically. Find $\int_1^2 \ln x\,dx$ using antiderivatives. Check that your answers agree.

In Problems 23–25, find the exact area.

23. Under $y = te^{-t}$ for $0 \le t \le 2$.

24. Between $y = \ln x$ and $y = \ln(x^2)$ for $1 \le x \le 2$.

25. Between $f(t) = \ln(t^2 - 1)$ and $g(t) = \ln(t - 1)$ for $2 \le t \le 3$.

26. Estimate $\int_0^{10} f(x)g'(x)\,dx$ if $f(x) = x^2$ and g has the values in the following table.

x	0	2	4	6	8	10
$g(x)$	2.3	3.1	4.1	5.5	5.9	6.1

27. The concentration, C, in ng/ml, of a drug in the blood as a function of the time, t, in hours since the drug was administered is given by $C = 15te^{-0.2t}$. The area under the concentration curve is a measure of the overall effect of the drug on the body, called the bioavailability. Find the bioavailability of the drug between $t = 0$ and $t = 3$.

28. During a surge in the demand for electricity, the rate, r, at which energy is used can be approximated by

$$r = te^{-at},$$

where t is the time in hours and a is a positive constant.

(a) Find the total energy, E, used in the first T hours. Give your answer as a function of a.

(b) What happens to E as $T \to \infty$?

29. Derive the formula (called a reduction formula):

$$\int x^n e^x\,dx = x^n e^x - n\int x^{n-1}e^x\,dx.$$

7.5 ANALYZING ANTIDERIVATIVES GRAPHICALLY AND NUMERICALLY

In this section, we use the Fundamental Theorem to approximate values of F when the rate of change, F', and one value of the function, $F(a)$, are known.

Example 1 Suppose $F'(t) = (1.8)^t$ and $F(0) = 2$. Find the value of $F(b)$ for $b = 0, 0.1, 0.2, \ldots, 1.0$.

Solution Apply the Fundamental Theorem with $F'(t) = (1.8)^t$ and $a = 0$ to get values for $F(b)$. Since

$$F(b) - F(0) = \int_0^b F'(t)\,dt = \int_0^b (1.8)^t\,dt$$

and $F(0) = 2$, we have

$$F(b) = 2 + \int_0^b (1.8)^t\,dt.$$

Use a calculator or computer to estimate the definite integral $\int_0^b (1.8)^t\,dt$ for each value of b. For example, when $b = 0.1$, we find that $\int_0^b (1.8)^t\,dt = 0.103$. Thus $F(0.1) = 2.103$. Continuing in this way gives the values in Table 7.1.

Table 7.1 *Approximate values of F*

b	0	0.1	0.2	0.3	0.4	0.5	0.6	0.7	0.8	0.9	1.0
$F(b)$	2	2.103	2.212	2.328	2.451	2.581	2.719	2.866	3.021	3.186	3.361

Notice from the table that the function $F(b)$ is increasing between $b = 0$ and $b = 1$. This is because the derivative $F'(t) = (1.8)^t$ is positive for t between 0 and 1.

Example 2 The graph of the derivative F' of a function F is shown in Figure 7.3. Assuming that $F(20) = 150$, estimate the maximum value attained by F.

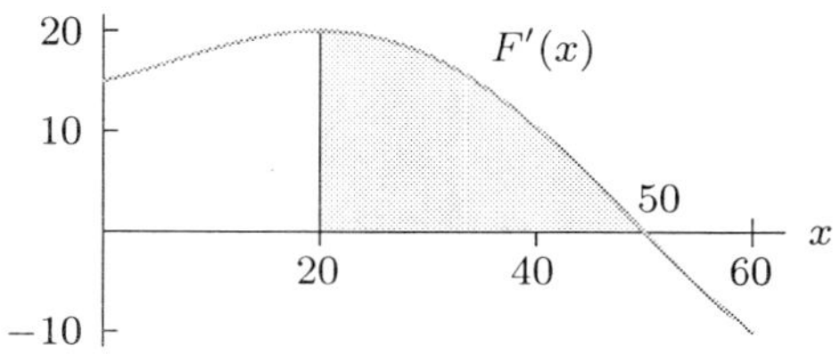

Figure 7.3: Graph of the derivative F' of some function F

Solution We know that $F(x)$ increases for $x < 50$ because the derivative of F is positive for $x < 50$. Similarly, $F(x)$ decreases for $x > 50$ because $F'(x)$ is negative for $x > 50$. Therefore, the graph of F rises until the point at which $x = 50$, and then it begins to fall. So the highest point on the graph of F is at $x = 50$ and the maximum value attained by F is $F(50)$. By the Fundamental Theorem:

$$F(50) - F(20) = \int_{20}^{50} F'(x)\,dx.$$

Since $F(20) = 150$, we have

$$F(50) = F(20) + \int_{20}^{50} F'(x)\,dx = 150 + \int_{20}^{50} F'(x)\,dx.$$

The definite integral is the area of the shaded region under the graph of F', which is roughly a triangle of base 30 and height 20. Therefore, the shaded area is about 300 and the maximum value attained by F is $F(50) \approx 150 + 300 = 450$.

Graphing a Function Given a Graph of its Derivative

Suppose we have the graph of f' and we want to sketch the graph of f. We know that when f' is positive, f is increasing, and when f' is negative, f is decreasing. If we want to know how much f increases or decreases, we compute a definite integral.

Example 3 Figure 7.4 shows the rate of change of the concentration of adrenaline, in micrograms per milliliter per minute, in a person's body. Sketch a graph of the concentration of adrenaline, in micrograms per milliliter, in the body as a function of time, in minutes.

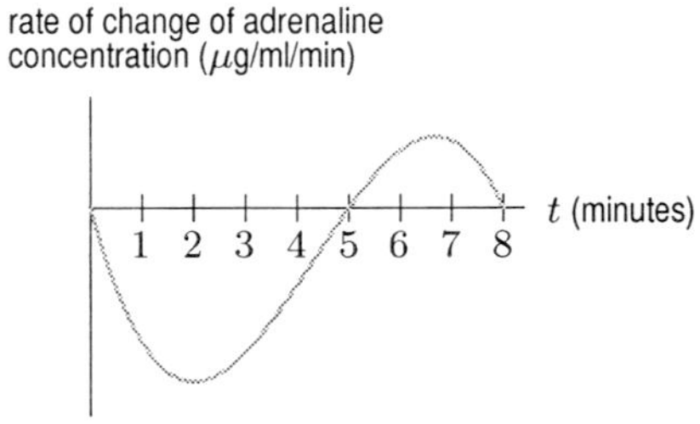

Figure 7.4: Rate of change of adrenaline concentration

Solution Since the initial concentration of the drug is not given, we can start the graph anywhere on the positive vertical axis. The rate of change is negative for $t < 5$ and positive for $t > 5$, so the concentration of adrenaline decreases until $t = 5$ and then increases. Since the area below the t-axis is greater than the area above the t-axis, the concentration of adrenaline decreases more than it increases. Thus, the concentration at $t = 8$ is less than the concentration at $t = 0$. Since the rate of change of concentration is zero at $t = 0$, 5, and 8, the graph of concentration is horizontal at these points. See Figure 7.5.

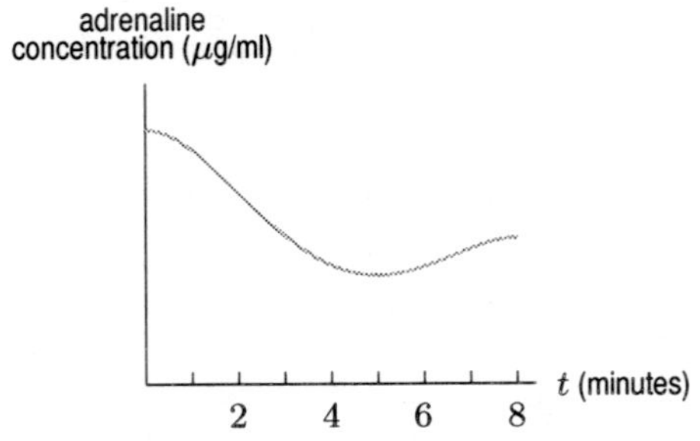

Figure 7.5: Adrenaline concentration

Example 4 Figure 7.6 shows the derivative $f'(x)$ of a function $f(x)$ and the values of some areas. If $f(0) = 10$, sketch a graph of the function $f(x)$. Give the coordinates of the local maxima and minima.

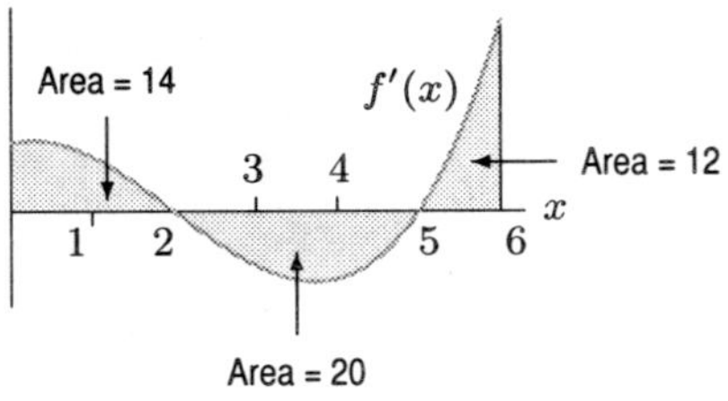

Figure 7.6: The graph of a derivative f'

Solution Figure 7.6 shows that the derivative f' is positive between 0 and 2, negative between 2 and 5, and positive between 5 and 6. Therefore, the function f is increasing between 0 and 2, decreasing between 2 and 5, and increasing between 5 and 6. (See Figure 7.7.) There is a local maximum at $x = 2$ and a local minimum at $x = 5$.

Notice that we can sketch the general shape of the graph of f without knowing any areas. The areas are used to make the graph more precise. We are told that $f(0) = 10$, so we plot the point $(0, 10)$ on the graph of f in Figure 7.8. The Fundamental Theorem and Figure 7.6 show that

$$f(2) - f(0) = \int_0^2 f'(x)\,dx = 14.$$

Therefore, the total change in f between $x = 0$ and $x = 2$ is 14. Since $f(0) = 10$, we have

$$f(2) = 10 + 14 = 24.$$

The point $(2, 24)$ is on the graph of f. See Figure 7.8.

Figure 7.6 shows that the area between $x = 2$ and $x = 5$ is 20. Since this area lies entirely below the x-axis, the Fundamental Theorem gives

$$f(5) - f(2) = \int_2^5 f'(x)\,dx = -20.$$

The total change in f is -20 between $x = 2$ and $x = 5$. Since $f(2) = 24$, we have

$$f(5) = 24 - 20 = 4.$$

Thus, the point $(5, 4)$ lies on the graph of f. Finally,

$$f(6) = f(5) + \int_5^6 f'(x)\,dx = 4 + 12 = 16,$$

so the point $(6, 16)$ is on the graph of f.

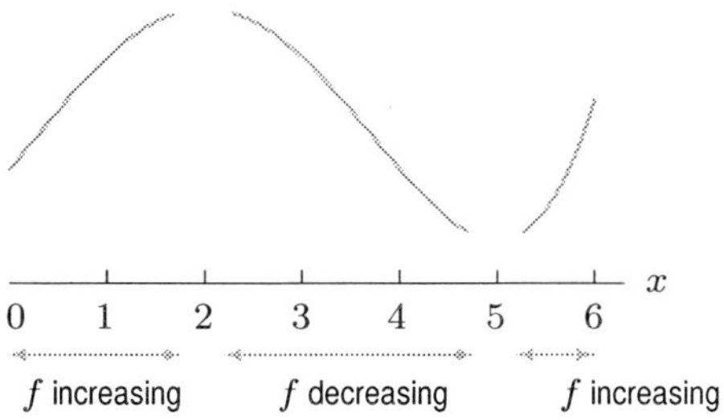

Figure 7.7: The shape of f

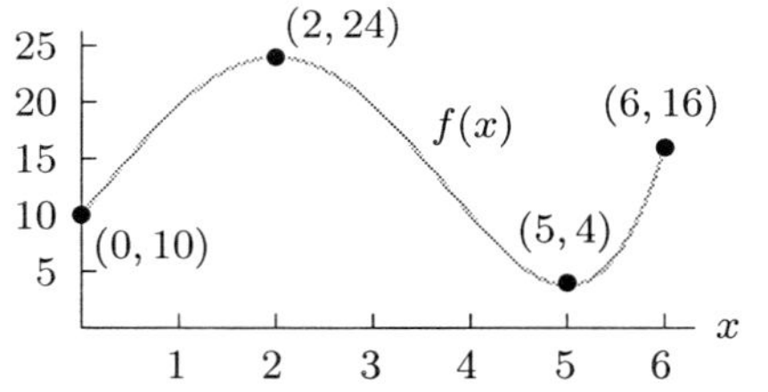

Figure 7.8: The graph of f

In this section we have seen how antiderivatives can by analyzed using the Fundamental Theorem of Calculus in the form

$$F(b) = F(0) + \int_0^b F'(t)\,dt.$$

In Section 8.2 we will see how antiderivatives can be constructed using the Second Fundamental Theorem introduced in Chapter 5's Focus on Theory section. This version of the theorem says that if f is a continuous function on an interval, and if a is any number in that interval, then the function F defined by

$$F(x) = \int_a^x f(t)\,dt$$

is an antiderivative of f; that is $F' = f$.

Problems for Section 7.5

1. The derivative $F'(t)$ is graphed in Figure 7.9. Given that $F(0) = 5$, calculate $F(t)$ for $t = 1, 2, 3, 4, 5$.

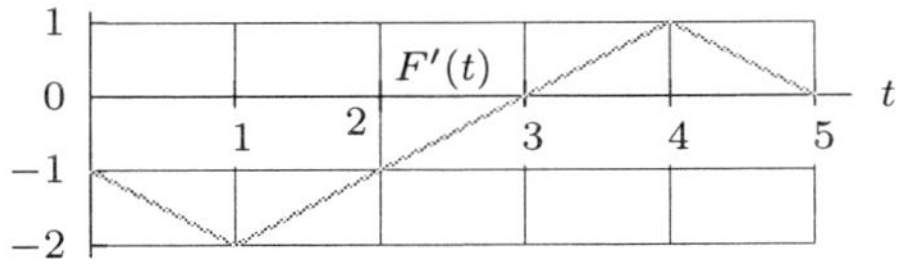

Figure 7.9

2. Figure 7.10 shows the derivative g'. If $g(0) = 0$, graph g. Give (x, y)-coordinates of all local maxima and minima.

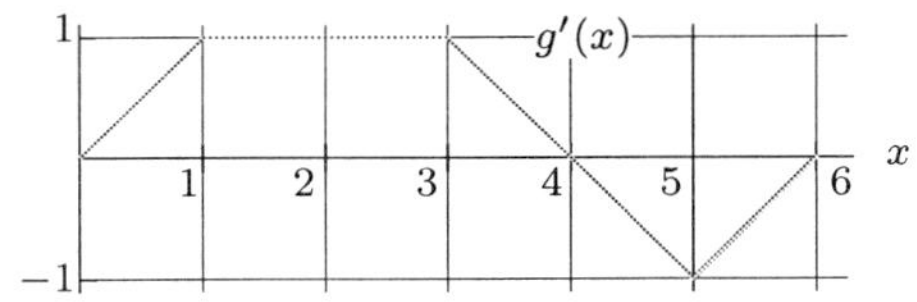

Figure 7.10

3. Figure 7.11 shows f. If $F' = f$ and $F(0) = 0$, find $F(b)$ for $b = 1, 2, 3, 4, 5, 6$.

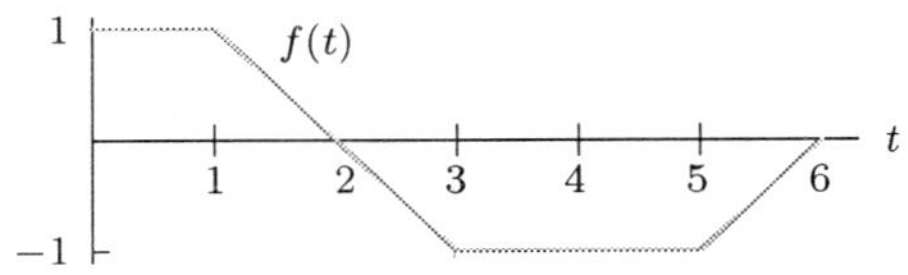

Figure 7.11

4. **(a)** Using Figure 7.12, estimate $\int_0^7 f(x)dx$.
(b) If F is an antiderivative of the same function f and $F(0) = 25$, estimate $F(7)$.

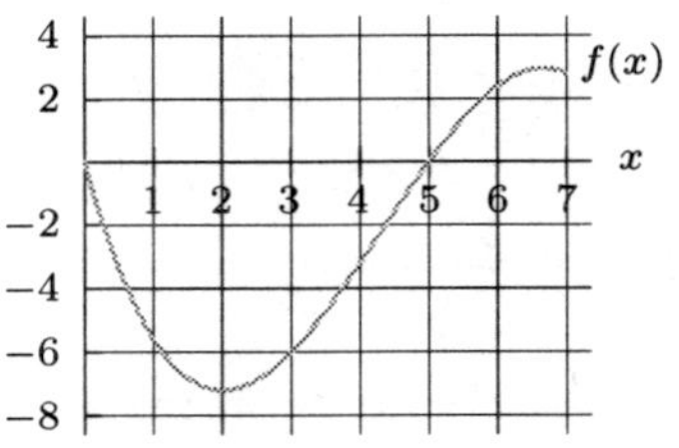

Figure 7.12

In Problems 5–8, sketch two functions F such that $F' = f$. In one case let $F(0) = 0$ and in the other, let $F(0) = 1$.

5.

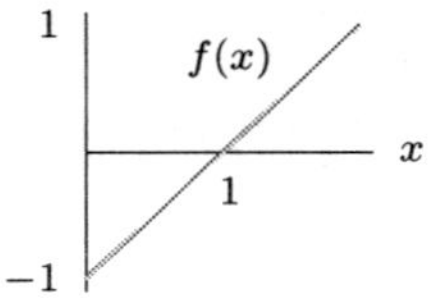

6.

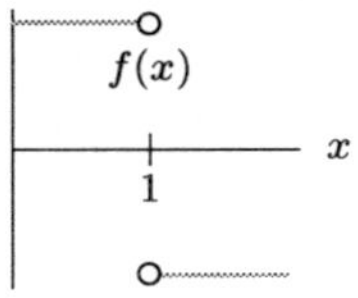

7.

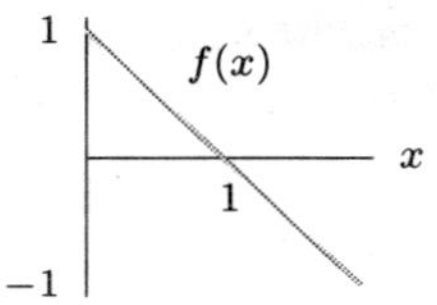

8.

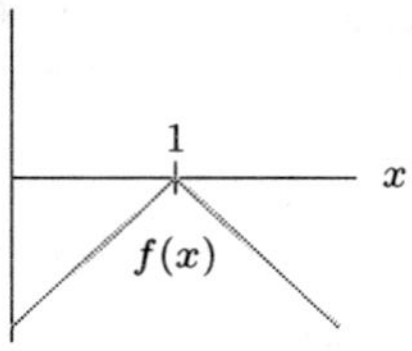

Problems 9–12 show the derivative f' of f.

(a) Where is f increasing and where is f decreasing? What are the x-coordinates of the local maxima and minima of f?

(b) Sketch a possible graph for f. (You don't need a scale on the vertical axis.)

9.

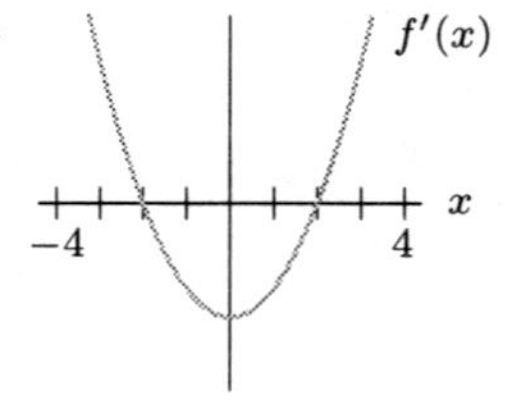

10.

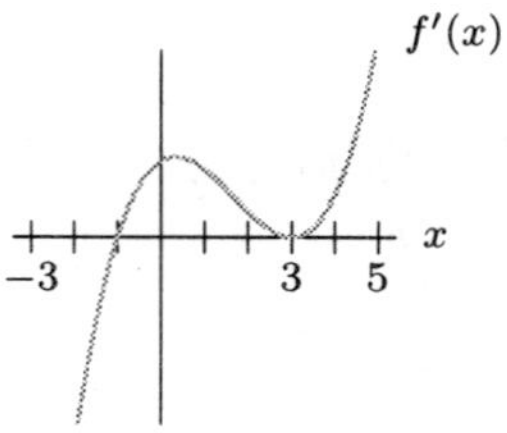

11.

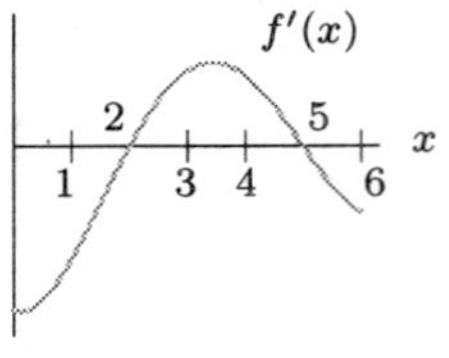

12.

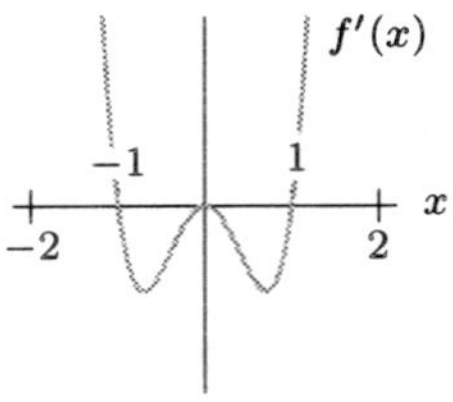

13. Figure 7.13 shows the rate at which photosynthesis is taking place in a leaf. The rate at which the leaf grows is approximately proportional to the rate of photosynthesis. Sketch the size of the leaf against time for 100 days.

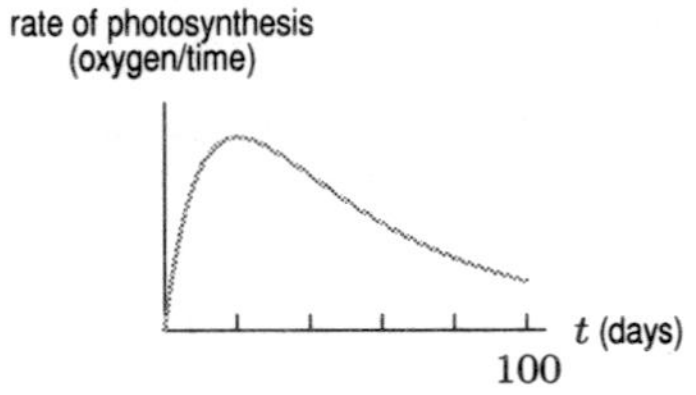

Figure 7.13

14. Figure 7.14 shows the derivative F' of F. Let $F(0) = 0$. Of the four numbers $F(1)$, $F(2)$, $F(3)$, and $F(4)$, which is largest? Which is smallest? How many of these numbers are negative?

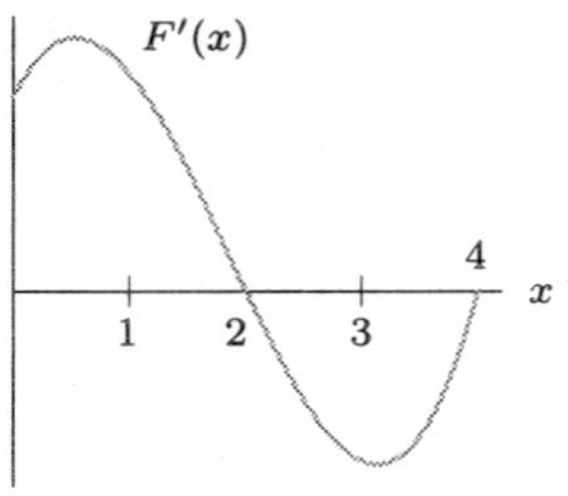

Figure 7.14

15. Urologists are physicians who specialize in the health of the bladder. In a common diagnostic test, urologists monitor the emptying of the bladder using a device that produces two graphs. In one of the graphs the flow rate (in milliliters per second) is measured as a function of time (in seconds). In the other graph, the volume emptied from the bladder is measured (in milliliters) as a function of time (in seconds). See Figure 7.15.

(a) Which graph is the flow rate and which is the volume?

(b) Which one of these graphs is an antiderivative of the other?

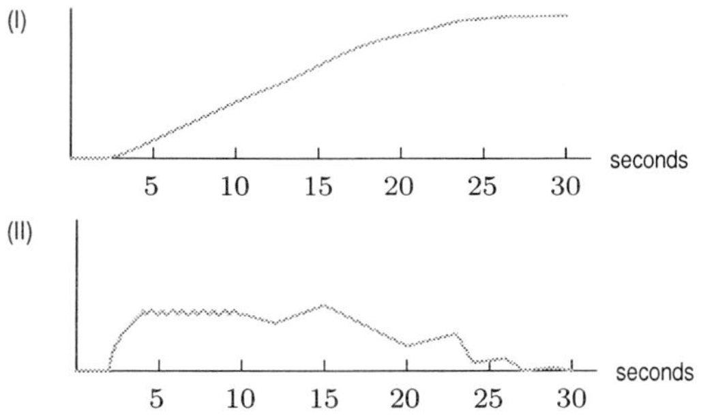

Figure 7.15

16. Use Figure 7.16 and the fact that $F(2) = 3$ to sketch the graph of $F(x)$. Label the values of at least four points.

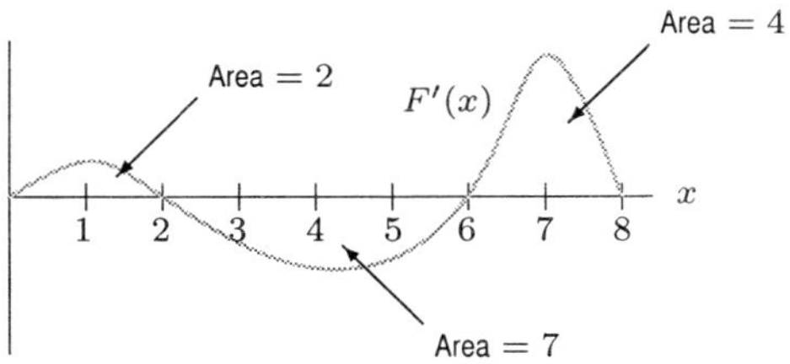

Figure 7.16

17. Figure 7.17 shows the derivative F'. If $F(0) = 14$, graph F. Give (x, y)-coordinates of all local maxima and minima.

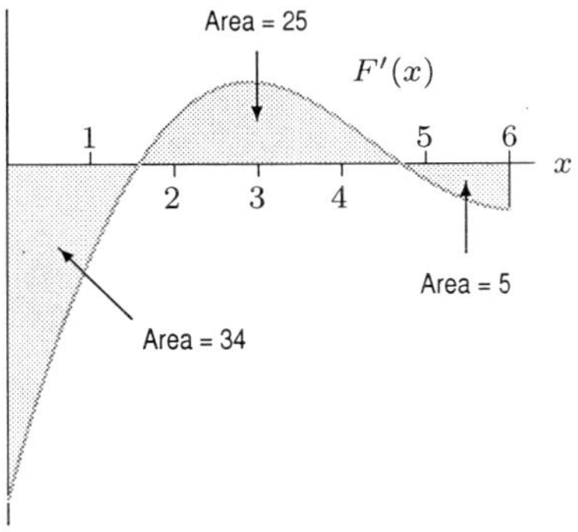

Figure 7.17

18. Figure 7.18 shows the derivative $F'(t)$. If $F(0) = 3$, find the values of $F(2)$, $F(5)$, $F(6)$. Graph $F(t)$.

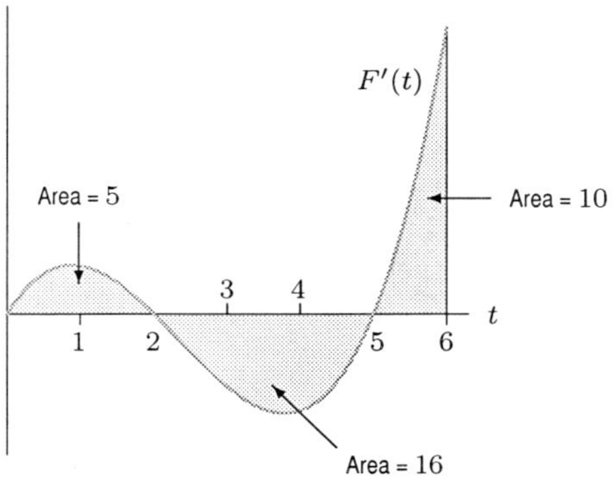

Figure 7.18

19. Using Figure 7.19, sketch a graph of an antiderivative $G(t)$ of $g(t)$ satisfying $G(0) = 5$. Label each critical point of $G(t)$ with its coordinates.

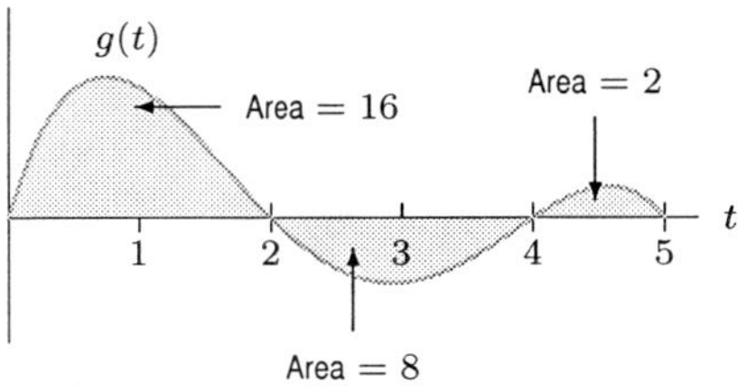

Figure 7.19

In Problems 20–21, a graph of f is given. Let $F'(x) = f(x)$.

(a) What are the critical points of $F(x)$?

(b) Which critical points are local maxima, which are local minima, and which are neither?

(c) Sketch a possible graph of $F(x)$.

20.

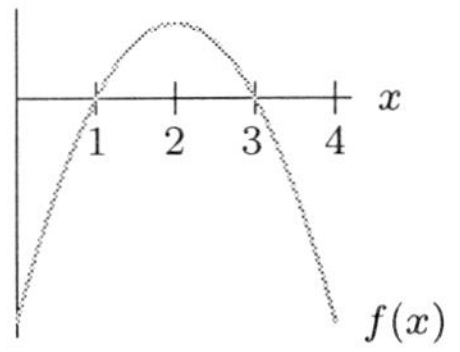

21.

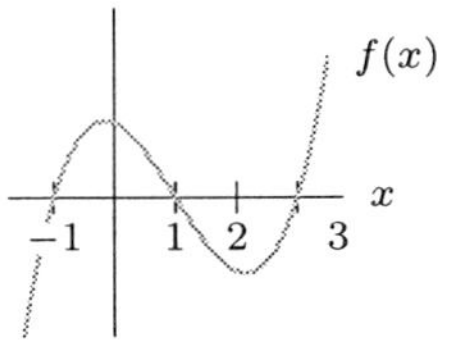

Problems 22–23 concern the graph of f' in Figure 7.20.

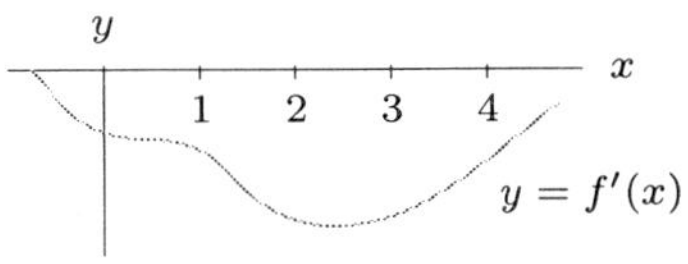

Figure 7.20: Note: Graph of f', not f

22. Which is greater, $f(0)$ or $f(1)$?

23. List the following in increasing order:

$$\frac{f(4) - f(2)}{2}, \quad f(3) - f(2), \quad f(4) - f(3).$$

For Problems 24–27, show the following quantities on Figure 7.21.

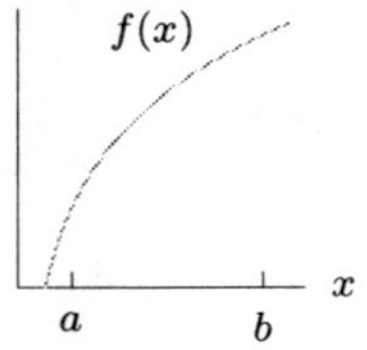

Figure 7.21

24. A length representing $f(b) - f(a)$.

25. A slope representing $\dfrac{f(b) - f(a)}{b - a}$.

26. An area representing $F(b) - F(a)$, where $F' = f$.

27. A length roughly approximating

$$\frac{F(b) - F(a)}{b - a}, \text{ where } F' = f.$$

CHAPTER SUMMARY

- **Computing antiderivatives**
  Powers and polynomials, e^{kx}, $\sin(kx)$, $\cos(kx)$
- Integration by substitution
- Integration by parts
- **Using antiderivatives to compute definite integrals analytically**: Fundamental Theorem of Calculus
- **Using definite integrals to analyze antiderivatives numerically and graphically**
- **Improper integrals**

REVIEW PROBLEMS FOR CHAPTER SEVEN

In Problems 1–10, find an antiderivative.

1. $f(t) = 2t^2 + 3t^3 + 4t^4$

2. $h(t) = 3t^2 + 7t + 1$

3. $f(x) = 6x^2 - 8x + 3$

4. $f(z) = e^z + 3$

5. $p(r) = 2\pi r$

6. $p(y) = \dfrac{1}{y} + y + 1$

7. $r(t) = \dfrac{1}{t^2}$

8. $g(t) = \sin t$

9. $g(x) = (x+1)^3$

10. $f(x) = (2x+1)^3$

Find the indefinite integrals in Problems 11–22.

11. $\displaystyle\int (4t + 7)\, dt$

12. $\displaystyle\int 3x\, dx$

13. $\displaystyle\int (x^3 - x)\, dx$

14. $\displaystyle\int e^{-0.05t}\, dt$

15. $\displaystyle\int \left(\frac{5}{t^2} + \frac{6}{t^3}\right) dt$

16. $\displaystyle\int \left(\frac{x+1}{x}\right) dx$

17. $\displaystyle\int (8t + 3)\, dt$

18. $\displaystyle\int \left(1 + \frac{1}{p}\right) dp$

19. $\displaystyle\int \left(8x^3 + \frac{1}{x}\right) dx$

20. $\displaystyle\int (5\cos x - 3\sin x) dx$

21. $\displaystyle\int \left(\frac{2}{x} + \pi \sin x\right) dx$

22. $\displaystyle\int (4x + 2e^x) dx$

Find the definite integrals in Problems 23–28 using the Fundamental Theorem.

23. $\displaystyle\int_0^2 (12x^2 + 1)\, dx$

24. $\displaystyle\int_{-3}^{-1} \frac{2}{r^3}\, dr$

25. $\displaystyle\int_0^1 \sin\theta\, d\theta$

26. $\displaystyle\int_1^2 \frac{1}{x}\, dx$

27. $\displaystyle\int_1^2 \frac{1}{x^2}\, dx$

28. $\displaystyle\int_0^2 \left(\frac{x^3}{3} + 2x\right) dx$

In Problems 29–31, find an antiderivative $F(x)$ with $F'(x) = f(x)$ and $F(0) = 0$. Is there only one possible solution?

29. $f(x) = 2x$

30. $f(x) = \dfrac{1}{4}x$

31. $\displaystyle\int \frac{2x}{\sqrt{x^2+1}} dx$

In Problems 32–41, integrate by substitution.

32. $\displaystyle\int 2x(x^2+1)^5 dx$

33. $\displaystyle\int x e^{-x^2}\, dx$

34. $\displaystyle\int \frac{4x^3}{x^4+1} dx$

35. $\displaystyle\int \frac{1}{(3x+1)^2} dx$

36. $\displaystyle\int \frac{x}{\sqrt{x^2+4}}dx$

37. $\displaystyle\int (5x-7)^{10}dx$

38. $\displaystyle\int 100e^{-0.2t}dt$

39. $\displaystyle\int x\sqrt{x^2+1}dx$

40. $\displaystyle\int x\sin(x^2)dx$

41. $\displaystyle\int t\cos(t^2)\,dt$

In Problems 42–43, find the exact area.

42. Under $f(x) = xe^{x^2}$ between $x = 0$ and $x = 2$.

43. Under $f(x) = 1/(x+1)$ between $x = 0$ and $x = 2$.

44. Find the exact average value of $f(x) = 1/(x+1)$ on the interval $x = 0$ to $x = 2$. Sketch a graph showing the function and the average value.

45. If t is in years, and $t = 0$ is January 1, 2005, worldwide energy consumption, r, in quadrillion (10^{15}) BTUs per year, is modeled by

$$r = 462e^{0.019t}.$$

(a) Write a definite integral for the total energy use between the start of 2005 and the start of 2010.
(b) Use the Fundamental Theorem of Calculus to evaluate the integral. Give units with your answer.

46. Use Figure 7.22 and the fact that $P = 2$ when $t = 0$ to find values of P when $t = 1, 2, 3, 4$ and 5.

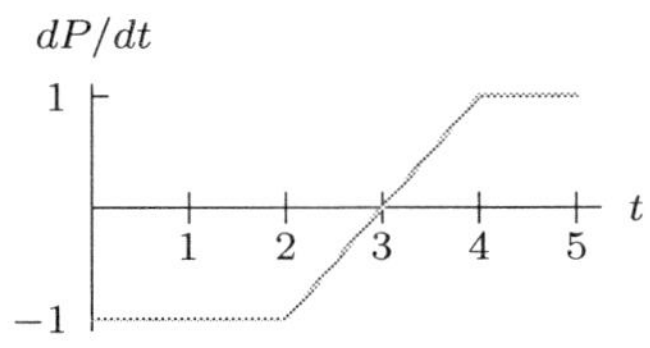

Figure 7.22

Problems 47–48 give a graph of $f'(x)$. Graph $f(x)$. Mark the points $x_1, \ldots, x_4$ on your graph and label local maxima, local minima and points of inflection.

47.

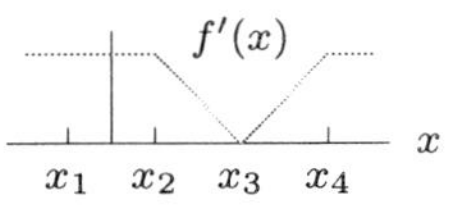

48.

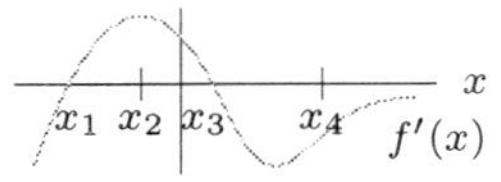

49. Figure 7.23 shows the derivative $F'(x)$. If $F(0) = 5$, find the value of $F(1)$, $F(3)$, and $F(4)$.

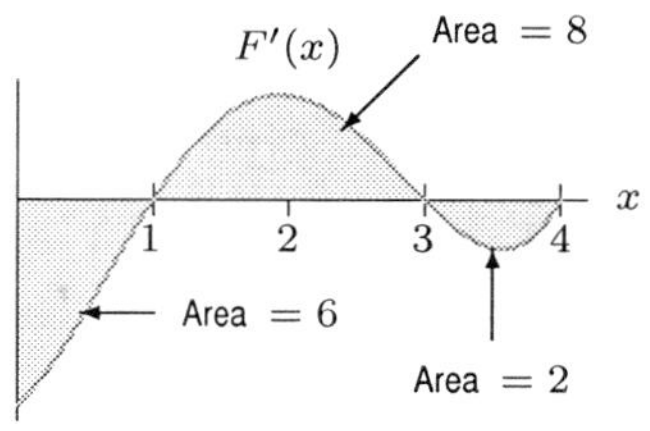

Figure 7.23

50. Suppose $\int_0^2 g(t)\,dt = 5$. Calculate the following:

(a) $\int_0^4 g(t/2)\,dt$ **(b)** $\int_0^2 g(2-t)\,dt$

51. **(a)** Sketch the area represented by the improper integral $\int_0^\infty xe^{-x}\,dx$.
(b) Calculate $\int_0^b xe^{-x}\,dx$ for $b = 5, 10, 20$.
(c) The improper integral in part (a) converges. Use your answers to part (b) to estimate its value.

52. Consider the improper integral

$$\int_1^\infty \frac{1}{\sqrt{x}}\,dx.$$

(a) Use a calculator or computer to find $\int_1^b 1/(\sqrt{x})\,dx$ for $b = 100, 1000, 10{,}000$. What do you notice?
(b) Find $\int_1^b 1/(\sqrt{x})\,dx$ using the Fundamental Theorem of Calculus. Your answer will contain b.
(c) Now take the limit as $b \to \infty$. What does this tell you about the improper integral?

53. At a time t hours after taking a tablet, the rate at which a drug is being eliminated is

$$r(t) = 50\left(e^{-0.1t} - e^{-0.2t}\right)\text{ mg/hr.}$$

Assuming that all the drug is eventually eliminated, calculate the original dose.

Use integration by parts to solve Problems 54–56.

54. $\displaystyle\int x\cos x\,dx$

55. $\displaystyle\int y\sin y\,dy$

56. $\displaystyle\int \ln(x^2)\,dx$

CHECK YOUR UNDERSTANDING

In Problems 1–50, indicate whether the statement is true or false.

1. $F(t)=\dfrac{t^3}{3}+5$ is an antiderivative of $f(t)=t^2$.
2. $F(x)=\dfrac{x^{-3}}{-3}$ is an antiderivative of $f(x)=x^{-2}$.
3. $F(x)=e^{3x}$ is an antiderivative of $f(x)=e^{3x}$.
4. $\displaystyle\int t^3\,dt=\frac{t^4}{4}+C$.
5. $\displaystyle\int \frac{1}{\sqrt{z}}\,dz=2\sqrt{z}+C$.
6. $\displaystyle\int e^x\,dx=\frac{e^{x+1}}{x+1}+C$.
7. $\displaystyle\int \ln|t|\,dt=\frac{1}{t}+C$.
8. $\displaystyle\int 2^x\,dx=\frac{1}{\ln 2}2^x+C$.
9. $\displaystyle\int 5\cos x\,dx=5\sin x+C$.
10. If $F(x)$ and $G(x)$ are both antiderivatives of $h(x)$, then $F(x)-G(x)$ is a constant.
11. The substitution $w=q^3+3q^2-q$ converts the integral $\int(q^3+3q^2-q)^{10}\,dq$ into $\int w^{10}\,dw$.
12. The substitution $w=\ln x$ converts the integral $\displaystyle\int\frac{(\ln x)^2}{x}\,dx$ into $\int w^2\,dw$.
13. The substitution $w=x^2$ converts the integral $\int e^{x^2}\,dx$ into $\int e^w\,dw$.
14. The substitution $w=x^2$ converts the integral $\int xe^{x^2}\,dx$ into $\displaystyle\int\frac{1}{2}e^w\,dw$.
15. The substitution $w=1-s$ converts the integral $\displaystyle\int\frac{1}{\sqrt{1-s}}\,ds$ into $\displaystyle\int\frac{1}{\sqrt{w}}\,dw$.
16. The substitution $w=t^2+1$ converts the integral $\displaystyle\int\frac{t}{\sqrt{t^2+1}}\,dt$ into $\displaystyle\int\frac{1}{2\sqrt{w}}\,dw$.
17. The substitution $w=e^x+e^{-x}$ converts the integral $\displaystyle\int\frac{e^x-e^{-x}}{e^x+e^{-x}}\,dx$ into $\displaystyle\int\frac{1}{w}\,dw$.
18. The substitution $w=q^3$ converts the integral $\int q^2(q^3+5)^{10}\,dq$ into $\displaystyle\int\frac{1}{3}w^{10}\,dw$.
19. The substitution $w=\sin\alpha$ converts the integral $\int\sin^3\alpha\cos\alpha\,d\alpha$ into $\int w^3\,dw$.
20. The substitution $w=\cos x$ converts the integral $\displaystyle\int\frac{\sin x}{\cos x}\,dx$ into $\displaystyle\int\frac{1}{w}\,dw$.
21. $\displaystyle\int_2^3\frac{1}{x}\,dx=\frac{1}{3}-\frac{1}{2}$.
22. $\displaystyle\int_2^3\frac{1}{x}\,dx=\ln 3-\ln 2$.
23. $\displaystyle\int_1^5 2x\,dx=5^2-1^2$.
24. $\displaystyle\int_1^2 3x^2\,dx=3(2^2)-3(1^2)$.
25. $\displaystyle\int_0^3 t^4\,dt=\frac{t^5}{5}+C$.
26. $\displaystyle\int_1^2 e^t\,dt=e^2-e$.
27. The substitution $w=x^2$ converts $\int_0^5 2xe^{x^2}\,dx$ into $\int_0^5 e^w\,dw$.
28. The substitution $w=\ln x$ converts $\int_1^e(\ln x)^3/x\,dx$ into $\int_0^1 w^3\,dw$.
29. If $f(x)$ is any function then $\int_{10}^{20} f(x)\,dx = 10\int_1^2 f(x)\,dx$.
30. The integral $\int_1^2 e^{-kx}\,dx$ is positive for any $k\neq 0$.
31. If a function f is positive on an interval, then its antiderivatives are all increasing on that interval.
32. If a function f is increasing on an interval, then its antiderivatives are all concave up on that interval.
33. For any function F with continuous derivative, $F(5)=F(0)+\int_0^5 F'(t)\,dt$.
34. For any function F with continuous derivative, $F(3)=F(0)+\int_1^3 F'(t)\,dt$.
35. If f' is in Figure 7.24, then $f(4)>f(3)$.

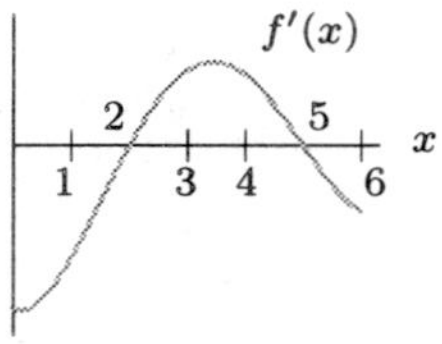

Figure 7.24

36. If f' is in Figure 7.24, then $f(2)>f(1)$.
37. If f' is in Figure 7.24, then $f(3)>f(2)$.
38. If f' is in Figure 7.24, then $f(5)>f(6)$.
39. If f' is in Figure 7.24, then $f(1)>f(0)$.
40. If f' is in Figure 7.24, then $f(1)\approx f(3)$.

41. Setting $u = x^3/3$ and $dv = e^x\,dx$ is a reasonable assignment of the parts for the integral $\int x^2 e^x\,dx$.

42. Setting $u = x$ and $dv = \sqrt{3x+1}\,dx$ is a reasonable assignment of the parts for the integral $\int x\sqrt{3x+1}\,dx$.

43. If $u = x^2$ and $dv = e^{3x}\,dx$ in integration by parts, then $du = 2x\,dx$ and $v = 3e^{3x}$.

44. If $u = x^2$ and $dv = \ln x\,dx$ in integration by parts, then $du = 2x\,dx$ and $v = 1/x$.

45. Integration by parts is an appropriate method to use in evaluating $\int xe^{3x^2}\,dx$.

46. Integration by parts is an appropriate method to use in evaluating $\int xe^{3x}\,dx$.

47. Integration by parts is an appropriate method to use in evaluating $\int \frac{x}{\sqrt{5x+1}}\,dx$.

48. Integration by parts is an appropriate method to use in evaluating $\int \frac{x}{\sqrt{5x^2+1}}\,dx$.

49. Integration by parts is an appropriate method to use in evaluating $\int x^2\sqrt{2x^3+9}\,dx$.

50. Integration by parts is an appropriate method to use in evaluating $\int x^3 \ln x\,dx$.

PROJECTS FOR CHAPTER SEVEN

1. **Quabbin Reservoir** The Quabbin Reservoir in the western part of Massachusetts provides most of Boston's water. The graph in Figure 7.25 represents the flow of water in and out of the Quabbin Reservoir throughout 2007.

   (a) Sketch a graph of the quantity of water in the reservoir, as a function of time.

   (b) When, in the course of 2007, was the quantity of water in the reservoir largest? Smallest? Mark and label these points on the graph you drew in part (a).

   (c) When was the quantity of water increasing most rapidly? Decreasing most rapidly? Mark and label these times on both graphs.

   (d) By July 2008 the quantity of water in the reservoir was about the same as in January 2007. Draw plausible graphs for the flow into and the flow out of the reservoir for the first half of 2008.

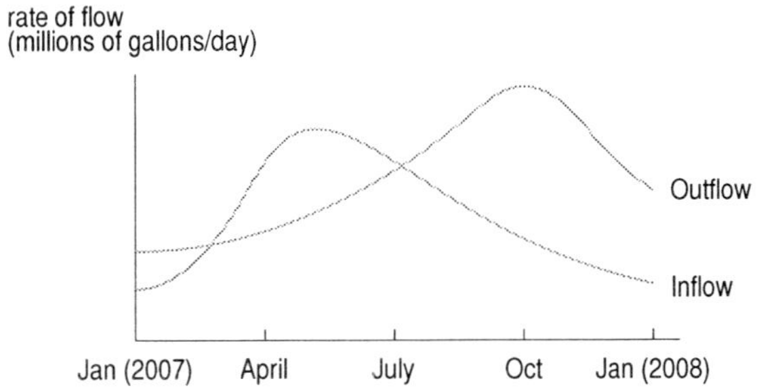

Figure 7.25

FOCUS ON PRACTICE

For Problems 1–48, evaluate the integrals. Assume a, b, A, B, P_0, h, and k are constants.

1. $\int (q^2 + 5q + 2)dq$

2. $\int (u^4 + 5)\, du$

3. $\int (x^2 + 1)\, dx$

4. $\int e^{-3t}\, dt$

5. $\int (6\sqrt{x})dx$

6. $\int (ax^2 + b)\, dx$

7. $\int (x^3 + 4x + 8)\, dx$

8. $\int 100e^{-0.5t}\, dt$

9. $\int (w^4 - 12w^3 + 6w^2 - 10)\, dw$

10. $\int \left(\frac{4}{x} + \frac{5}{x^2}\right)\, dx$

11. $\int \frac{dq}{\sqrt{q}}$

12. $\int 3\sin\theta\, d\theta$

13. $\int (p^2 + \frac{5}{p})\, dp$

14. $\int P_0 e^{kt}\, dt$

15. $\int (q^3 + 8q + 15)\, dq$

16. $\int 1000e^{0.075t}\, dt$

17. $\int (5\sin x + 3\cos x)\, dx$

18. $\int (10 + 5\sin x)\, dx$

19. $\int \frac{5}{w}\, dw$

20. $\int \pi r^2 h\, dr$

21. $\int \left(q + \frac{1}{q^3}\right)\, dq$

22. $\int 15p^2q^4\, dp$

23. $\int 15p^2q^4\, dq$

24. $\int (3x^2 + 6e^{2x})\, dx$

25. $\int 5e^{2q}\, dq$

26. $\int \left(p^3 + \frac{1}{p}\right)\, dp$

27. $\int (Ax^3 + Bx)\, dx$

28. $\int (6\sqrt{x} + 15)\, dx$

29. $\int (x^2 + 8 + e^x)\, dx$

30. $\int 30e^{-0.2t} dt$

31. $\int (t^2 - 6t + 5)\, dt$

32. $\int \left(\frac{a}{x} + \frac{b}{x^2}\right)\, dx$

33. $\int (Aq + B)dq$

34. $\int \left(\frac{6}{\sqrt{x}} + 8\sqrt{x}\right)\, dx$

35. $\int (e^{2t} + 5)dt$

36. $\int \sin(3x)dx$

37. $\int 12\cos(4x)dx$

38. $\int \frac{1}{y+2} dy$

39. $\int y(y^2 + 5)^8\, dy$

40. $\int \cos(4x)dx$

41. $\int A\sin(Bt)dt$

42. $\int \sqrt{3x+1} dx$

43. $\int \frac{e^x}{2 + e^x}\, dx$

44. $\int \sin^6(5\theta)\cos(5\theta)\, d\theta$

45. $\int \frac{\cos x}{\sqrt{1 + \sin x}}\, dx$

46. $\int x\ln x\, dx$

47. $\int xe^x\, dx$

48. $\int_0^{10} ze^{-z}\, dz$

Chapter Eight

PROBABILITY

Contents

8.1 DENSITY FUNCTIONS

Understanding the distribution of various quantities through the population can be important to decision makers. For example, the income distribution gives useful information about the economic structure of a society. In this section we look at the distribution of ages in the US. To allocate funding for education, health care, and social security, the government needs to know how many people are in each age group. We see how to represent such information by a density function.

US Age Distribution

Table 8.1 *Distribution of ages in the US in 2000*

Age group	Percentage of total population
0 – 20	29%
20 – 40	29%
40 – 60	26%
60 – 80	13%
80 – 100	3%

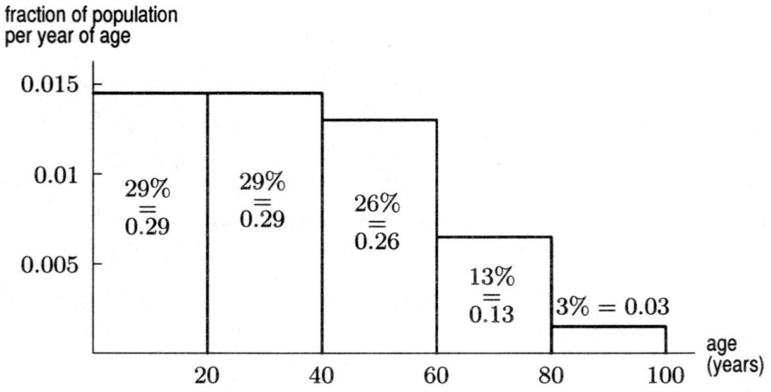

Figure 8.1: How ages were distributed in the US in 2000

Suppose we have the data in Table 8.1 showing how the ages of the US population[1] were distributed in 2000. To represent this information graphically we use a type of *histogram*[2] putting a vertical bar above each age group in such a way that the *area* of each bar represents the percentage in that age group. The total area of all the rectangles is $100\% = 1$. We only consider people who are less than 100 years old.[3] For the 0–20 age group, the base of the rectangle is 20, and we want the area to be 29%, so the height must be $29\%/20 = 1.45\%$. We treat ages as though they were continuously distributed. The category 0–20, for example, contains people who are just one day short of their twentieth birthday. Notice that the vertical axis is measured in percent/year. (See Figure 8.1.)

Example 1 In 2000, estimate what percentage of the US population was:

(a) Between 20 and 60 years old?
(b) Less than 10 years old?
(c) Between 75 and 80 years old?
(d) Between 80 and 85 years old?

Solution

(a) We add the percentages, so $29\% + 26\% = 55\%$.

(b) To find the percentage less than 10 years old, we could assume, for example, that the population was distributed evenly over the 0–20 group. (This means we are assuming that babies were born at a fairly constant rate over the last 20 years, which is probably reasonable.) If we make this assumption, then we can say that the population less than 10 years old was about half that in the 0–20 group, that is, 14.5%. Notice that we get the same result by computing the area of the rectangle from 0 to 10. (See Figure 8.2.)

(c) To find the population between 75 and 80 years old, since 13% of Americans in 2000 were in the 60-80 group, we might apply the same reasoning and say that $\frac{1}{4}(13\%) = 3.25\%$ of the population was in this age group. This result is represented as an area in Figure 8.2. The assumption that the population was evenly distributed is not a good one here; certainly there were more people between the ages of 60 and 65 than between 75 and 80. Thus, the estimate of 3.25% is certainly too high.

[1] www.censusscope.org/us/chart_age.html, accessed May 10, 2005.
[2] There are other types of histograms which have frequency on the vertical axis.
[3] In fact, 0.02% of the population is over 100, but this is too small to be visible on the histogram.

(d) Again using the (faulty) assumption that ages in each group were distributed uniformly, we would find that the percentage between 80 and 85 was $\frac{1}{4}(3\%) = 0.75\%$. (See Figure 8.2.) This estimate is also poor—there were certainly more people in the 80–85 group than, say, the 95–100 group, and so the 0.75% estimate is too low.

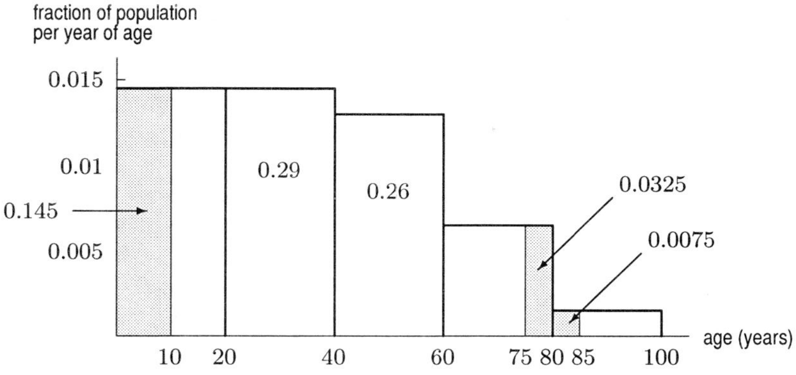

Figure 8.2: Ages in the US in 2000—various subgroups (for Example 1)

Smoothing Out the Histogram

We could get better estimates if we had smaller age groups (each age group in Figure 8.1 is 20 years, which is quite large) or if the histogram were smoother. Suppose we have the more detailed data in Table 8.2, which leads to the new histogram in Figure 8.3.

As we get more detailed information, the upper silhouette of the histogram becomes smoother, but the area of any of the bars still represents the percentage of the population in that age group. Imagine, in the limit, replacing the upper silhouette of the histogram by a smooth curve in such a way that area under the curve above one age group is the same as the area in the corresponding rectangle. The total area under the whole curve is again 100% = 1. (See Figure 8.3.)

The Age Density Function

If t is age in years, we define $p(t)$, the age *density function*, to be a function which "smooths out" the age histogram. This function has the property that

$$\text{Fraction of population between ages } a \text{ and } b = \text{Area under graph of } p \text{ between } a \text{ and } b = \int_a^b p(t)dt.$$

Table 8.2 *Ages in the US in 2000 (more detailed)*

Age group	Percentage of total population
0 – 10	14%
10 – 20	15%
20 – 30	14%
30 – 40	15%
40 – 50	15%
50 – 60	11%
60 – 70	7%
70 – 80	6%
80 – 90	2%
90 – 100	1%

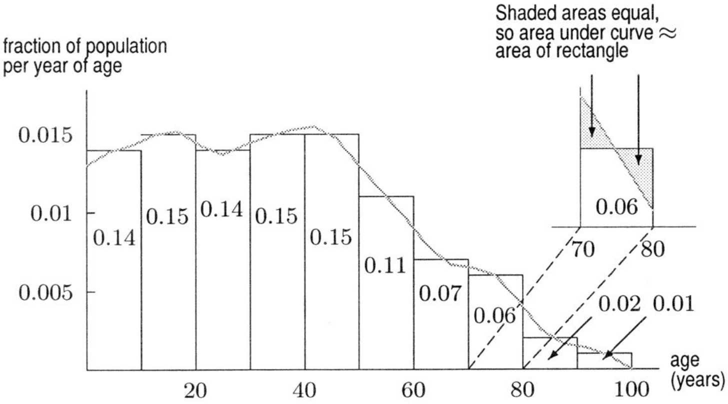

Figure 8.3: Smoothing out the age histogram

If a and b are the smallest and largest possible ages (say, $a = 0$ and $b = 100$), so that the ages of all of the population are between a and b, then

$$\int_a^b p(t)dt = \int_0^{100} p(t)dt = 1.$$

What does the age density function p tell us? Notice that we have not talked about the meaning of $p(t)$ itself, but *only* of the integral $\int_a^b p(t)\,dt$. Let's look at this in a bit more detail. Suppose, for example, that $p(10) = 0.014 = 1.4\%$ per year. This is *not* telling us that 1.4% of the population is precisely 10 years old (where 10 years old means exactly 10, not $10\frac{1}{2}$, not $10\frac{1}{4}$, not 10.1). However, $p(10) = 0.014$ does tell us that for some small interval Δt around 10, the fraction of the population with ages in this interval is approximately $p(10)\,\Delta t = 0.014\,\Delta t$. Notice also that the units of $p(t)$ are % *per year*, so $p(t)$ must be multiplied by years to give a percentage of the population.

The Density Function

Suppose we are interested in how a certain numerical characteristic, x, is distributed through a population. For example, x might be height or age if the population is people, or might be wattage for a population of light bulbs. Then we define a general density function with the following properties:

The function, $p(x)$, is a **density function** if

$$\begin{array}{ccccc} \text{Fraction of population} & & \text{Area under} & & \\ \text{for which } x \text{ is} & = & \text{graph of } p & = & \displaystyle\int_a^b p(x)dx. \\ \text{between } a \text{ and } b & & \text{between } a \text{ and } b & & \end{array}$$

$$\int_{-\infty}^{\infty} p(x)\,dx = 1 \quad \text{and} \quad p(x) \geq 0 \quad \text{for all } x.$$

The density function must be nonnegative if its integral always gives a fraction of the population. The fraction of the population with x between $-\infty$ and ∞ is 1 because the entire population has the characteristic x between $-\infty$ and ∞. The function p that was used to smooth out the age histogram satisfies this definition of a density function. Notice that we do not assign a meaning to the value $p(x)$ directly, but rather interpret $p(x)\,\Delta x$ as the fraction of the population with the characteristic in a short interval of length Δx around x.

Example 2 Figure 8.4 gives the density function for the amount of time spent waiting at a doctor's office.

(a) What is the longest time anyone has to wait?
(b) Approximately what fraction of patients wait between 1 and 2 hours?
(c) Approximately what fraction of patients wait less than an hour?

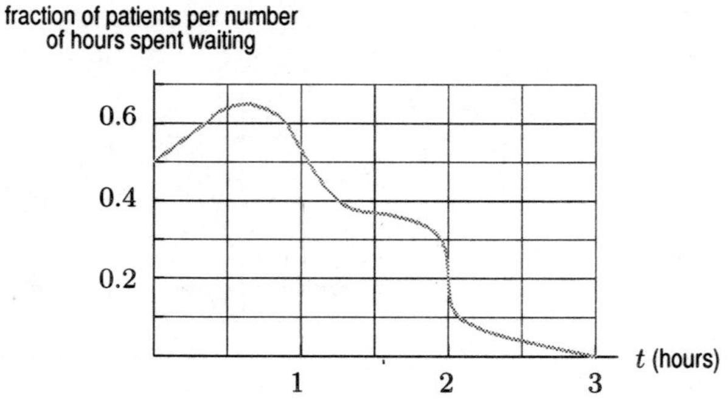

Figure 8.4: Distribution of waiting time at a doctor's office

Solution

(a) The density function is zero for all $t > 3$, so no one waits more than 3 hours. The longest time anyone has to wait is 3 hours.

(b) The fraction of patients who wait between 1 and 2 hours is equal to the area under the density curve between $t = 1$ and $t = 2$. We can estimate this area by counting squares: There are about 7.5 squares in this region, each of area $(0.5)(0.1) = 0.05$. The area is approximately $(7.5)(0.05) = 0.375$. Thus about 37.5% of patients wait between 1 and 2 hours.

(c) This fraction is equal to the area under the density function for $t < 1$. There are about 12 squares in this area, and each has area 0.05 as in part (b), so our estimate for the area is $(12)(0.05) = 0.60$. Therefore, about 60% of patients see the doctor in less than an hour.

Problems for Section 8.1

In Problems 1–4, the distribution of the heights, x, in meters, of trees is represented by the density function $p(x)$. In each case, calculate the fraction of trees which are:

(a) Less than 5 meters high
(b) More than 6 meters high
(c) Between 2 and 5 meters high

1.

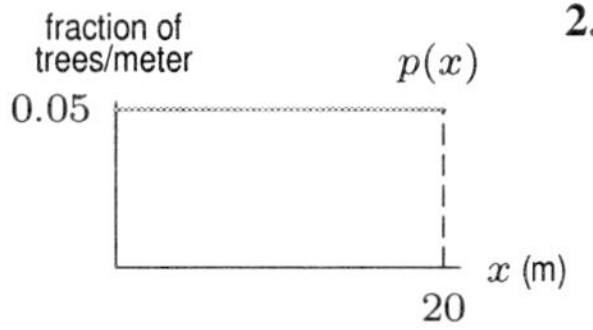

2.

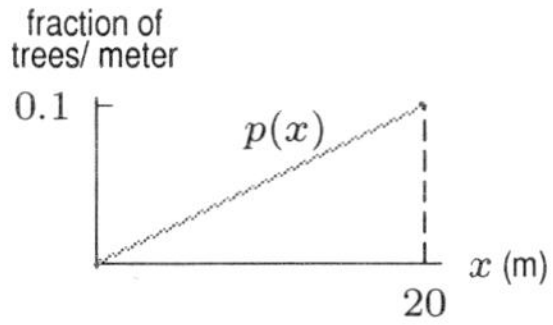

3.

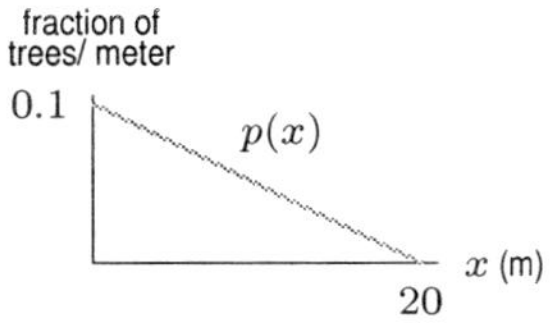

4.

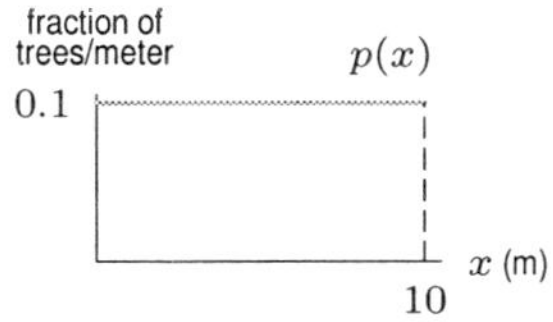

5. Suppose that $p(x)$ is the density function for heights of American men, in inches. What is the meaning of the statement $p(68) = 0.2$?

6. The density function $p(t)$ for the length of the larval stage, in days, for a breed of insect is given in Figure 8.5. What fraction of these insects are in the larval stage for between 10 and 12 days? For less than 8 days? For more than 12 days? In which one-day interval is the length of a larval stage most likely to fall?

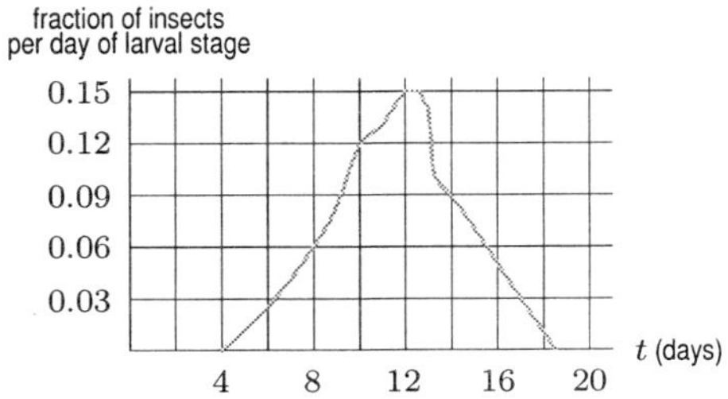

Figure 8.5

7. A machine lasts up to 10 years. Figure 8.6 shows the density function, $p(t)$, for the length of time it lasts.

(a) What is the value of C?
(b) Is a machine more likely to break in its first year or in its tenth year? In its first or second year?
(c) What fraction of the machines lasts 2 years or less? Between 5 and 7 years? Between 3 and 6 years?

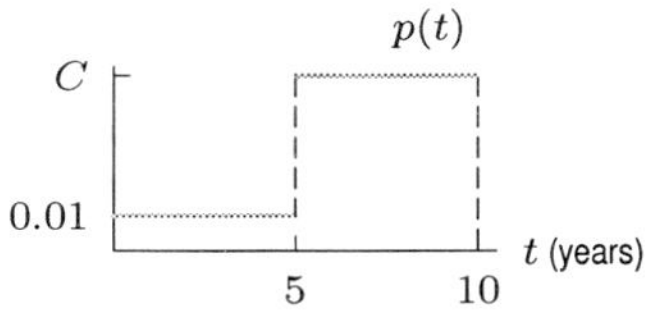

Figure 8.6

8. Figure 8.7 shows the distribution of the number of years of education completed by adults in a population. What does the shape of the graph tell you? Estimate the percentage of adults who have completed less than 10 years of education.

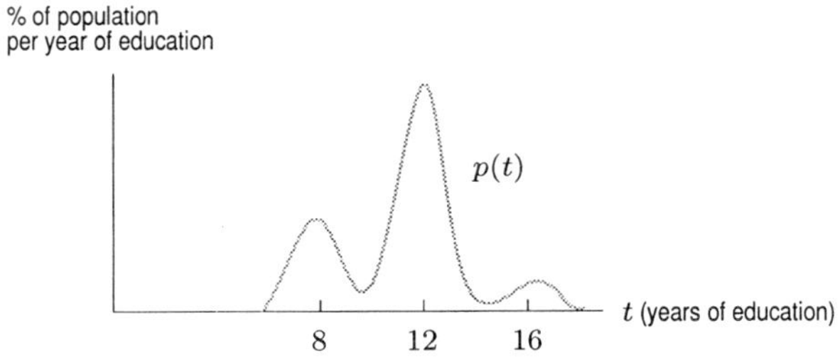

Figure 8.7

In Problems 9–12, given that $p(x)$ is a density function, find the value of a.

9.

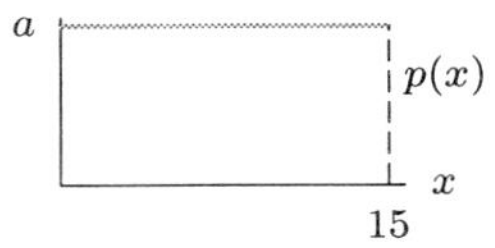

10.

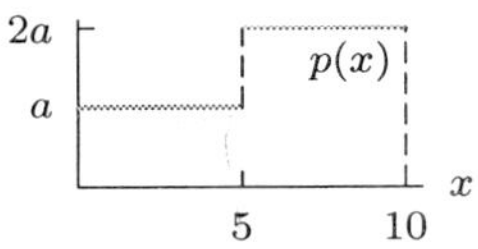

11. 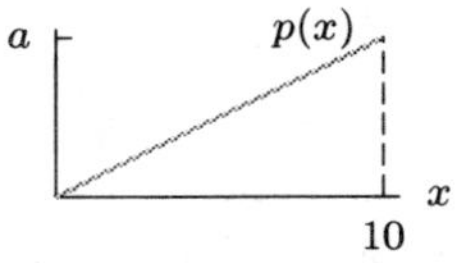

12. 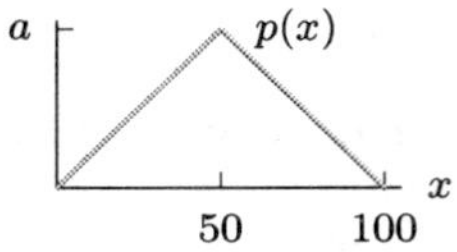

In Problems 13–15, graph a possible density function representing crop yield (in kilograms) from a field under the given circumstance.

13. All yields from 0 to 100 kg are equally likely; the field never yields more than 100 kg.

14. High yields are more likely than low. The maximum yield is 200 kg.

15. A drought makes low yields most common, and there is no yield greater than 30 kg.

16. Which of the following functions makes the most sense as a model for the probability density representing the time (in minutes, starting from $t = 0$) that the next customer walks into a store?

(a) $p(t) = \begin{cases} \cos t & 0 \le t \le 2\pi \\ e^{t-2\pi} & t \ge 2\pi \end{cases}$
(b) $p(t) = 3e^{-3t}$ for $t \ge 0$
(c) $p(t) = e^{-3t}$ for $t \ge 0$
(d) $p(t) = 1/4$ for $0 \le t \le 4$

8.2 CUMULATIVE DISTRIBUTION FUNCTIONS AND PROBABILITY

Section 8.1 introduced density functions which describe the way in which a numerical characteristic is distributed through a population. In this section we study another way to present the same information.

Cumulative Distribution Function for Ages

An alternative way of showing how ages are distributed in the US is by using the *cumulative distribution function* $P(t)$, defined by

$$P(t) = \frac{\text{Fraction of population}}{\text{of age less than } t} = \int_0^t p(x)dx.$$

Thus, P is the antiderivative of p with $P(0) = 0$, so $P(t)$ is the area under the density curve between 0 and t. See the left-hand part of Figure 8.8.

Notice that the cumulative distribution function is nonnegative and increasing (or at least nondecreasing), since the number of people younger than age t increases as t increases. Another way of seeing this is to notice that $P' = p$, and p is positive (or nonnegative). Thus the cumulative age distribution is a function which starts with $P(0) = 0$ and increases as t increases. We have $P(t) = 0$ for $t < 0$ because, when $t < 0$, there is no one whose age is less than t. The limiting value of P, as $t \to \infty$, is 1 since as t becomes very large (100 say), everyone is younger than age t, so the fraction of people with age less than t tends toward 1.

We want to find the cumulative distribution function for the age density function shown in Figure 8.3. We see that $P(10)$ is equal to 0.14, since Figure 8.3 shows that 14% of the population is between 0 and 10 years of age. Also,

$$P(20) = \begin{array}{c}\text{Fraction of the population}\\ \text{between 0 and 20 years old}\end{array} = 0.14 + 0.15 = 0.29$$

and similarly

$$P(30) = 0.14 + 0.15 + 0.14 = 0.43.$$

Continuing in this way gives the values for $P(t)$ in Table 8.3. These values were used to graph $P(t)$ in the right-hand part of Figure 8.8.

Table 8.3 *Cumulative distribution function, $P(t)$, giving fraction of US population of age less than t years*

t	0	10	20	30	40	50	60	70	80	90	100
$P(t)$	0	0.14	0.29	0.43	0.58	0.73	0.84	0.91	0.97	0.99	1.00

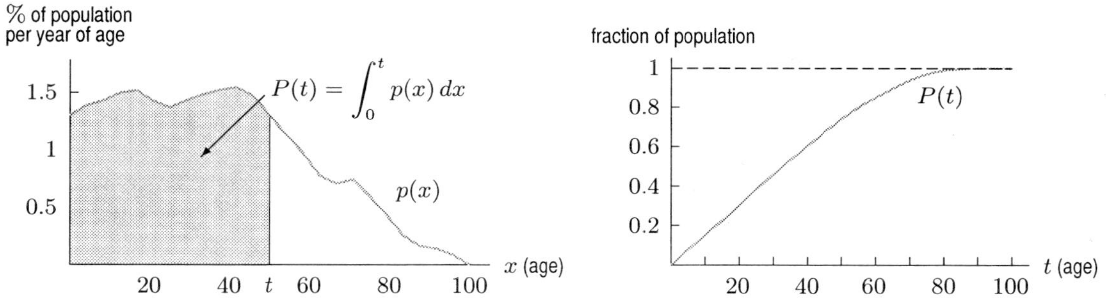

Figure 8.8: Graph of $p(x)$, the age density function, and its relation to $P(t)$, the cumulative age distribution function

Cumulative Distribution Function

A **cumulative distribution function**, $P(t)$, of a density function p, is defined by

$$P(t) = \int_{-\infty}^{t} p(x)\,dx = \text{Fraction of population having values of } x \text{ below } t.$$

Thus, P is an antiderivative of p, that is, $P' = p$.
Any cumulative distribution function has the following properties:

- P is increasing (or nondecreasing).
- $\lim_{t\to\infty} P(t) = 1$ and $\lim_{t\to-\infty} P(t) = 0$.
- $$\text{Fraction of population having values of } x \text{ between } a \text{ and } b = \int_a^b p(x)\,dx = P(b) - P(a).$$

Example 1 The time to conduct a routine maintenance check on a machine has a cumulative distribution function $P(t)$, which gives the fraction of maintenance checks completed in time less than or equal to t minutes. Values of $P(t)$ are given in Table 8.4.

Table 8.4 *Cumulative distribution function for time to conduct maintenance checks*

t (minutes)	0	5	10	15	20	25	30
$P(t)$ (fraction completed)	0	0.03	0.08	0.21	0.38	0.80	0.98

(a) What fraction of maintenance checks are completed in 15 minutes or less?
(b) What fraction of maintenance checks take longer than 30 minutes?
(c) What fraction take between 10 and 15 minutes?
(d) Draw a histogram showing how times for maintenance checks are distributed.
(e) In which of the given 5-minute intervals is the length of a maintenance check most likely to fall?
(f) Give a rough sketch of the density function.
(g) Sketch a graph of the cumulative distribution function.

Solution

(a) The fraction of maintenance checks completed in 15 minutes is $P(15) = 0.21$, or 21%.
(b) Since $P(30) = 0.98$, we see that 98% of maintenance checks take 30 minutes or less. Therefore, only 2% take more than 30 minutes.
(c) Since 8% take 10 minutes or less and 21% take 15 minutes or less, the fraction taking between 10 and 15 minutes is $0.21 - 0.08 = 0.13$, or 13%.
(d) We begin by making a table showing how the times are distributed. Table 8.4 shows that the fraction of checks completed between 0 and 5 minutes is 0.03, and the fraction completed between 5 and 10 minutes is 0.05, and so on. See Table 8.5.

Table 8.5 *Distribution of time to conduct maintenance checks*

t (minutes)	0 – 5	5 – 10	10 – 15	15 – 20	20 – 25	25 – 30	> 30
Fraction completed	0.03	0.05	0.13	0.17	0.42	0.18	0.02

The histogram in Figure 8.9 is drawn so that the area of each bar is the fraction of checks completed in the corresponding time period. For instance, the first bar has area 0.03 and width 5 minutes, so its height is $0.03/5 = 0.006$.

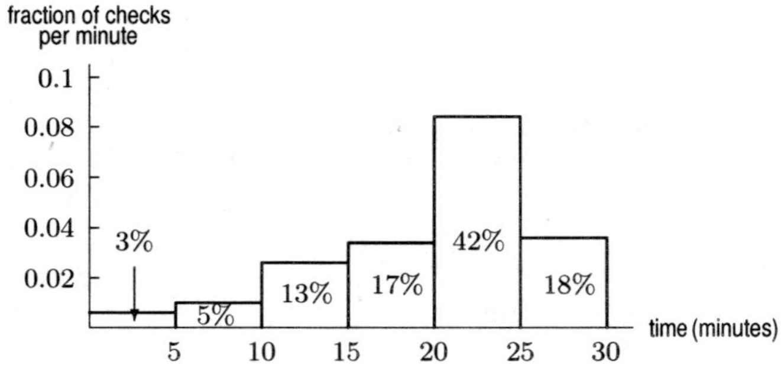

Figure 8.9: Histogram of times for maintenance checks

(e) From Figure 8.9, we see that more of the checks take between 20 and 25 minutes to complete, so this is the most likely length of time.
(f) The density function, $p(t)$, is a smoothed version of the histogram in Figure 8.9. A reasonable sketch is given in Figure 8.10.
(g) A graph of $P(t)$ is given in Figure 8.11. Since $P(t)$ is a cumulative distribution function, $P(t)$ is approaching 1 as t gets large, but is never larger than 1.

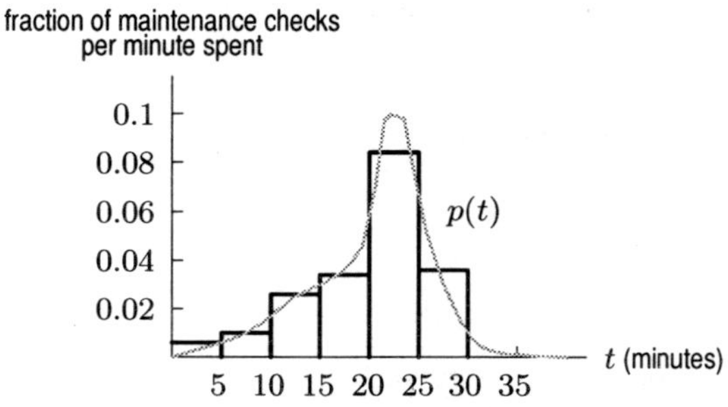

Figure 8.10: Density function for time to conduct maintenance checks

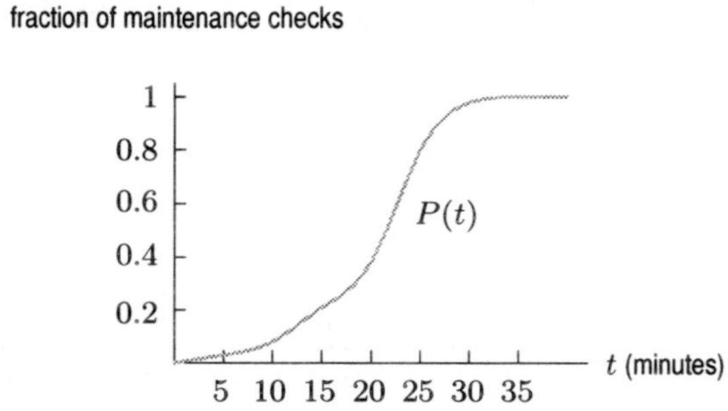

Figure 8.11: Cumulative distribution function for time to conduct maintenance checks

Suppose we pick a member of the US population at random. What is the probability that we pick a person who is between, say, the ages of 70 and 80? We saw in Table 8.2 on page 329 that 6% of the population is in this age group. We say that the probability, or chance, that the person is between 70 and 80 is 0.06. Using any age density function $p(t)$, we define probabilities as follows:

$$\begin{array}{c}\text{Probability that a person is}\\ \text{between ages } a \text{ and } b\end{array} = \begin{array}{c}\text{Fraction of population}\\ \text{between ages } a \text{ and } b\end{array} = \int_a^b p(t)\,dt.$$

Since the cumulative distribution gives the fraction of the population younger than age t, the cumulative distribution function can also be used to calculate the probability that a randomly selected person is in a given age group.

$$\begin{array}{c}\text{Probability that a person is}\\ \text{younger than age } t\end{array} = \begin{array}{c}\text{Fraction of population}\\ \text{younger than age } t\end{array} = P(t) = \int_0^t p(x)\,dx.$$

In the next example, both a density function and a cumulative distribution function are used to describe the same situation.

Example 2 Suppose you want to analyze the fishing industry in a small town. Each day, the boats bring back at least 2 tons of fish, but never more than 8 tons.

(a) Using the density function describing the daily catch in Figure 8.12, find and graph the corresponding cumulative distribution function and explain its meaning.
(b) What is the probability that the catch is between 5 and 7 tons?

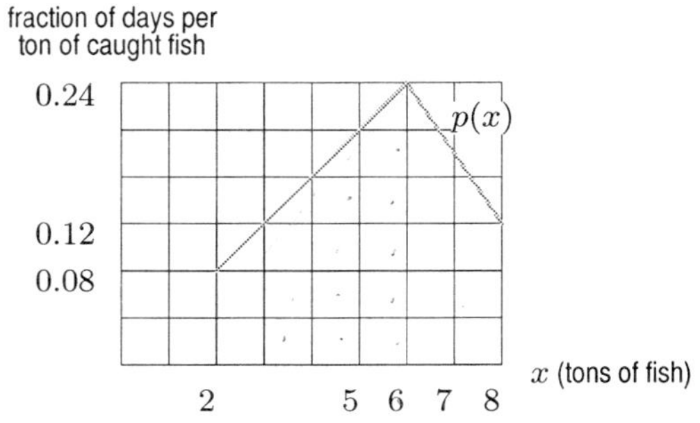

Figure 8.12: Density function of daily catch

Solution (a) The cumulative distribution function $P(t)$ is equal to the fraction of days on which the catch is less than t tons of fish. Since the catch is never less than 2 tons, we have $P(t) = 0$ for $t \leq 2$. Since the catch is always less than 8 tons, we have $P(t) = 1$ for $t \geq 8$. For t in the range $2 < t < 8$ we must evaluate the integral

$$P(t) = \int_{-\infty}^{t} p(x)dx = \int_2^t p(x)dx.$$

This integral equals the area under the graph of $p(x)$ between $x = 2$ and $x = t$. It can be computed by counting grid squares in Figure 8.12; each square has area 0.04. For example,

$$P(3) = \int_2^3 p(x)dx \approx \text{Area of 2.5 squares} = 2.5(0.04) = 0.10.$$

Table 8.6 contains values of $P(t)$; the graph is shown in Figure 8.13.

Table 8.6 *Estimates for* $P(t)$ *of daily catch*

t (tons of fish)	$P(t)$ (fraction of fishing days)
2	0
3	0.10
4	0.24
5	0.42
6	0.64
7	0.85
8	1

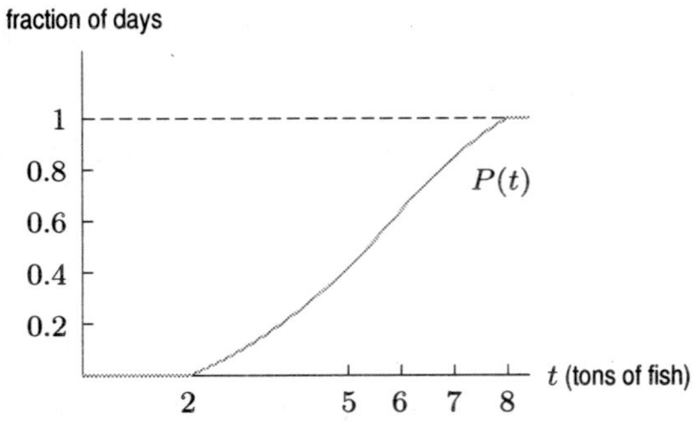

Figure 8.13: Cumulative distribution, $P(t)$, of daily catch

(b) The probability that the catch is between 5 and 7 tons can be found using either the density function p or the cumulative distribution function P. When we use the density function, this probability is represented by the shaded area in Figure 8.14, which is about 10.75 squares, so

$$\begin{matrix}\text{Probability catch is} \\ \text{between 5 and 7 tons}\end{matrix} = \int_5^7 p(x)\,dx \approx \text{Area of 10.75 squares} = 10.75(0.04) = 0.43.$$

The probability can be found from the cumulative distribution function as follows:

$$\begin{matrix}\text{Probability catch is} \\ \text{between 5 and 7 tons}\end{matrix} = P(7) - P(5) = 0.85 - 0.42 = 0.43.$$

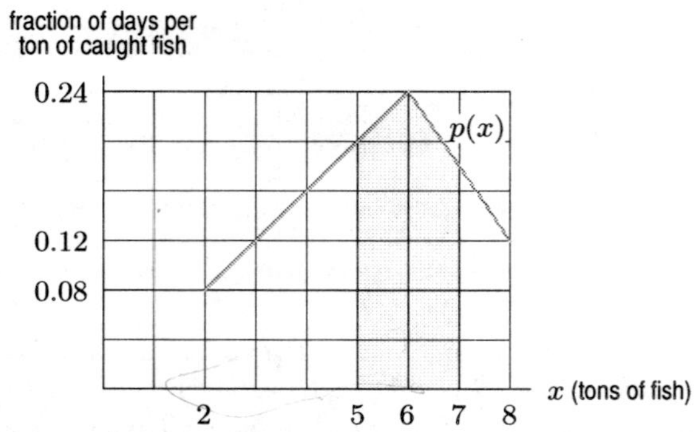

Figure 8.14: Shaded area represents the probability the catch is between 5 and 7 tons

Problems for Section 8.2

1. (a) Using the density function in Example 2 on page 330, fill in values for the cumulative distribution function $P(t)$ for the length of time people wait in the doctor's office.

t (hours)	0	1	2	3	4
$P(t)$ (fraction of people waiting)					

(b) Graph $P(t)$.

2. In an agricultural experiment, the quantity of grain from a given size field is measured. The yield can be anything from 0 kg to 50 kg. For each of the following situations, pick the graph that best represents the:
(i) Probability density function
(ii) Cumulative distribution function.

(a) Low yields are more likely than high yields.
(b) All yields are equally likely.
(c) High yields are more likely than low yields.

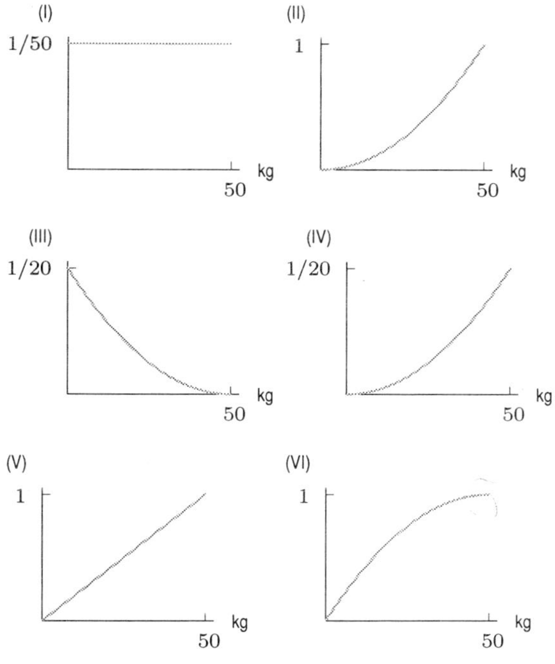

3. A person who travels regularly on the 9:00 am bus from Oakland to San Francisco reports that the bus is almost always a few minutes late but rarely more than five minutes late. The bus is never more than two minutes early, although it is on very rare occasions a little early.

(a) Sketch a density function, $p(t)$, where t is the number of minutes that the bus is late. Shade the region under the graph between $t = 2$ minutes and $t = 4$ minutes. Explain what this region represents.

(b) Now sketch the cumulative distribution function $P(t)$. What measurement(s) on this graph correspond to the area shaded? What do the inflection point(s) on your graph of P correspond to on the graph of p? Interpret the inflection points on the graph of P without referring to the graph of p.

4. The density function and cumulative distribution function of heights of grass plants in a meadow are in Figures 8.15 and 8.16, respectively.

(a) There are two species of grass in the meadow, a short grass and a tall grass. Explain how the graph of the density function reflects this fact.
(b) Explain how the graph of the cumulative distribution function reflects the fact that there are two species of grass in the meadow.
(c) About what percentage of the grasses in the meadow belong to the short grass species?

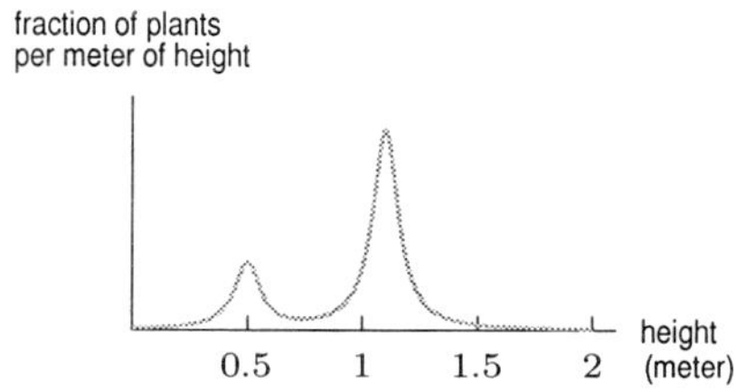

Figure 8.15

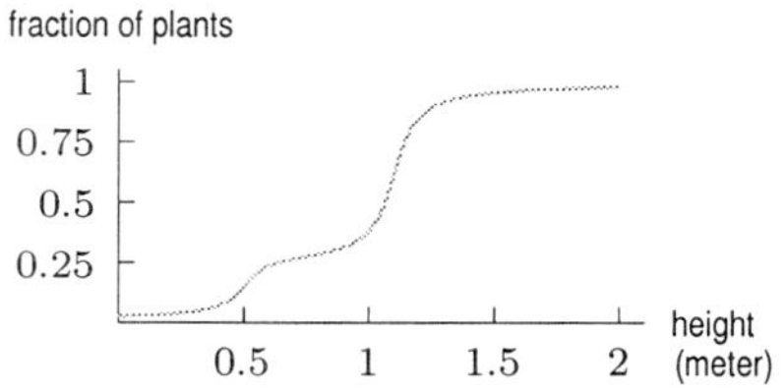

Figure 8.16

5. Show that the area under the fishing density function in Figure 8.12 on page 335 is 1. Why is this to be expected?

6. Figure 8.17 shows a density function and the corresponding cumulative distribution function.[4]

(a) Which curve represents the density function and which represents the cumulative distribution function? Give a reason for your choice.
(b) Put reasonable values on the tick marks on each of the axes.

[4] Adapted from *Calculus*, by David A. Smith and Lawerence C. Moore (Lexington, D.C. Heath, 1994).

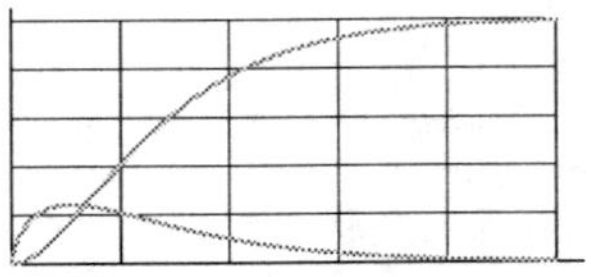

Figure 8.17

7. Students at the University of California were surveyed and asked their grade point average. (The GPA ranges from 0 to 4, where 2 is just passing.) The distribution of GPAs is shown in Figure 8.18.[5]

(a) Roughly what fraction of students are passing?
(b) Roughly what fraction of the students have honor grades (GPAs above 3)?
(c) Why do you think there is a peak around 2?
(d) Sketch the cumulative distribution function.

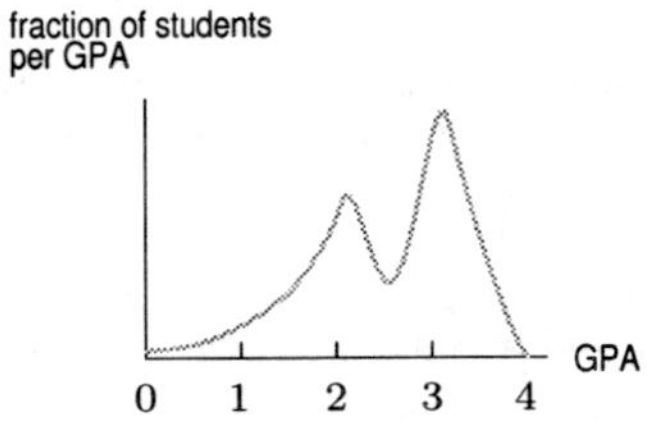

Figure 8.18

8. Suppose $F(x)$ is the cumulative distribution function for heights (in meters) of trees in a forest.

(a) Explain in terms of trees the meaning of the statement $F(7) = 0.6$.
(b) Which is greater, $F(6)$ or $F(7)$? Justify your answer in terms of trees.

9. Figure 8.19 shows $P(t)$, the percentage of inventory of an item that has sold by time t, where t is in days and day 1 is January 1.

(a) When did the first item sell? The last item?
(b) On May 1 (day 121), what percentage of the inventory had been sold?
(c) Approximately what percentage of the inventory sold during May and June (days 121–181)?
(d) What percentage of the inventory remained after half of the year had passed (at day 181)?
(e) Estimate when items went on sale and sold quickly.

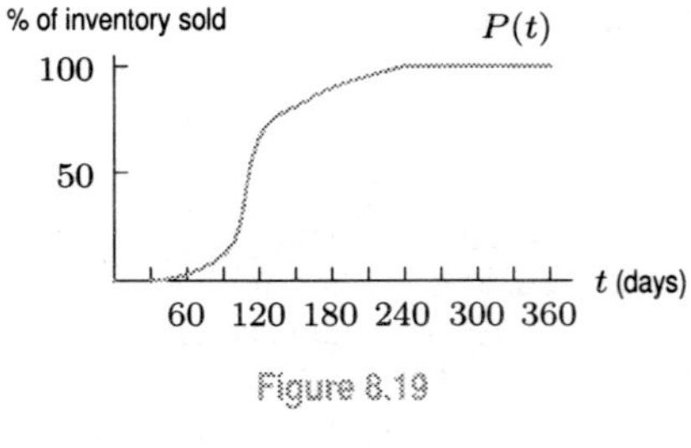

Figure 8.19

10. An experiment is done to determine the effect of two new fertilizers A and B on the growth of a species of peas. The cumulative distribution functions of the heights of the mature peas without treatment and treated with each of A and B are graphed in Figure 8.20.

(a) About what height are most of the unfertilized plants?
(b) Explain in words the effect of the fertilizers A and B on the mature height of the plants.

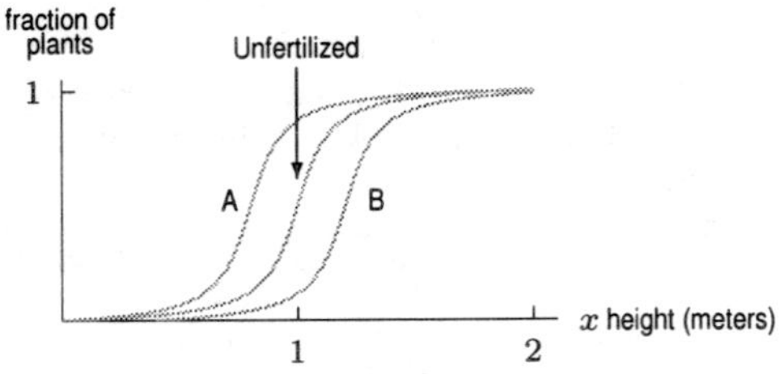

Figure 8.20

For Problems 11–12, let $p(t) = -0.0375t^2 + 0.225t$ be the density function for the shelf life of a brand of banana, with t in weeks and $0 \le t \le 4$. See Figure 8.21.

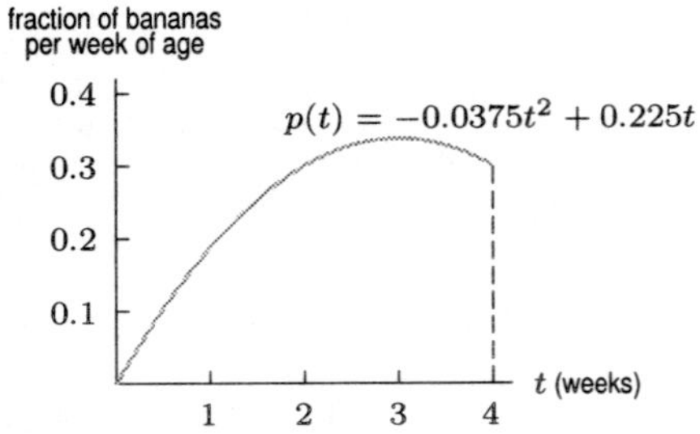

Figure 8.21

11. Find the probability that a banana will last

(a) Between 1 and 2 weeks.
(b) More than 3 weeks. (c) More than 4 weeks.

12. (a) Sketch the cumulative distribution function for the shelf life of bananas. [Note: The domain of your function should be all real numbers, including to the left of $t = 0$ and to the right of $t = 4$.]
(b) Use the cumulative distribution function to estimate the probability that a banana lasts between 1 and 2 weeks. Check with Problem 11(a).

13. A group of people have received treatment for cancer. Let t be the survival time, the number of years a person lives after the treatment. The density function giving the distribution of t is $p(t) = Ce^{-Ct}$ for some positive constant C. What is the practical meaning of the cumulative distribution function $P(t) = \int_0^t p(x)\,dx$?

[5] Adapted from *Statistics*, by Freedman, Pisani, Purves, and Adikhari (New York: Norton, 1991).

14. The probability of a transistor failing between $t = a$ months and $t = b$ months is given by $c\int_a^b e^{-ct}dt$, for some constant c.

(a) If the probability of failure within the first six months is 10%, what is c?

(b) Given the value of c in part (a), what is the probability the transistor fails within the second six months?

15. While taking a walk along the road where you live, you accidentally drop your glove, but you don't know where. The probability density $p(x)$ for having dropped the glove x kilometers from home (along the road) is

$$p(x) = 2e^{-2x} \quad \text{for } x \geq 0.$$

(a) What is the probability that you dropped it within 1 kilometer of home?

(b) At what distance y from home is the probability that you dropped it within y km of home equal to 0.95?

8.3 THE MEDIAN AND THE MEAN

It is often useful to be able to give an "average" value for a distribution. Two measures that are in common use are the *median* and the *mean.*

The Median

A **median** of a quantity x distributed through a population is a value T such that half the population has values of x less than (or equal to) T, and half the population has values of x greater than (or equal to) T. Thus, if p is the density function, a median T satisfies

$$\int_{-\infty}^{T} p(x)\,dx = 0.5.$$

In other words, half the area under the graph of p lies to the left of T. Equivalently, if P is the cumulative distribution function,

$$P(T) = 0.5.$$

Example 1 Let t days be the length of time a pair of jeans remains in a shop before it is sold. The density function of t is graphed in Figure 8.22 and given by

$$p(t) = 0.04 - 0.0008t.$$

(a) What is the longest time a pair of jeans remains unsold?
(b) Would you expect the median time till sale to be less than, equal to, or greater than 25 days?
(c) Find the median time required to sell a pair of jeans.

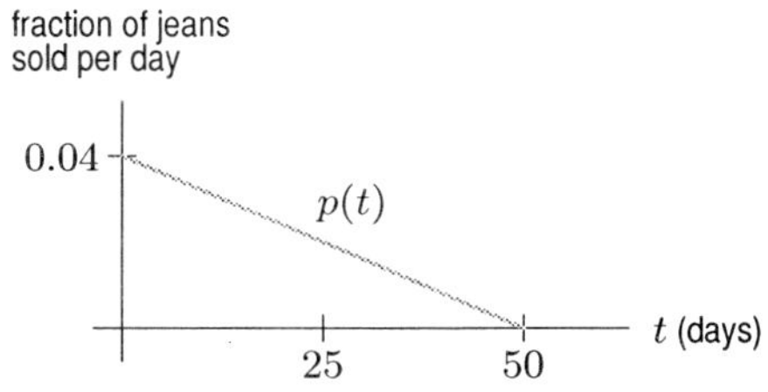

Figure 8.22: Density function for time till sale of a pair of jeans

Solution (a) The density function is 0 for all times $t > 50$, so all jeans are sold within 50 days.
(b) The area under the graph of the density function in the interval $0 \leq t \leq 25$ is greater than the area under the graph in the interval $25 \leq t \leq 50$. So more than half the jeans are sold before their 25th day in the shop. The median time till sale is less than 25 days.

(c) Let P be the cumulative distribution function. We want to find the value of T such that

$$P(T) = \int_{-\infty}^{T} p(t)\,dt = \int_{0}^{T} p(t)\,dt = 0.5.$$

Using a calculator to evaluate the integrals, we obtain the values for P in Table 8.7.

Table 8.7 *Cumulative distribution for selling time*

T (days)	0	5	10	15	20	25
$P(T)$ (fraction of jeans sold by day T)	0	0.19	0.36	0.51	0.64	0.75

Since about half the jeans are sold within 15 days, the median time to sale is about 15 days. See Figures 8.23 and 8.24. We could also use the Fundamental Theorem of Calculus to find the median exactly. See Problem 19.

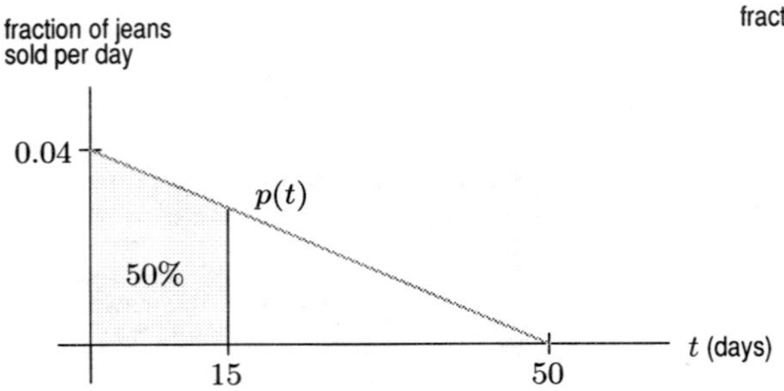

Figure 8.23: Median and density function

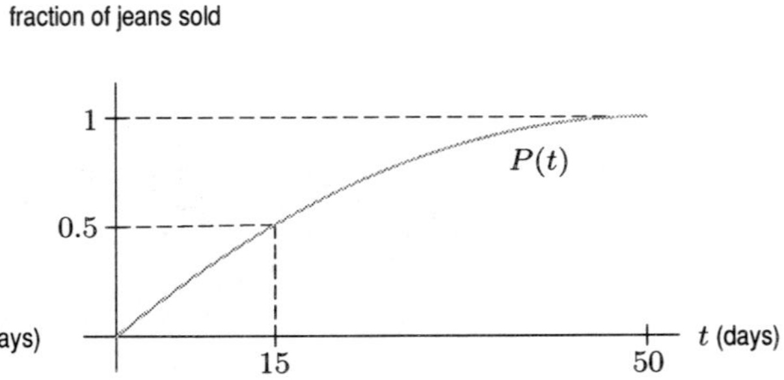

Figure 8.24: Median and cumulative distribution function

The Mean

Another commonly used average value is the *mean*. To find the mean of N numbers, we add the numbers and divide the sum by N. For example, the mean of the numbers 1, 2, 7, and 10 is $(1 + 2 + 7 + 10)/4 = 5$. The mean age of the entire US population is therefore defined as

$$\text{Mean age} = \frac{\sum \text{Ages of all people in the US}}{\text{Total number of people in the US}}.$$

Calculating the sum of all the ages directly would be an enormous task; we approximate the sum by an integral. We consider the people whose age is between t and $t+\Delta t$. How many are there?

The fraction of the population with age between t and $t + \Delta t$ is the area under the graph of p between these points, which is approximated by the area of the rectangle, $p(t)\Delta t$. (See Figure 8.25.) If the total number of people in the population is N, then

$$\begin{array}{c}\text{Number of people with age}\\ \text{between } t \text{ and } t+\Delta t\end{array} \approx p(t)\Delta t N.$$

The age of each of these people is approximately t, so

$$\begin{array}{c}\text{Sum of ages of people}\\ \text{between age } t \text{ and } t+\Delta t\end{array} \approx tp(t)\Delta t N.$$

Therefore, adding and factoring out an N gives us

$$\text{Sum of ages of all people} \approx \left(\sum tp(t)\Delta t\right) N.$$

In the limit, as Δt shrinks to 0, the sum becomes an integral. Assuming no one is over 100 years old, we have

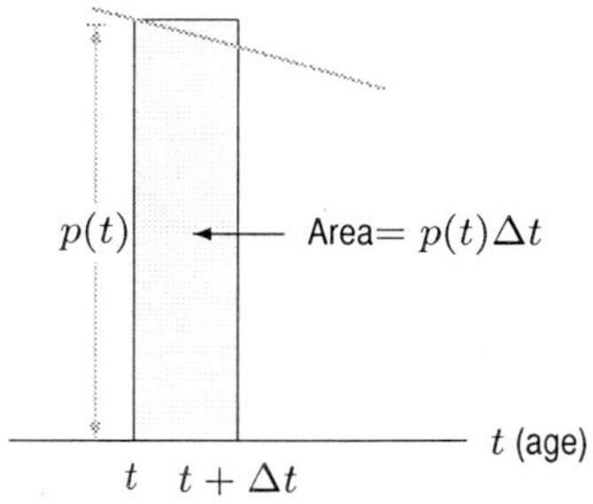

Figure 8.25: Shaded area is percentage of population with age between t and $t+\Delta t$

$$\text{Sum of ages of all people} = \left(\int_0^{100} tp(t)dt\right) N.$$

Since N is the total number of people in the US,

$$\text{Mean age} = \frac{\text{Sum of ages of all people in US}}{N} = \int_0^{100} tp(t)dt.$$

We can give the same argument for any[6] density function $p(x)$.

> If a quantity has density function $p(x)$,
>
> $$\textbf{Mean value} \text{ of the quantity} = \int_{-\infty}^{\infty} xp(x)\,dx.$$

It can be shown that the mean is the point on the horizontal axis where the region under the graph of the density function, if it were made out of cardboard, would balance.

Example 2 Find the mean time for jeans sales, using the density function of Example 1.

Solution The formula for p is $p(t) = 0.04 - 0.0008t$. We compute

$$\text{Mean time} = \int_0^{50} tp(t)\,dt = \int_0^{50} t(0.04 - 0.0008t)\,dt = 16.67 \text{ days}.$$

The mean is represented by the balance point in Figure 8.26. Notice that the mean is different from the median computed in Example 1.

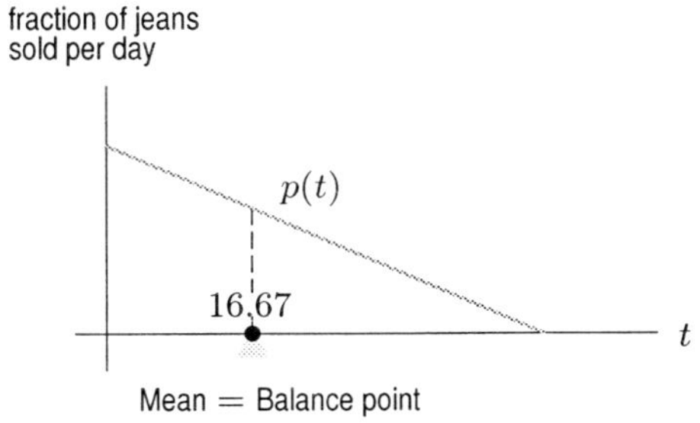

Figure 8.26: Mean sale time for jeans

[6]Provided all the relevant improper integrals converge.

Normal Distributions

How much rain do you expect to fall in your home town this year? If you live in Anchorage, Alaska, the answer is something close to 15 inches (including the snow). Of course, you don't expect exactly 15 inches. Some years there are more than 15 inches, and some years there are less. Most years, however, the amount of rainfall is close to 15 inches; only rarely is it well above or well below 15 inches. What does the density function for the rainfall look like? To answer this question, we look at rainfall data over many years. Records show that the distribution of rainfall is well-approximated by a *normal distribution*. The graph of its density function is a bell-shaped curve which peaks at 15 inches and slopes downward approximately symmetrically on either side.

Normal distributions are frequently used to model real phenomena, from grades on an exam to the number of airline passengers on a particular flight. A normal distribution is characterized by its *mean*, μ, and its *standard deviation*, σ. The mean tells us the location of the central peak. The standard deviation tells us how closely the data is clustered around the mean. A small value of σ tells us that the data is close to the mean; a large σ tells us the data is spread out. In the following formula for a normal distribution, the factor of $1/(\sigma\sqrt{2\pi})$ makes the area under the graph equal to 1.

> A **normal distribution** has a density function of the form
>
> $$p(x) = \frac{1}{\sigma\sqrt{2\pi}} e^{-(x-\mu)^2/(2\sigma^2)},$$
>
> where μ is the mean of the distribution and σ is the standard deviation, with $\sigma > 0$.

To model the rainfall in Anchorage, we use a normal distribution with $\mu = 15$ and $\sigma = 1$. (See Figure 8.27.)

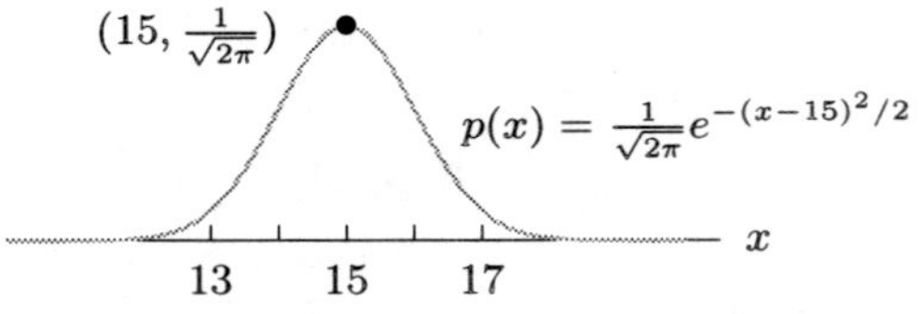

Figure 8.27: Normal distribution with $\mu = 15$ and $\sigma = 1$

Example 3 For Anchorage's rainfall, use the normal distribution with the density function with $\mu = 15$ and $\sigma = 1$ to compute the fraction of the years with rainfall between

(a) 14 and 16 inches, (b) 13 and 17 inches, (c) 12 and 18 inches.

Solution (a) The fraction of the years with annual rainfall between 14 and 16 inches is $\int_{14}^{16} \frac{1}{\sqrt{2\pi}} e^{-(x-15)^2/2}\,dx$. Since there is no elementary antiderivative for $e^{-(x-15)^2/2}$, we find the integral numerically. Its value is about 0.68.

$$\text{Fraction of years with rainfall between 14 and 16 inches} = \int_{14}^{16} \frac{1}{\sqrt{2\pi}} e^{-(x-15)^2/2}\,dx \approx 0.68.$$

(b) Finding the integral numerically again:

$$\text{Fraction of years with rainfall between 13 and 17 inches} = \int_{13}^{17} \frac{1}{\sqrt{2\pi}} e^{-(x-15)^2/2}\,dx \approx 0.95.$$

(c)

$$\text{Fraction of years with rainfall between 12 and 18 inches} = \int_{12}^{18} \frac{1}{\sqrt{2\pi}} e^{-(x-15)^2/2}\, dx \approx 0.997.$$

Since 0.95 is so close to 1, we expect that most of the time the rainfall will be between 13 and 17 inches a year.

Among the normal distributions, the one having $\mu = 0$, $\sigma = 1$ is called the *standard normal distribution*. Values of the corresponding cumulative distribution function are published in tables.

Problems for Section 8.3

1. Estimate the median daily catch for the fishing data given in Example 2 of Section 8.2.

2. **(a)** Use the cumulative distribution function in Figure 8.28 to estimate the median.
(b) Describe the density function: For what values is it positive? Increasing? Decreasing? Identify all local maxima and minima values.

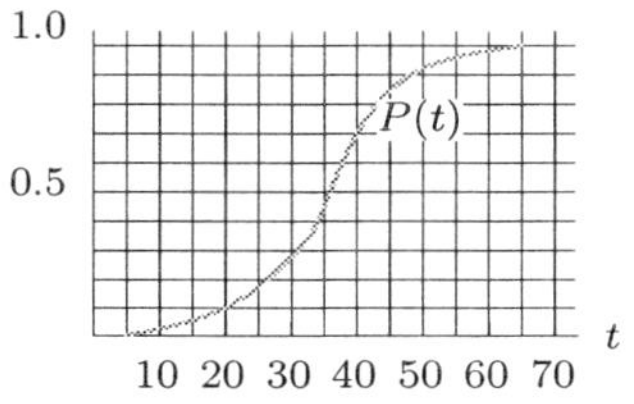

Figure 8.28

3. The speeds of cars on a road are approximately normally distributed with a mean $\mu = 58$ km/hr and standard deviation $\sigma = 4$ km/hr.

(a) What is the probability that a randomly selected car is going between 60 and 65 km/hr?
(b) What fraction of all cars are going slower than 52 km/hr?

4. The distribution of IQ scores can be modeled by a normal distribution with mean 100 and standard deviation 15.

(a) Write the formula for the density function of IQ scores.
(b) Estimate the fraction of the population with IQ between 115 and 120.

For Problems 5–6, let $p(t) = -0.0375t^2 + 0.225t$ be the density function for the shelf life of a brand of banana which lasts up to 4 weeks. Time, t, is measured in weeks and $0 \le t \le 4$.

5. Find the median shelf life of a banana using $p(t)$. Plot the median on a graph of $p(t)$. Does it look like half the area is to the right of the median and half the area is to the left?

6. Find the mean shelf life of a banana using $p(t)$. Plot the mean on a graph of $p(t)$. Does it look like the mean is the place where the density function balances?

7. Suppose that x measures the time (in hours) it takes for a student to complete an exam. All students are done within two hours and the density function for x is

$$p(x) = \begin{cases} x^3/4 & \text{if } 0 < x < 2 \\ 0 & \text{otherwise.} \end{cases}$$

(a) What proportion of students take between 1.5 and 2.0 hours to finish the exam?
(b) What is the mean time for students to complete the exam?
(c) Compute the median of this distribution.

8. Let $p(t) = 0.1e^{-0.1t}$ be the density function for the waiting time at a subway stop, with t in minutes, $0 \le t \le 60$.

(a) Graph $p(t)$. Use the graph to estimate visually the median and the mean.
(b) Calculate the median and the mean. Plot both on the graph of $p(t)$.
(c) Interpret the median and mean in terms of waiting time.

9. Let $P(x)$ be the cumulative distribution function for the household income distribution in the US in 2006. Values of $P(x)$ are in the following table:

Income x (thousand \$)	20	40	60	80	100	150
$P(x)$ (%)	21.7	45.4	63.0	75.8	84.0	94.0

(a) What percent of the households made between \$40,000 and \$60,000? More than \$150,000?
(b) Approximately what was the median income?
(c) Is the statement "More than one-third of households made between \$40,000 and \$80,000" true or false?

CHAPTER SUMMARY

- **Density function**
- **Cumulative distribution function**
- **Probability**
- **Median**
- **Mean**
- **Normal distribution**

REVIEW PROBLEMS FOR CHAPTER EIGHT

1. Match the graphs of the density functions (a), (b), and (c) with the graphs of the cumulative distribution functions I, II, and III.

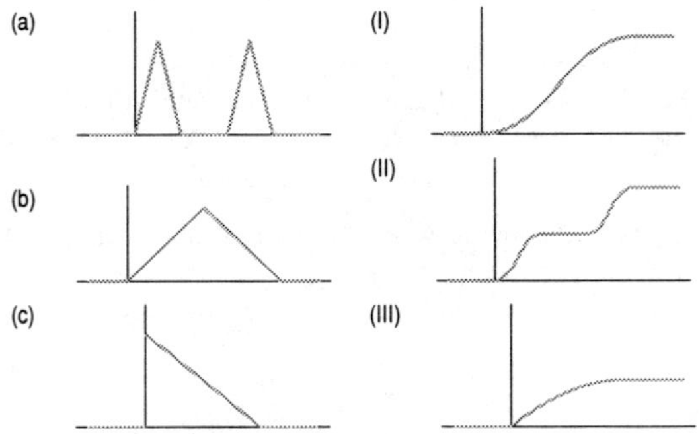

In Problems 2–3, graph a density function representing the given distribution.

2. The age at which a person dies in a society with high infant mortality and in which adults usually die between age 40 and age 60.

3. The heights of the people in an elementary school.

In Problems 4–6, graph a density function and a cumulative distribution function which could represent the distribution of income through a population with the given characteristics.

4. A large middle class.

5. Small middle and upper classes and many poor people.

6. Small middle class, many poor and many rich people.

In Problems 7–10, calculate the value of c if p is a density function.

7.

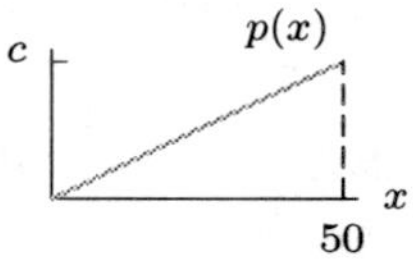

8.

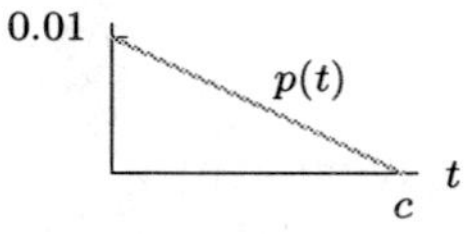

9.

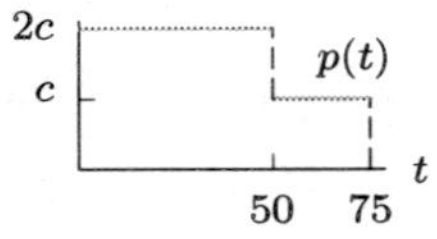

10.

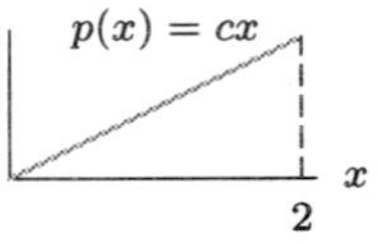

11. An insect has a life-span of no more than one year. Figure 8.29 shows the density function, $p(t)$, for the life-span.

(a) Do more insects die in the first month of their life or the twelfth month of their life?

(b) What fraction of the insects live no more than 6 months?

(c) What fraction of the insects live more than 9 months?

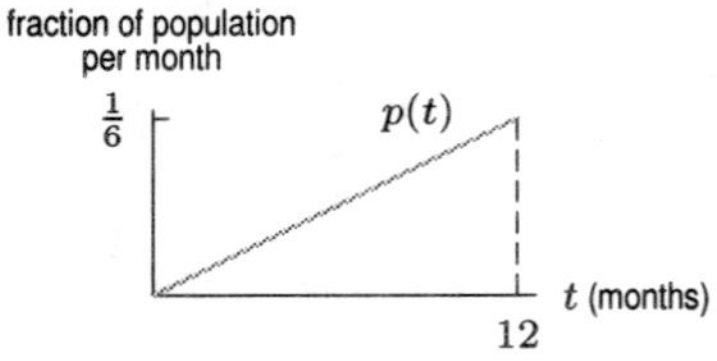

Figure 8.29

12. A large number of people take a standardized test, receiving scores described by the density function p graphed in Figure 8.30. Does the density function imply that most people receive a score near 50? Explain why or why not.

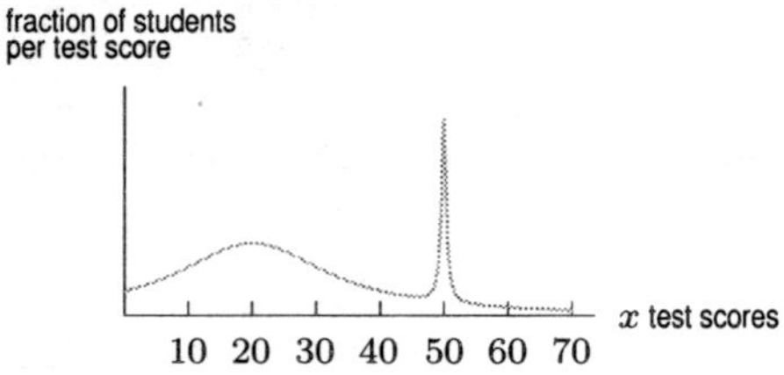

Figure 8.30

13. Figure 8.31[7] shows the distribution of elevation, in miles, across the earth's surface. Positive elevation denotes land above sea level; negative elevation shows land below sea level (i.e., the ocean floor).

(a) Describe in words the elevation of most of the earth's surface.

[7] Adapted from *Statistics*, by Freedman, Pisani, Purves, and Adikhari (New York: Norton).

(b) Approximately what fraction of the earth's surface is below sea level?

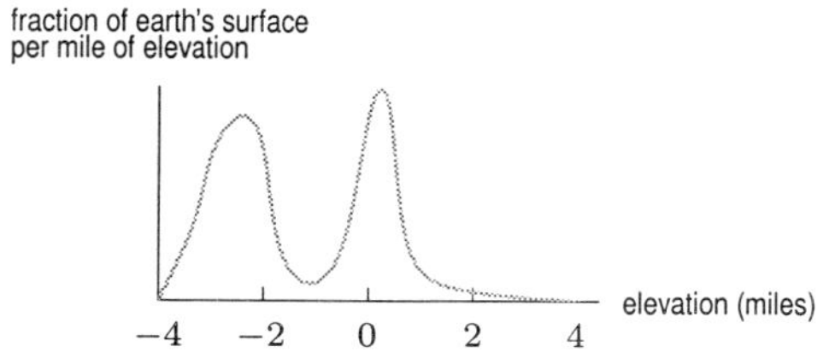

Figure 8.31

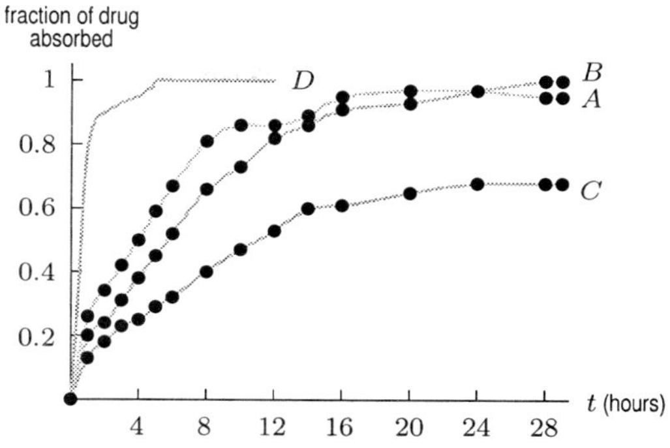

Figure 8.32

14. A congressional committee is investigating a defense contractor whose projects often incur cost overruns. The data in Table 8.8 show y, the fraction of the projects with an overrun of at most $C\%$.

(a) Plot the data with C on the horizontal axis. Is this a density function or a cumulative distribution function? Sketch a curve through these points.
(b) If you think you drew a density function in part (a), sketch the corresponding cumulative distribution function on another set of axes. If you think you drew a cumulative distribution function in part (a), sketch the corresponding density function.
(c) Based on the table, what is the probability that there will be a cost overrun of 50% or more? Between 20% and 50%? Near what percent is the cost overrun most likely to be?

Table 8.8 *Fraction, y, of overruns that are at most $C\%$*

C	-20%	-10%	0%	10%	20%	30%	40%	50%
y	0.01	0.08	0.19	0.32	0.50	0.80	0.94	0.99

15. The density function for radii r (mm) of spherical raindrops during a storm is constant over the range $0 < r < 5$ and zero elsewhere.

(a) Find the density function $f(r)$ for the radii.
(b) Find the cumulative distribution function $F(r)$.

16. Major absorption differences have been reported for different commercial versions of the same drug. One study compared three commercial versions of timed-release theophylline capsules.[8] A theophylline solution was included for comparison. Figure 8.32 shows the cumulative distribution functions $P(t)$, which represent the fraction of the drug absorbed by time t. Which curve represents the solution? Compare absorption rates of the four versions of the drug.

17. After measuring the duration of many telephone calls, the telephone company found their data was well-approximated by the density function $p(x) = 0.4e^{-0.4x}$, where x is the duration of a call, in minutes.

(a) What percentage of calls last between 1 and 2 minutes?
(b) What percentage of calls last 1 minute or less?
(c) What percentage of calls last 3 minutes or more?
(d) Find the cumulative distribution function.

18. In southern Switzerland, most rain falls in the spring and fall; summers and winters are relatively dry. Sketch possible graphs for the density function and the cumulative distribution function of the rain distribution over the course of one year. Put the date on the horizontal axis and fraction of the year's rainfall on the vertical axis.

19. Find the median of the density function given by $p(t) = 0.04 - 0.0008t$ for $0 \leq t \leq 50$ using the Fundamental Theorem of Calculus.

20. In 1950 an experiment was done observing the time gaps between successive cars on the Arroyo Seco Freeway. The data[9] show that, if x is time in seconds and $0 \leq x \leq 40$, the density function of these time gaps is

$$p(x) = 0.122e^{-0.122x}.$$

Find the median and mean time gap. Interpret them in terms of cars on the freeway.

In Problems 21–25, a quantity x is distributed through a population with probability density function $p(x)$ and cumulative distribution function $P(x)$. Decide if the statements in Problems 21–25 are true or false. Give an explanation for your answer.

21. If $p(10) = 1/2$, then half the population has $x < 10$.

22. If $P(10) = 1/2$, then half the population has $x < 10$.

23. If $p(10) = 1/2$, then the fraction of the population lying between $x = 9.98$ and $x = 10.04$ is about 0.03.

24. If $p(10) = p(20)$, then none of the population has x values lying between 10 and 20.

25. If $P(10) = P(20)$, then none of the population has x values lying between 10 and 20.

[8] *Progress in Drug Metabolism*, Bridges and Chasseaud (eds.) (New York: Wiley, 1980).
[9] Reported by Daniel Furlough and Frank Barnes.

CHECK YOUR UNDERSTANDING

In Problems 1–30, indicate whether the statement is true or false.

1. If $\int_{-\infty}^{\infty} p(x)\,dx = 1$ then $p(x)$ is a density function.
2. The function given by $p(x) = 0.25$ when $0 \le x \le 4$ and $p(x) = 0$ when $x < 0$ or $x > 4$ is a density function.
3. If $p(x)$ is a density function for the age in years of a population, then $\int_0^{10} p(x)\,dx$ is the fraction of the population with ages between 0 and 10 years.
4. If $p(t)$ is a density function for waiting times in minutes at a bank teller, then $\int_{10}^{15} p(x)\,dx$ is the fraction of the people waiting in line for that teller who wait between 10 and 15 minutes.
5. If $p(x)$ is the density function for age in years of the US population, then $\int_0^{70} p(x)\,dx \approx 0.5$.
6. If $p(x)$ is the density function for age in years of a population, then $p(50)$ is the fraction of the population exactly 50 years old.
7. If $p(x)$ is the density function for age in years of a population, and $p(10) = 0.014$, then about 1.4 percent of the population have ages between 9.5 and 10.5 years.
8. If $p(x)$ is a density function, then $\int_0^{\infty} p(x)\,dx = 0.5$.
9. If $p(x)$ is a density function, then $\int_0^1 p(x)\,dx \le \int_0^2 p(x)\,dx$.
10. If $p(x)$ is a density function, then it could never happen that $\int_0^{100} p(x)\,dx = \int_0^{200} p(x)\,dx$.
11. A cumulative distribution function $P(t)$ can never be decreasing.
12. If $P(t)$ is the cumulative distribution function corresponding to the density function $p(t)$, then $p' = P$.
13. If $P(t)$ is a cumulative distribution function for a population, then $P(30)$ is the fraction of the population with values above 30.
14. If $P(t)$ is a cumulative distribution function for a population, then $P(20) - P(10)$ is the fraction of the population with values between 10 and 20.
15. If $P(t)$ is a cumulative distribution function for a population, then $P(30) - P(10) = P(20)$.
16. If $P(t)$ is a cumulative distribution function for age in years of a population and $P(10) = 0.5$, then one half of the population is less than 10 years old.
17. The units of a density function $p(x)$ and its associated cumulative distribution function $P(x)$ are the same.
18. A cumulative distribution function $P(t)$ satisfies $0 \le P(t) \le 1$ for all t.
19. If $P(t)$ is a cumulative distribution function such that $P(10) = 0.05$ and $P(15) = 0.08$, then $P(25) = 0.13$.
20. If $P(t)$ is the cumulative distribution function for a density function $p(x)$, then $P(50) - P(25) = \int_{25}^{50} p(x)\,dx$.
21. If $p(x)$ is the probability density function for a quantity, then the mean value of the quantity is $\int_{-\infty}^{\infty} xp(x)\,dx$.
22. If $P(t)$ is the cumulative distribution function for a quantity, then the mean value of the quantity is given by $\int_{-\infty}^{\infty} tP(t)\,dt$.
23. If $P(t)$ is the cumulative distribution function for a quantity, then the median value T of the quantity is given by $T = P(0.5)$.
24. If $p(x)$ is the density function for a quantity x, then the median T satisfies $\int_{-\infty}^{T} p(x)\,dx = 0.5$.
25. If $p(x)$ is the density function for a quantity x, then the median T satisfies $\int_{T}^{\infty} p(x)\,dx = 0.5$.
26. If a quantity has density function $p(x)$, then the mean value and median are never equal.
27. The normal distribution with density function $p(x) = (1/\sqrt{2\pi})e^{-x^2/2}$ has mean $\mu = 0$ and standard deviation $\sigma = 1$.
28. The normal distribution with density function $p(x) = (1/\sqrt{2\pi})e^{-(x-5)^2/2}$ has mean $\mu = -5$ and standard deviation $\sigma = 1$.
29. The integral $\int_{-\infty}^{\infty} (1/\sqrt{2\pi})e^{-(x-7)^2/2}\,dx$ equals 1.
30. The mean value of a quantity x distributed through a population is the value T such that half of the population has values x less than or equal to T.

PROJECTS FOR CHAPTER EIGHT

1. Triangular Probability Distribution

Triangular probability distributions, such as the one with density function graphed in Figure 8.33, are used in business to model uncertainty. Such a distribution can be used to model a variable where only three pieces of information are available: a lower bound ($x = a$), a most likely value ($x = c$), and an upper bound ($x = b$).

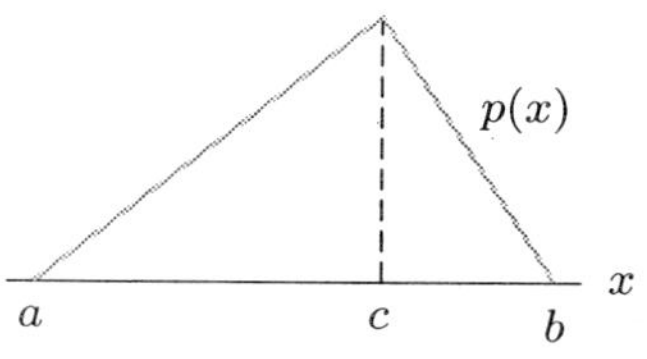

Figure 8.33

Thus, we can write the function $p(x)$ as two linear functions:

$$p(x) = \begin{cases} m_1x + b_1 & a \leq x \leq c \\ m_2x + b_2 & c < x \leq b. \end{cases}$$

(a) Find the value of $p(c)$ geometrically, using the criterion that the probability that x takes on some value between a and b is 1.

Suppose a new product costs between \$6 and \$10 per unit to produce, with a most likely cost of \$9.

(b) Find $p(9)$.

(c) Use the fact that $p(6) = p(10) = 0$ and the value of $p(9)$ you found in part (b) to find m_1, m_2, b_1, and b_2.

(d) What is the probability that the production cost per unit will be less than \$8?

(e) What is the median cost?

(f) Write a formula for the cumulative probability distribution function $P(x)$ for

(i) $6 \leq x \leq 9$, (ii) $9 < x \leq 10$.

Sketch the graph of $P(x)$.

Chapter Nine

FUNCTIONS OF SEVERAL VARIABLES

9.1 UNDERSTANDING FUNCTIONS OF TWO VARIABLES

In business, science, and politics, outcomes are rarely determined by a single variable. For example, consider an airline's ticket pricing. To avoid flying planes with many empty seats, it sells some tickets at full price and some at a discount. For a particular route, the airline's revenue, R, earned in a given time period is determined by the number of full-price tickets, x, and the number of discount tickets, y, sold. We say that R is a function of x and y, and we write

$$R = f(x, y).$$

This is just like the function notation of one-variable calculus. The variable R is the dependent variable and the variables x and y are the independent variables. The letter f stands for the *function* or rule that gives the value, or output, of R corresponding to given values of x and y. The collection of all possible inputs, (x, y), is called the *domain* of f. We say a function is an *increasing* (*decreasing*) function of one of its variables if it increases (decreases) as that variable increases while the other independent variables are held constant.

A function of two variables can be represented numerically by a table of values, algebraically by a formula, or pictorially by a contour diagram. In this section we give numerical and algebraic examples; contour diagrams are introduced in Section 9.2.

Functions Given Numerically

The revenue, R, (in dollars) from a particular airline route is shown in Table 9.1 as a function of the number of full-price tickets and the number of discount tickets sold.

Table 9.1 *Revenue from ticket sales as a function of x and y*

		Number of full-price tickets, x			
		100	200	300	400
Number of discount tickets, y	200	75,000	110,000	145,000	180,000
	400	115,000	150,000	185,000	220,000
	600	155,000	190,000	225,000	260,000
	800	195,000	230,000	265,000	300,000
	1000	235,000	270,000	305,000	340,000

Values of x are shown across the top, values of y are down the left side, and the corresponding values of $f(x, y)$ are in the table. For example, to find the value of $f(300, 600)$, we look in the column corresponding to $x = 300$ at the row $y = 600$, where we find the number 225,000. Thus,

$$f(300, 600) = 225{,}000.$$

This means that the revenue from 300 full-price tickets and 600 discount tickets is \$225,000. We see in Table 9.1 that f is an increasing function of x and an increasing function of y.

Notice how this differs from the table of values of a one-variable function, where one row or one column is enough to list the values of the function. Here many rows and columns are needed because the function has an output for every pair of values of the independent variables.

Functions Given Algebraically

The function given in Table 9.1 can be represented by a formula. Looking across the rows, we see that each additional 100 full-price tickets sold raises the revenue by \$35,000, so each full-price ticket must cost \$350. Similarly, looking down a column shows that an additional 200 discount tickets sold increases the revenue by \$40,000, so each discount ticket must cost \$200. Thus, the revenue function is given by the formula

$$R = 350x + 200y.$$

Example 1 Give a formula for the function $M = f(B, t)$ where M is the amount of money in a bank account t years after an initial investment of B dollars, if interest accrues at a rate of 5% per year compounded
(a) Annually (b) Continuously.

Solution (a) Annual compounding means that M increases by a factor of 1.05 every year, so

$$M = f(B,t) = B(1.05)^t.$$

(b) Continuous compounding means that M grows according to the function e^{kt}, with $k = 0.05$, so

$$M = f(B,t) = Be^{0.05t}.$$

Example 2 A car rental company charges \$40 a day and 15 cents a mile for its cars.

(a) Write a formula for the cost, C, of renting a car as a function of the number of days, d, and the number of miles driven, m.
(b) If $C = f(d,m)$, find $f(5,300)$ and interpret it.

Solution (a) The total cost in dollars of renting a car is 40 times the numbers of days plus 0.15 times the number of miles, so

$$C = 40d + 0.15m.$$

(b) We have

$$\begin{aligned} f(5,300) &= 40(5) + 0.15(300) \\ &= 200 + 45 \\ &= 245. \end{aligned}$$

We see that $f(5,300) = 245$. This tells us that if we rent a car for 5 days and drive it 300 miles, it costs us \$245.

Strategy to Investigate Functions of Two Variables: Vary One Variable at a Time

We can learn a great deal about a function of two variables by letting one variable vary while holding the other fixed. This gives a function of one variable, called a *cross-section* of the original function.

Concentration of a Drug in the Blood

When a drug is injected into muscle tissue, it diffuses into the bloodstream. The concentration of the drug in the blood increases until it reaches a maximum, and then decreases. The concentration, C (in mg per liter), of the drug in the blood is a function of two variables: x, the amount (in mg) of the drug given in the injection, and t, the time (in hours) since the injection was administered. We are told that

$$C = f(x,t) = te^{-t(5-x)} \qquad \text{for } 0 \le x \le 4 \text{ and } t \ge 0.$$

Example 3 In terms of the drug concentration in the blood, explain the significance of the cross-sections:

(a) $f(4,t)$ (b) $f(x,1)$

Solution (a) Holding x fixed at 4 means that we are considering an injection of 4 mg of the drug; letting t vary means we are watching the effect of this dose as time passes. Thus the function $f(4,t)$ describes the concentration of the drug in the blood resulting from a 4-mg injection as a function of time. Figure 9.1 shows the graph of $f(4,t) = te^{-t}$. Notice that the concentration in the blood from this dose is at a maximum at 1 hour after injection, and that the concentration in the blood eventually approaches zero.

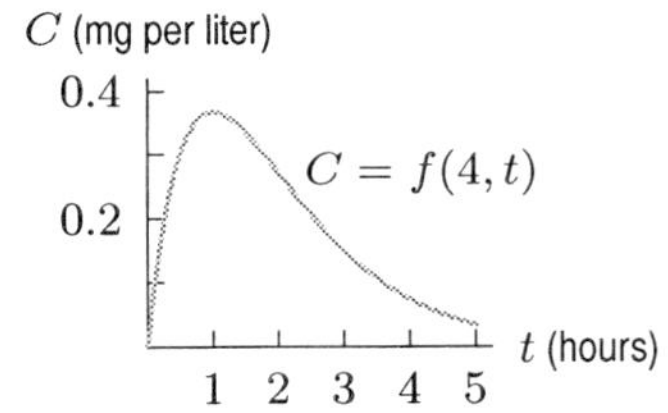

Figure 9.1: The function $f(4,t)$ shows the concentration in the blood resulting from a 4-mg injection

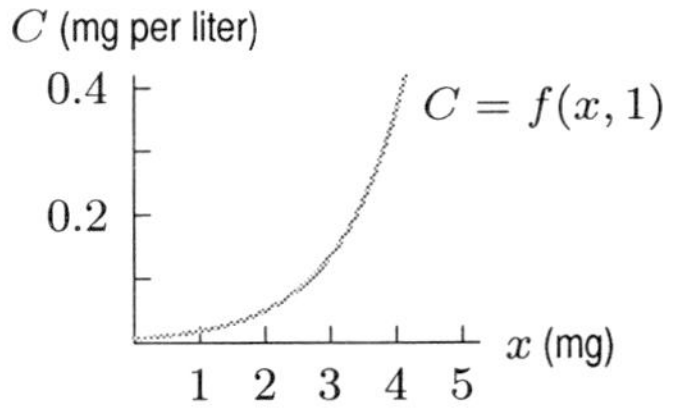

Figure 9.2: The function $f(x,1)$ shows the concentration in the blood 1 hour after the injection

(b) Holding t fixed at 1 means that we are focusing on the blood 1 hour after the injection; letting x vary means we are considering the effect of different doses at that instant. Thus, the function $f(x,1)$ gives the concentration of the drug in the blood 1 hour after injection as a function of the amount injected. Figure 9.2 shows the graph of $f(x,1) = e^{-(5-x)} = e^{x-5}$. Notice that $f(x,1)$ is an increasing function of x. This makes sense: If we administer more of the drug, the concentration in the bloodstream is higher.

Example 4 Continue with $C = f(x,t) = te^{-t(5-x)}$. Graph the cross-sections of $f(a,t)$ for $a = 1, 2, 3$, and 4 on the same axes. Describe how the graph changes for larger values of a and explain what this means in terms of drug concentration in the blood.

Solution The one-variable function $f(a,t)$ represents the effect of an injection of a mg at time t. Figure 9.3 shows the graphs of the four functions $f(1,t) = te^{-4t}$, $f(2,t) = te^{-3t}$, $f(3,t) = te^{-2t}$, and $f(4,t) = te^{-t}$ corresponding to injections of 1, 2, 3, and 4 mg of the drug. The general shape of the graph is the same in every case: The concentration in the blood is zero at the time of injection $t = 0$, then increases to a maximum value, and then decreases toward zero again. We see that if a larger dose of the drug is administered, the peak of the graph is later and higher. This makes sense, since a larger dose will take longer to diffuse fully into the bloodstream and will produce a higher concentration when it does.

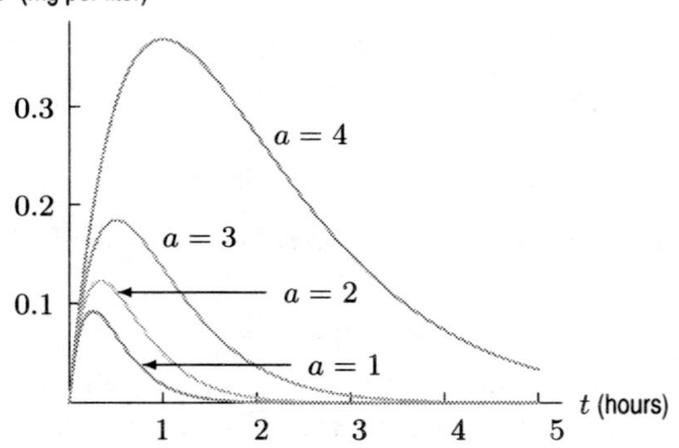

Figure 9.3: Concentration $C = f(a,t)$ of the drug resulting from an a-mg injection

Problems for Section 9.1

Problems 1–2 concern the cost, C, of renting a car from a company which charges \$40 a day and 15 cents a mile, so $C = f(d,m) = 40d + 0.15m$, where d is the number of days, and m is the number of miles.

1. Make a table of values for C, using $d = 1, 2, 3, 4$ and $m = 100, 200, 300, 400$. You should have 16 values in your table.

2. (a) Find $f(3,200)$ and interpret it.
   (b) Explain the significance of $f(3,m)$ in terms of rental car costs. Graph this function, with C as a function of m.
   (c) Explain the significance of $f(d,100)$ in terms of rental car costs. Graph this function, with C as a function of d.

Problems 3–7 refer to Table 9.2, which shows[1] the weekly beef consumption, C, (in lbs) of an average household as a function of p, the price of beef (in \$/lb) and I, annual household income (in \$1000s).

Table 9.2 *Quantity of beef bought (lbs/household/week)*

		p			
		3.00	3.50	4.00	4.50
I	20	2.65	2.59	2.51	2.43
	40	4.14	4.05	3.94	3.88
	60	5.11	5.00	4.97	4.84
	80	5.35	5.29	5.19	5.07
	100	5.79	5.77	5.60	5.53

3. Give tables for beef consumption as a function of p, with I fixed at $I = 20$ and $I = 100$. Give tables for beef consumption as a function of I, with p fixed at $p = 3.00$ and $p = 4.00$. Comment on what you see in the tables.

4. How does beef consumption vary as a function of household income if the price of beef is held constant?

5. Make a table showing the amount of money, M, that the average household spends on beef (in dollars per household per week) as a function of the price of beef and household income.

[1]From Richard G. Lipsey, *An Introduction to Positive Economics*, 3rd Ed. (London: Weidenfeld and Nicolson, 1971).

6. Make a table of the proportion, P, of household income spent on beef per week as a function of price and income. (Note that P is the fraction of income spent on beef.)
7. Express P, the proportion of household income spent on beef per week, in terms of the original function $f(I, p)$ which gave consumption as a function of p and I.
8. Graph the bank-account function f in Example 1 on page 350, holding B fixed at $B = 10, 20, 30$ and letting t vary. Then graph f, holding t fixed at $t = 0, 5, 10$ and letting B vary. Explain what you see.
9. The total sales of a product, S, can be expressed as a function of the price p charged for the product and the amount, a, spent on advertising, so $S = f(p, a)$. Do you expect f to be an increasing or decreasing function of p? Do you expect f to be an increasing or decreasing function of a? Why?
10. The number, n, of new cars sold in a year is a function of the price of new cars, c, and the average price of gas, g.
    (a) If c is held constant, is n an increasing or decreasing function of g? Why?
    (b) If g is held constant, is n an increasing or decreasing function of c? Why?
11. The heat index is a temperature which tells you how hot it feels as a result of the combination of temperature and humidity. See Table 9.3. Heat exhaustion is likely to occur when the heat index reaches 105°F.
    (a) If the temperature is 80°F and the humidity is 50%, how hot does it feel?
    (b) At what humidity does 90°F feel like 90°F?
    (c) Make a table showing the approximate temperature at which heat exhaustion becomes a danger, as a function of humidity.
    (d) Explain why the heat index is sometimes above the actual temperature and sometimes below it.

Table 9.3 *Heat index (°F) as a function of humidity (H%) and temperature (T°F)*

H \ T	70	75	80	85	90	95	100	105	110	115
0	64	69	73	78	83	87	91	95	99	103
10	65	70	75	80	85	90	95	100	105	111
20	66	72	77	82	87	93	99	105	112	120
30	67	73	78	84	90	96	104	113	123	135
40	68	74	79	86	93	101	110	123	137	151
50	69	75	81	88	96	107	120	135	150	
60	70	76	82	90	100	114	132	149		

12. Using Table 9.3, graph heat index as a function of humidity with temperature fixed at 70°F and at 100°F. Explain the features of each graph and the difference between them in common-sense terms.
13. The monthly payments, P dollars, on a mortgage in which A dollars were borrowed at an annual interest rate of $r\%$ for t years is given by $P = f(A, r, t)$. Is f an increasing or decreasing function of A? Of r? Of t?
14. You are planning a trip whose principal cost is gasoline.
    (a) Make a table showing how the daily fuel cost varies as a function of the price of gasoline (in dollars per gallon) and the number of gallons you buy each day.
    (b) If your car goes 30 miles on each gallon of gasoline, make a table showing how your daily fuel cost varies as a function of your daily travel distance and the price of gas.

In Problems 15–16, the fallout, V (in kilograms per square kilometer), from a volcanic explosion depends on the distance, d, from the volcano and the time, t, since the explosion:

$$V = f(d, t) = (\sqrt{t})e^{-d}.$$

15. On the same axes, graph cross-sections of f with $t = 1$, and $t = 2$. As distance from the volcano increases, how does the fallout change? Look at the relationship between the graphs: how does the fallout change as time passes? Explain your answers in terms of volcanoes.
16. On the same axes, graph cross-sections of f with $d = 0$, $d = 1$, and $d = 2$. As time passes since the explosion, how does the fallout change? Look at the relationship between the graphs: how does fallout change as a function of distance? Explain your answers in terms of volcanoes.
17. An airport can be cleared of fog by heating the air. The amount of heat required, $H(T, w)$ (in calories per cubic meter of fog), depends on the temperature of the air, T (in °C), and the wetness of the fog, w (in grams per cubic meter of fog). Figure 9.4 shows several graphs of H against T with w fixed.
    (a) Estimate $H(20, 0.3)$ and explain what information it gives us.
    (b) Make a table of values for $H(T, w)$. Use $T = 0$, 10, 20, 30, 40, and $w = 0.1, 0.2, 0.3, 0.4$.

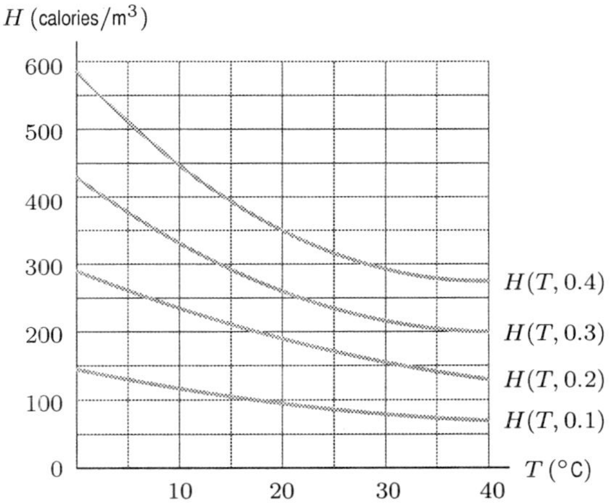

Figure 9.4

9.4 COMPUTING PARTIAL DERIVATIVES ALGEBRAICALLY

The partial derivative $f_x(x, y)$ is the ordinary derivative of the function $f(x, y)$ with respect to x with y fixed, and the partial derivative $f_y(x, y)$ is the ordinary derivative of $f(x, y)$ with respect to y with x fixed. Thus, we can use all the techniques for differentiation from single-variable calculus to find partial derivatives.

Example 1 Let $f(x, y) = x^2 + 5y^2$. Find $f_x(3, 2)$ and $f_y(3, 2)$ algebraically.

Solution We use the fact that $f_x(3, 2)$ is the derivative of $f(x, 2)$ at $x = 3$. To find f_x, we fix y at 2:

$$f(x, 2) = x^2 + 5(2^2) = x^2 + 20.$$

Differentiating with respect to x gives

$$f_x(x, 2) = 2x \quad \text{so} \quad f_x(3, 2) = 2(3) = 6.$$

Similarly, $f_y(3, 2)$ is the derivative of $f(3, y)$ at $y = 2$. To find f_y, we fix x at 3:

$$f(3, y) = 3^2 + 5y^2 = 9 + 5y^2.$$

Differentiating with respect to y, we have

$$f_y(3, y) = 10y \quad \text{so} \quad f_y(3, 2) = 10(2) = 20.$$

Example 2 Let $f(x, y) = x^2 + 5y^2$ as in Example 1. Find f_x and f_y as functions of x and y.

Solution To find f_x, we treat y as a constant. Thus $5y^2$ is a constant and the derivative with respect to x of this term is 0. We have

$$f_x(x, y) = 2x + 0 = 2x.$$

To find f_y, we treat x as a constant and so the derivative of x^2 with respect to y is zero. We have

$$f_y(x, y) = 0 + 10y = 10y.$$

Example 3 Find both partial derivatives of each of the following functions:

(a) $f(x, y) = 3x + e^{-5y}$ (b) $f(x, y) = x^2y$ (c) $f(u, v) = u^2e^{2v}$

Solution (a) To find f_x, we treat y as a constant, so the term e^{-5y} is a constant, and the derivative of this term is zero. Likewise, to find f_y, we treat x as a constant. We have

$$f_x(x, y) = 3 + 0 = 3 \quad \text{and} \quad f_y(x, y) = 0 + (-5)e^{-5y} = -5e^{-5y}.$$

(b) To find f_x, we treat y as a constant, so the function is treated as a constant times x^2. The derivative of a constant times x^2 is the constant times $2x$, and so we have

$$f_x(x, y) = (2x)y = 2xy \quad \text{Similarly,} \quad f_y(x, y) = (x^2)(1) = x^2.$$

(c) To find f_u, we treat v as a constant, and to find f_v, we treat u as a constant. We have

$$f_u(u, v) = (2u)(e^{2v}) = 2ue^{2v} \quad \text{and} \quad f_v(u, v) = u^2(2e^{2v}) = 2u^2e^{2v}.$$

Example 4 The concentration C of bacteria in the blood (in millions of bacteria/ml) following the injection of an antibiotic is a function of the dose x (in gm) injected and the time t (in hours) since the injection. Suppose we are told that $C = f(x, t) = te^{-xt}$. Evaluate the following quantities and explain what each one means in practical terms: (a) $f_x(1, 2)$ (b) $f_t(1, 2)$

Solution (a) To find f_x, we treat t as a constant and differentiate with respect to x, giving

$$f_x(x, t) = -t^2e^{-xt}.$$

Substituting $x = 1, t = 2$ gives

$$f_x(1, 2) = -4e^{-2} \approx -0.54.$$

To see what $f_x(1, 2)$ means, think about the function $f(x, 2)$ of which it is the derivative. The graph of $f(x, 2)$ in Figure 9.39 gives the concentration of bacteria as a function of the dose two hours after the injection. The derivative $f_x(1, 2)$ is the slope of this graph at the point $x = 1$; it is negative because a larger dose reduces the bacteria population. More precisely, the partial derivative $f_x(1, 2)$ gives the rate of change of bacteria concentration with respect to the dose injected, namely a decrease in bacteria concentration of 0.54 million/ml per gram of additional antibiotic injected.

(b) To find f_t, treat x as a constant and differentiate using the product rule:

$$f_t(x, t) = 1 \cdot e^{-xt} - xte^{-xt}.$$

Substituting $x = 1, t = 2$ gives

$$f_t(1, 2) = e^{-2} - 2e^{-2} \approx -0.14.$$

To see what $f_t(1, 2)$ means, think about the function $f(1, t)$ of which it is the derivative. The graph of $f(1, t)$ in Figure 9.40 gives the concentration of bacteria at time t if the dose of antibiotic is 1 gm. The derivative $f_t(1, 2)$ is the slope of the graph at the point $t = 2$; it is negative because after 2 hours the concentration of bacteria is decreasing. More precisely, the partial derivative $f_t(1, 2)$ gives the rate at which the bacteria concentration is changing with respect to time, namely a decrease in bacteria concentration of 0.14 million/ml per hour.

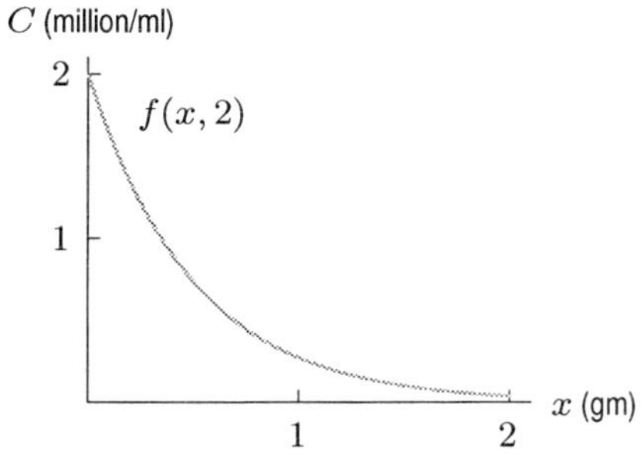

Figure 9.39: Bacteria concentration after 2 hours as a function of the quantity of antibiotic injected

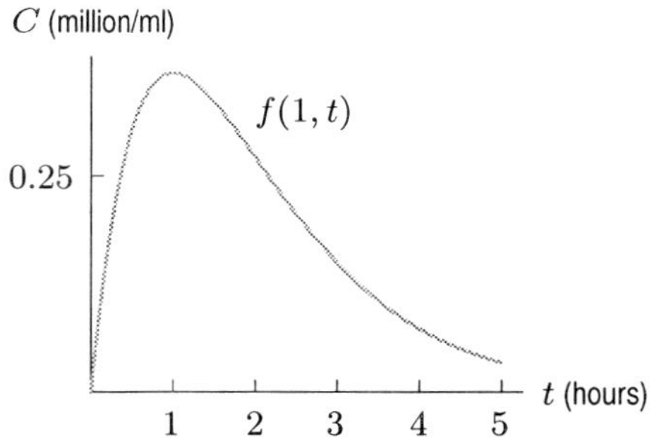

Figure 9.40: Bacteria concentration as a function of time if 1 unit of antibiotic is injected

Example 5 Let's consider a small printing business where N is the number of workers, V is the value of the equipment (in units of \$25,000), and P is the production, measured in thousands of pages per day. Suppose the production function for this company is given by

$$P = f(N, V) = 2N^{0.6}V^{0.4}.$$

(a) If this company has a labor force of 100 workers and 200 units' worth of equipment, what is the production output of the company?
(b) Find $f_N(100, 200)$ and $f_V(100, 200)$. Interpret your answers in terms of production.

Solution (a) We have $N = 100$ and $V = 200$, so

$$\text{Production } = 2(100)^{0.6}(200)^{0.4} = 263.9 \text{ thousand pages per day.}$$

(b) To find f_N, we treat V as a constant and differentiate with respect to N:

$$f_N(N, V) = 2(0.6)N^{-0.4}V^{0.4}.$$

Substituting $N = 100$, $V = 200$ gives

$$f_N(100, 200) = 1.2(100^{-0.4})(200^{0.4}) \approx 1.583 \text{ thousand pages/worker.}$$

This tells us that if we have 200 units of equipment and increase the number of workers by 1 from 100 to 101, the production output will go up by about 1.58 units, or 1580 pages per day.

Similarly, to find $f_V(100, 200)$, we treat N as a constant and differentiate with respect to V:

$$f_V(N, V) = 2(0.4)N^{0.6}V^{-0.6}.$$

Substituting $N = 100$, $V = 200$ gives

$$f_V(100, 200) = 0.8(100^{0.6})(200^{-0.6}) \approx 0.53 \text{ thousand pages/unit of equipment.}$$

This tells us that if we have 100 workers and increase the value of the equipment by 1 unit (\$25,000) from 200 units to 201 units, the production goes up by about 0.53 units, or 530 pages per day.

Second-Order Partial Derivatives

Since the partial derivatives of a function are themselves functions, we can usually differentiate them, giving *second-order partial derivatives.* A function $z = f(x, y)$ has two first-order partial derivatives, f_x and f_y, and four second-order partial derivatives.

The Second-Order Partial Derivatives of $z = f(x, y)$

$$\frac{\partial^2 z}{\partial x^2} = f_{xx} = (f_x)_x, \qquad \frac{\partial^2 z}{\partial x \partial y} = f_{yx} = (f_y)_x,$$

$$\frac{\partial^2 z}{\partial y \partial x} = f_{xy} = (f_x)_y, \qquad \frac{\partial^2 z}{\partial y^2} = f_{yy} = (f_y)_y.$$

It is usual to omit the parentheses, writing f_{xy} instead of $(f_x)_y$ and $\frac{\partial^2 z}{\partial y\, \partial x}$ instead of $\frac{\partial}{\partial y}\left(\frac{\partial z}{\partial x}\right)$.

Example 6 Use the values of the function $f(x, y)$ in Table 9.7 to estimate $f_{xy}(1, 2)$ and $f_{yx}(1, 2)$.

Table 9.7 *Values of $f(x, y)$*

		x		
		0.9	1.0	1.1
	1.8	4.72	5.83	7.06
y	2.0	6.48	8.00	9.60
	2.2	8.62	10.65	12.88

Solution Since $f_{xy} = (f_x)_y$, we first estimate f_x:

$$f_x(1, 2) \approx \frac{f(1.1, 2) - f(1, 2)}{0.1} = \frac{9.60 - 8.00}{0.1} = 16.0,$$
$$f_x(1, 2.2) \approx \frac{f(1.1, 2.2) - f(1, 2.2)}{0.1} = \frac{12.88 - 10.65}{0.1} = 22.3.$$

Thus,

$$f_{xy}(1, 2) \approx \frac{f_x(1, 2.2) - f_x(1, 2)}{0.2} = \frac{22.3 - 16.0}{0.2} = 31.5.$$

Similarly,

$$f_{yx}(1, 2) \approx \frac{f_y(1.1, 2) - f_y(1, 2)}{0.1} \approx \frac{1}{0.1}\left(\frac{f(1.1, 2.2) - f(1.1, 2)}{0.2} - \frac{f(1, 2.2) - f(1, 2)}{0.2}\right)$$
$$= \frac{1}{0.1}\left(\frac{12.88 - 9.60}{0.2} - \frac{10.65 - 8.00}{0.2}\right) = 31.5.$$

Observe that in this example, $f_{xy} = f_{yx}$ at the point $(1, 2)$.

Example 7 Compute the four second-order partial derivatives of $f(x, y) = xy^2 + 3x^2e^y$.

Solution From $f_x(x, y) = y^2 + 6xe^y$ we get

$$f_{xx}(x, y) = \frac{\partial}{\partial x}(y^2 + 6xe^y) = 6e^y \quad \text{and} \quad f_{xy}(x, y) = \frac{\partial}{\partial y}(y^2 + 6xe^y) = 2y + 6xe^y.$$

From $f_y(x, y) = 2xy + 3x^2e^y$ we get

$$f_{yx}(x, y) = \frac{\partial}{\partial x}(2xy + 3x^2e^y) = 2y + 6xe^y \quad \text{and} \quad f_{yy}(x, y) = \frac{\partial}{\partial y}(2xy + 3x^2e^y) = 2x + 3x^2e^y.$$

Observe that $f_{xy} = f_{yx}$ in this example.

The Mixed Partial Derivatives Are Equal

It is not an accident that the estimates for $f_{xy}(1, 2)$ and $f_{yx}(1, 2)$ are equal in Example 6, because the same values of the function are used to calculate each one. The fact that $f_{xy} = f_{yx}$ in Example 7 corroborates the following general result:

> If f_{xy} and f_{yx} are continuous at (a, b), then
> $$f_{xy}(a, b) = f_{yx}(a, b).$$

Most of the functions we will encounter not only have f_{xy} and f_{yx} continuous, but all their higher-order partial derivatives (such as f_{xxy} or f_{xyyy}) will be continuous. We call such functions *smooth*.

Problems for Section 9.4

Find the partial derivatives in Problems 1–13. The variables are restricted to a domain on which the function is defined.

1. f_x and f_y if $f(x, y) = 2x^2 + 3y^2$
2. f_x and f_y if $f(x, y) = 100x^2y$
3. f_x and f_y if $f(x, y) = x^2 + 2xy + y^3$
4. z_x if $z = x^2y + 2x^5y$
5. f_u and f_v if $f(u, v) = u^2 + 5uv + v^2$
6. $\frac{\partial z}{\partial x}$ if $z = x^2e^y$
7. $\frac{\partial Q}{\partial p}$ if $Q = 5a^2p - 3ap^3$
8. f_t if $f(t, a) = 5a^2t^3$
9. f_x and f_y if $f(x, y) = 5x^2y^3 + 8xy^2 - 3x^2$
10. f_x and f_y if $f(x, y) = 10x^2e^{3y}$
11. $\frac{\partial P}{\partial r}$ if $P = 100e^{rt}$
12. $\frac{\partial A}{\partial h}$ if $A = \frac{1}{2}(a + b)h$
13. $\frac{\partial}{\partial m}\left(\frac{1}{2}mv^2\right)$
14. If $f(x, y) = x^3 + 3y^2$, find $f(1, 2)$, $f_x(1, 2)$, $f_y(1, 2)$.
15. If $f(u, v) = 5uv^2$, find $f(3, 1)$, $f_u(3, 1)$, and $f_v(3, 1)$.
16. (a) Let $f(x, y) = x^2 + y^2$. Estimate $f_x(2, 1)$ and $f_y(2, 1)$ using the contour diagram for f in Figure 9.41.
    (b) Estimate $f_x(2, 1)$ and $f_y(2, 1)$ from a table of values for f with $x = 1.9, 2, 2.1$ and $y = 0.9, 1, 1.1$.
    (c) Compare your estimates in parts (a) and (b) with the exact values of $f_x(2, 1)$ and $f_y(2, 1)$ found algebraically.

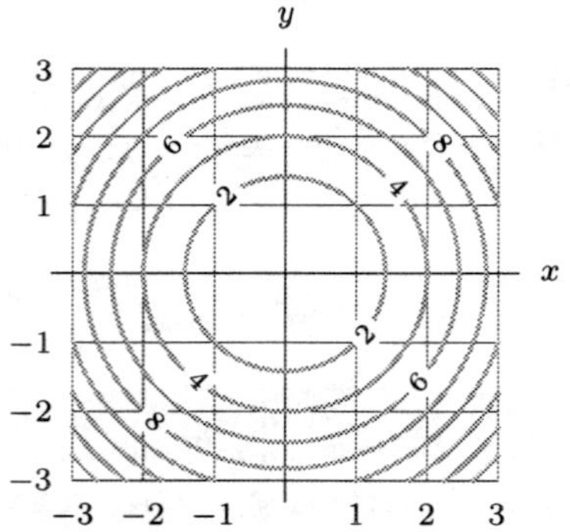

Figure 9.41

17. The amount of money, $\$B$, in a bank account earning interest at a continuous rate, r, depends on the amount deposited, $\$P$, and the time, t, it has been in the bank, where

$$B = Pe^{rt}.$$

Find $\partial B/\partial t$, $\partial B/\partial r$ and $\partial B/\partial P$ and interpret each in financial terms.

18. The cost of renting a car from a certain company is \$40 per day plus 15 cents per mile, and so we have

$$C = 40d + 0.15m.$$

Find $\partial C/\partial d$ and $\partial C/\partial m$. Give units and explain why your answers make sense.

19. A company's production output, P, is given in tons, and is a function of the number of workers, N, and the value of the equipment, V, in units of \$25,000. The production function for the company is

$$P = f(N, V) = 5N^{0.75}V^{0.25}.$$

The company currently employs 80 workers, and has equipment worth \$750,000. What are N and V? Find the values of f, f_N, and f_V at these values of N and V. Give units and explain what each answer means in terms of production.

For Problems 20–31, calculate all four second-order partial derivatives and confirm that the mixed partials are equal.

20. $f(x, y) = x^2y$
21. $f(x, y) = x^2+2xy+y^2$
22. $f(x, y) = xe^y$
23. $f(x, y) = \frac{2x}{y}, \quad y \neq 0$
24. $f = 5 + x^2y^2$
25. $f = e^{xy}$
26. $Q = 5p_1^2p_2^{-1}, \quad p_2 \neq 0$
27. $V = \pi r^2h$
28. $P = 2KL^2$
29. $B = 5xe^{-2t}$
30. $f(x, t) = t^3 - 4x^2t$
31. $f = 100e^{rt}$
32. Is there a function f which has the following partial derivatives? If so what is it? Are there any others?

$$f_x(x, y) = 4x^3y^2 - 3y^4,$$
$$f_y(x, y) = 2x^4y - 12xy^3.$$

33. Show that the Cobb-Douglas function

$$Q = bK^{\alpha}L^{1-\alpha} \quad \text{where} \quad 0 < \alpha < 1$$

satisfies the equation

$$K\frac{\partial Q}{\partial K} + L\frac{\partial Q}{\partial L} = Q.$$

Problems 34–36 are about the money supply, M, which is the total value of all the cash and checking account balances in an economy. It is determined by the value of all the cash, B, the ratio, c, of cash to checking deposits, and the fraction, r, of checking account deposits that banks hold as cash:

$$M = \frac{c + 1}{c + r}B.$$

(a) Find the partial derivative.
(b) Give its sign.
(c) Explain the significance of the sign in practical terms.

34. $\partial M/\partial B$
35. $\partial M/\partial r$
36. $\partial M/\partial c$

9.5 CRITICAL POINTS AND OPTIMIZATION

To optimize a function means to find the largest or smallest value of the function. If the function represents profit, we may want to find the conditions that maximize profit. On the other hand, if the function represents cost, we may want to find the conditions that minimize cost. In Chapter 4, we saw how to optimize a function of one variable by investigating critical points. In this section, we see how to extend the notions of critical points and local extrema to a function of more than one variable.

Local and Global Maxima and Minima for Functions of Two Variables

Functions of several variables, like functions of one variable, can have *local* and *global extrema* (that is, local and global maxima and minima). A function has a local extremum at a point where it takes on the largest or smallest value in a small region around the point. Global extrema are the largest or smallest value anywhere. For a function f defined on a domain R, we say:

- f has a **local maximum** at P_0 if $f(P_0) \geq f(P)$ for all points P near P_0
- f has a **local minimum** at P_0 if $f(P_0) \leq f(P)$ for all points P near P_0
- f has a **global maximum** at P_0 if $f(P_0) \geq f(P)$ for all points P in R
- f has a **global minimum** at P_0 if $f(P_0) \leq f(P)$ for all points P in R

Example 1 Table 9.8 gives a table of values for a function $f(x, y)$. Estimate the location and value of any global maxima or minima for $0 \leq x \leq 1$ and $0 \leq y \leq 20$.

Table 9.8 *Where are the extreme points of this function $f(x, y)$?*

y \ x	0	0.2	0.4	0.6	0.8	1.0
0	80	84	82	76	71	65
5	86	90	88	73	77	71
10	91	95	93	88	82	76
15	87	91	89	84	78	72
20	82	86	84	79	73	67

Solution The global maximum value of the function appears to be 95 at the point $(0.2, 10)$. Since the table only gives certain values, we cannot be sure that this is exactly the maximum. (The function might have a larger value at, for example, $(0.3, 11)$.) The global minimum value of this function on the points given is 65 at the point $(1, 0)$.

Example 2 Figure 9.42 gives a contour diagram for a function $f(x, y)$. Estimate the location and value of any local maxima or minima. Are any of these global maxima or minima on the square shown?

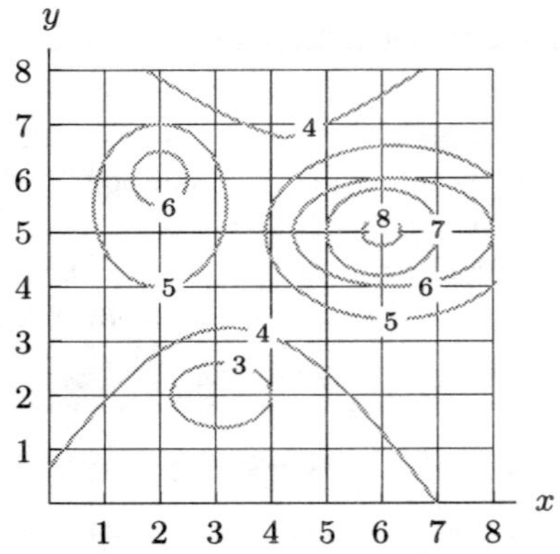

Figure 9.42: Where are the local and global extreme points of this function?

Solution There is a local maximum of above 8 near the point $(6, 5)$, a local maximum of above 6 near the point $(2, 6)$, and a local minimum of below 3 near the point $(3, 2)$. The value above 8 is the global maximum and the value below 3 is the global minimum on the given domain.

In Example 1 and Example 2, we can estimate the location and value of extreme points, but we do not have enough information to find them exactly. This is usually true when we are given a table of values or a contour diagram. To find local or global extrema exactly, we usually need to have a formula for the function.

Finding a Local Maximum or Minimum Analytically

In one-variable calculus, the local extrema of a function occur at points where the derivative is zero or undefined. How does this generalize to the case of functions of two or more variables? Suppose that a function $f(x, y)$ has a local maximum at a point (x_0, y_0) which is not on the boundary of the domain of f. If the partial derivative $f_x(x_0, y_0)$ were defined and positive, then we could increase f by increasing x. If $f_x(x_0, y_0) < 0$, then we could increase f by decreasing x. Since f has a local maximum at (x_0, y_0), there can be no direction in which f is increasing, so we must have $f_x(x_0, y_0) = 0$. Similarly, if $f_y(x_0, y_0)$ is defined, then $f_y(x_0, y_0) = 0$. The case in which $f(x, y)$ has a local minimum is similar. Therefore, we arrive at the following conclusion:

If a function $f(x, y)$ has a local maximum or minimum at a point (x_0, y_0) not on the boundary of the domain of f, then either

$$f_x(x_0, y_0) = 0 \quad \text{and} \quad f_y(x_0, y_0) = 0$$

or (at least) one partial derivative is undefined at the point (x_0, y_0). Points where each of the partial derivatives is either zero or undefined are called **critical points**.

As in the single-variable case, the fact that (x_0, y_0) is a critical point for f does not necessarily mean that f has a maximum or a minimum there.

How Do We Find Critical Points?

To find critical points of a function f, we find the points where both partial derivatives of f are zero or undefined.

Example 3 Find and analyze the critical points of $f(x,y) = x^2 - 2x + y^2 - 4y + 5$.

Solution To find the critical points, we set both partial derivatives equal to zero:

$$f_x(x,y) = 2x - 2 = 0,$$
$$f_y(x,y) = 2y - 4 = 0.$$

Solving these equations gives $x = 1$ and $y = 2$. Hence, f has only one critical point, namely $(1,2)$. What is the behavior of f near $(1,2)$? The values of the function in Table 9.9 suggest that the function has a local minimum value of 0 at the point $(1,2)$.

Table 9.9 *Values of $f(x,y)$ near the point $(1,2)$*

y \ x	0.8	0.9	1.0	1.1	1.2
1.8	0.08	0.05	0.04	0.05	0.08
1.9	0.05	0.02	0.01	0.02	0.05
2.0	0.04	0.01	0.00	0.01	0.04
2.1	0.05	0.02	0.01	0.02	0.05
2.2	0.08	0.05	0.04	0.05	0.08

Example 4 A manufacturing company produces two products which are sold in two separate markets. The company's economists analyze the two markets and determine that the quantities, q_1 and q_2, demanded by consumers and the prices, p_1 and p_2 (in dollars), of each item are related by the equations

$$p_1 = 600 - 0.3q_1 \quad \text{and} \quad p_2 = 500 - 0.2q_2.$$

Thus, if the price for either item increases, the demand for it decreases. The company's total production cost is given by

$$C = 16 + 1.2q_1 + 1.5q_2 + 0.2q_1q_2.$$

If the company wants to maximize its total profits, how much of each product should it produce? What is the maximum profit?[8]

Solution The total revenue R is the sum of the revenues, p_1q_1 and p_2q_2, from each market. Substituting for p_1 and p_2, we get

$$\begin{aligned} R &= p_1q_1 + p_2q_2 \\ &= (600 - 0.3q_1)q_1 + (500 - 0.2q_2)q_2 \\ &= 600q_1 - 0.3q_1^2 + 500q_2 - 0.2q_2^2. \end{aligned}$$

Thus the total profit π is given by

$$\begin{aligned} \pi &= R - C \\ &= 600q_1 - 0.3q_1^2 + 500q_2 - 0.2q_2^2 - (16 + 1.2q_1 + 1.5q_2 + 0.2q_1q_2) \\ &= -16 + 598.8q_1 - 0.3q_1^2 + 498.5q_2 - 0.2q_2^2 - 0.2q_1q_2. \end{aligned}$$

To maximize π, we compute partial derivatives:

$$\frac{\partial \pi}{\partial q_1} = 598.8 - 0.6q_1 - 0.2q_2,$$
$$\frac{\partial \pi}{\partial q_2} = 498.5 - 0.4q_2 - 0.2q_1.$$

[8] Adapted from M. Rosser, *Basic Mathematics for Economists*, p. 316 (New York: Routledge, 1993).

Since the partial derivatives are defined everywhere, the only critical points of π are those where the partial derivatives of π are both equal to zero. Thus, we solve the equations for q_1 and q_2,

$$598.8 - 0.6q_1 - 0.2q_2 = 0,$$
$$498.5 - 0.4q_2 - 0.2q_1 = 0,$$

giving

$$q_1 = 699.1 \approx 699 \quad \text{and} \quad q_2 = 896.7 \approx 897.$$

To see whether this is a maximum, we look at a table of values of profit π around this point. Table 9.10 suggests that profit is greatest at $(699, 897)$. So the company should produce 699 units of the first product priced at \$390.30 per unit, and 897 units of the second product priced at \$320.60 per unit. The maximum profit is then $\pi(699, 897) = \$432{,}797$.

Table 9.10 *Does this profit function have a maximum at* $(699, 897)$?

		Quantity, q_1		
		698	699	700
Quantity, q_2	896	432,796.4	432,796.9	432,796.8
	897	432,796.7	432,797.0	432,796.7
	898	432,796.6	432,796.7	432,796.2

Is a Critical Point a Local Maximum or a Local Minimum?

We can often see whether a critical point is a local maximum or minimum or neither by looking at a table or contour diagram. The following analytic method may also be useful in distinguishing between local maxima and minima.[9] It is analogous to the Second Derivative Test in Chapter 4.

Second Derivative Test for Functions of Two Variables

Suppose (x_0, y_0) is a critical point where $f_x(x_0, y_0) = f_y(x_0, y_0) = 0$. Let

$$D = f_{xx}(x_0, y_0)f_{yy}(x_0, y_0) - f_{xy}(x_0, y_0)^2.$$

- If $D > 0$ and $f_{xx}(x_0, y_0) > 0$, then f has a local minimum at (x_0, y_0).
- If $D > 0$ and $f_{xx}(x_0, y_0) < 0$, then f has a local maximum at (x_0, y_0).
- If $D < 0$, then f has neither a local maximum or minimum at (x_0, y_0).
- If $D = 0$, the test is inconclusive.

Example 5 Use the second derivative test to confirm that the critical point $q_1 = 699.1$, $q_2 = 896.7$ gives a local maximum of the profit function π of Example 4.

Solution To see whether or not we have found a maximum point, we compute the second-order partial derivatives:

$$\frac{\partial^2 \pi}{\partial q_1^2} = -0.6, \quad \frac{\partial^2 \pi}{\partial q_2^2} = -0.4, \quad \frac{\partial^2 \pi}{\partial q_1 \partial q_2} = -0.2.$$

Since

$$D = \frac{\partial^2 \pi}{\partial q_1^2}\frac{\partial^2 \pi}{\partial q_2^2} - \left(\frac{\partial^2 \pi}{\partial q_1 \partial q_2}\right)^2 = (-0.6)(-0.4) - (-0.2)^2 = 0.2 > 0,$$

the second derivative test implies that we have found a local maximum point.

[9] An explanation of this test can be found, for example, in *Multivariable Calculus* by W. McCallum et al. (New York: John Wiley, 1997).

Problems for Section 9.5

1. Figure 9.43 shows contours of $f(x, y)$. List x- and y-coordinates and the value of the function at any local maximum and local minimum points, and identify which is which. Are any of these local extrema also global extrema on the region shown? If so, which ones?

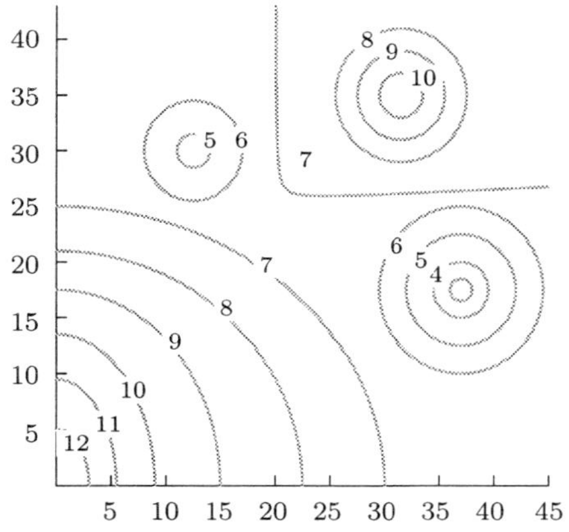

Figure 9.43

2. Figure 9.44 shows contours of $f(x, y)$. List the x- and y-coordinates and the value of the function at any local maximum and local minimum points, and identify which is which. Are any of these local extrema also global extrema on the region shown? If so, which ones?

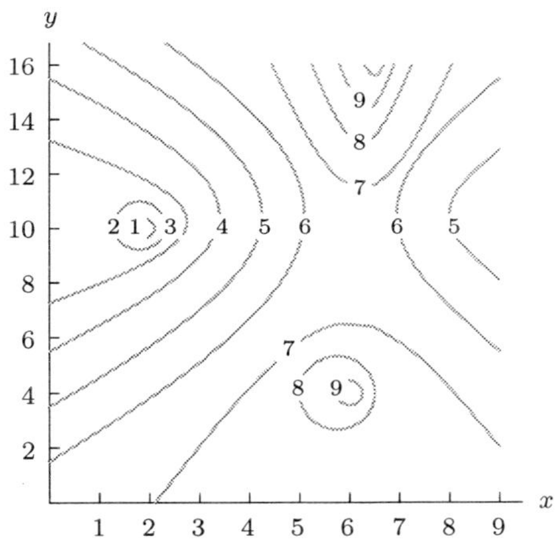

Figure 9.44

In Problems 3–5, estimate the position and approximate value of the global maxima and minima on the region shown.

3.

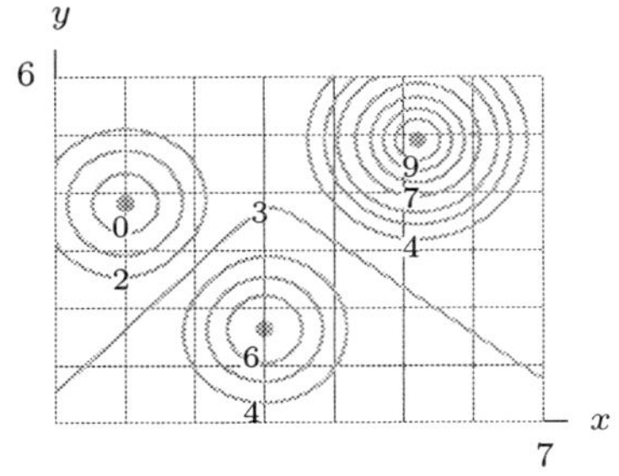

4.

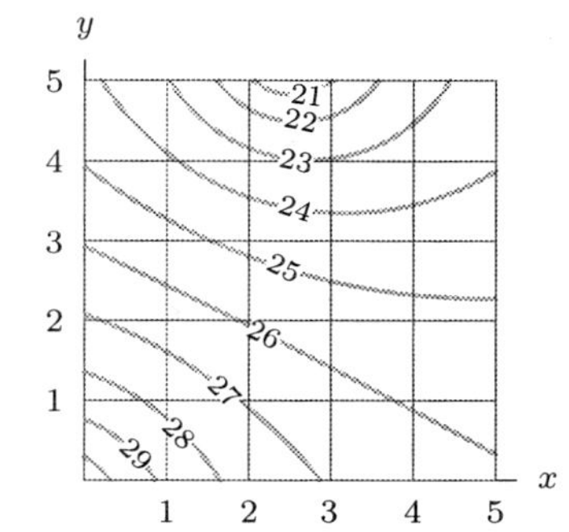

5.

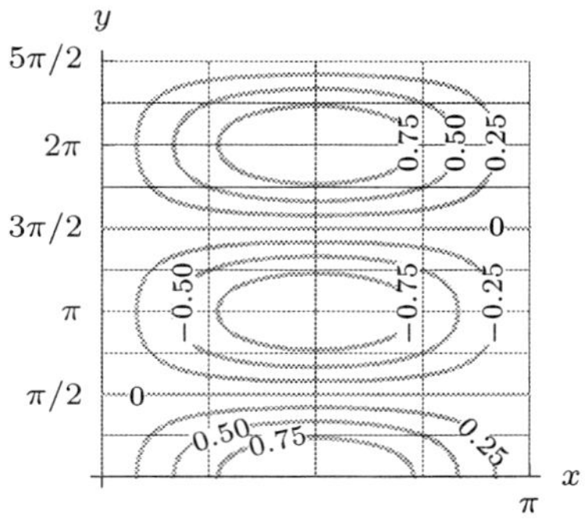

In Problems 6–15, find all the critical points and determine whether each is a local maximum, local minimum, or neither.

6. $f(x, y) = x^2 + 4x + y^2$

7. $f(x, y) = x^2 + xy + 3y$

8. $f(x, y) = x^2 + y^2 + 6x - 10y + 8$

9. $f(x, y) = y^3 - 3xy + 6x$

10. $f(x, y) = x^2 - 2xy + 3y^2 - 8y$

11. $f(x, y) = x^3 - 3x + y^3 - 3y$

12. $f(x, y) = x^3 + y^2 - 3x^2 + 10y + 6$

13. $f(x, y) = x^3 + y^3 - 6y^2 - 3x + 9$

14. $f(x, y) = x^3 + y^3 - 3x^2 - 3y + 10$

15. $f(x, y) = 400 - 3x^2 - 4x + 2xy - 5y^2 + 48y$

16. By looking at the weather map in Figure 9.5 on page 354, find the maximum and minimum daily high temperatures in the states of Mississippi, Alabama, Pennsylvania, New York, California, Arizona, and Massachusetts.

17. For $f(x, y) = A - (x^2 + Bx + y^2 + Cy)$, what values of A, B, and C give f a local maximum value of 15 at the point $(-2, 1)$?

18. A missile has a guidance device which is sensitive to both temperature, t°C, and humidity, h. The range in km over which the missile can be controlled is given by

$$\text{Range} = 27{,}800 - 5t^2 - 6ht - 3h^2 + 400t + 300h.$$

What are the optimal atmospheric conditions for controlling the missile?

19. The quantity of a product demanded by consumers is a function of its price. The quantity of one product demanded may also depend on the price of other products. For example, the demand for tea is affected by the price of coffee; the demand for cars is affected by the price of gas. The quantities demanded, q_1 and q_2, of two products depend on their prices, p_1 and p_2, as follows:

$$q_1 = 150 - 2p_1 - p_2$$

$$q_2 = 200 - p_1 - 3p_2.$$

(a) What does the fact that the coefficients of p_1 and p_2 are negative tell you? Give an example of two products that might be related this way.
(b) If one manufacturer sells both products, how should the prices be set to generate the maximum possible revenue? What is that maximum possible revenue?

20. Two products are manufactured in quantities q_1 and q_2 and sold at prices of p_1 and p_2, respectively. The cost of producing them is given by

$$C = 2q_1^2 + 2q_2^2 + 10.$$

(a) Find the maximum profit that can be made, assuming the prices are fixed.
(b) Find the rate of change of that maximum profit as p_1 increases.

21. A company operates two plants which manufacture the same item and whose total cost functions are

$$C_1 = 8.5 + 0.03q_1^2 \quad \text{and} \quad C_2 = 5.2 + 0.04q_2^2,$$

where q_1 and q_2 are the quantities produced by each plant. The company is a monopoly. The total quantity demanded, $q = q_1 + q_2$, is related to the price, p, by

$$p = 60 - 0.04q.$$

How much should each plant produce in order to maximize the company's profit?[10]

Chapter Ten

MATHEMATICAL MODELING USING DIFFERENTIAL EQUATIONS

10.2 SOLUTIONS OF DIFFERENTIAL EQUATIONS

What does it mean to "solve" a differential equation? A differential equation is an equation involving the derivative of an unknown function. The unknown is not a number but a function. A *solution* to a differential equation is any function that satisfies the differential equation.

In this section, we see how to solve a differential equation numerically and how to check whether or not a function is a solution to a differential equation. In the next section, we see how to visualize a solution.

Another Look at Marine Harvesting

Let's take another look at the fish population discussed in Section 10.1. Left alone, the population increases at a continuous rate of 20% per year. The fish are being harvested at a constant rate of 10 million fish per year. If P is the fish population, in millions, in year t, then we have

$$\frac{dP}{dt} = 0.20P - 10.$$

Solving this differential equation means to find a function giving P in terms of t. Combined with information about the initial population, we can use the equation to predict the population at any time in the future.

Solving the Differential Equation Numerically

Suppose at time $t = 0$, the fish population is 60 million. We can substitute $P = 60$ into the differential equation to compute the derivative, dP/dt:

$$\text{At time } t = 0, \qquad \frac{dP}{dt} = 0.20P - 10 = 0.20(60) - 10 = 12 - 10 = 2.$$

Since at $t = 0$, the fish population is changing at a rate of 2 million fish a year, at the end of the first year, the fish population will have increased by about 2 million fish. So:

$$\text{At } t = 1, \qquad \text{we estimate} \quad P = 60 + 2 = 62.$$

We use this new value of P to estimate dP/dt during the second year:

$$\text{At time } t = 1, \qquad \frac{dP}{dt} = 0.20P - 10 = 12.4 - 10 = 2.4.$$

During the second year, the fish population increased by about 2.4 million fish, so:

$$\text{At } t = 2, \qquad \text{we estimate} \quad P = 62 + 2.4 = 64.4.$$

We use this value of P to estimate the rate of change during the third year, and so on. Continuing in this fashion, we compute the approximate values of P in Table 10.1. This table gives approximate numerical values for P at future times.

Table 10.1 *Approximate values of the fish population as a function of time*

t (years)	0	1	2	3	4	5	...
P (millions)	60	62	64.4	67.28	70.74	74.89	...

A Formula for the Solution to the Differential Equation

A function $P = f(t)$ which satisfies the differential equation

$$\frac{dP}{dt} = 0.20P - 10$$

is called a *solution* of the differential equation. Table 10.1 shows approximate numerical values of a solution. It is sometimes (but not always) possible to find a formula for the solution. In this particular case, there is a formula; it is

$$P = 50 + Ce^{0.20t}, \qquad \text{where } C \text{ is any constant.}$$

We check that this is a solution to the differential equation by substituting it into the left and right sides of the differential equation separately. We find

$$\begin{aligned} \text{Left side} &= \frac{dP}{dt} = 0.20Ce^{0.20t} \\ \text{Right side} = 0.20P - 10 &= 0.20(50 + Ce^{0.20t}) - 10 \\ &= 10 + 0.20Ce^{0.20t} - 10 \\ &= 0.20Ce^{0.20t}. \end{aligned}$$

Since we get the same expression on both sides, we say that $P = 50 + Ce^{0.20t}$ is a solution of this differential equation. Any choice of C works, so the solutions form a family of functions with parameter C. Several members of the family of solutions are graphed in Figure 10.5.

Finding the Arbitrary Constant: Initial Conditions

To find a value for the constant C—in other words, to select a single solution from the family of solutions—we need an additional piece of information, usually the initial population. In this case, we know that $P = 60$ when $t = 0$, so substituting into

$$P = 50 + Ce^{0.20t}$$

gives

$$\begin{aligned} 60 &= 50 + Ce^{0.20(0)} \\ 60 &= 50 + C \cdot 1 \\ C &= 10. \end{aligned}$$

The function $P = 50 + 10e^{0.20t}$ satisfies the differential equation *and* the initial condition that $P = 60$ when $t = 0$.

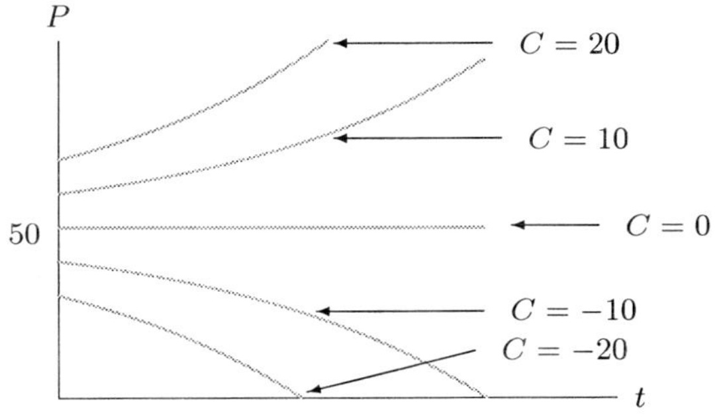

Figure 10.5: Solution curves for $dP/dt = 0.20P - 10$: Members of the family $P = 50 + Ce^{0.20t}$

General Solutions and Particular Solutions

For the differential equation $dP/dt = 0.20P - 10$, it can be shown that every solution is of the form $P = 50 + Ce^{0.20t}$ for some value of C. We say that the *general solution* of the differential equation $dP/dt = 0.20P - 10$ is the family of functions $P = 50 + Ce^{0.20t}$. The solution $P = 50 + 10e^{0.20t}$ that satisfies the differential equation together with the initial condition that $P = 60$ when $t = 0$ is calld a *particular solution.* The differential equation and the initial condition together are called an *initial-value problem.*

Example 1 (a) Check that $P = Ce^{2t}$ is a solution to the differential equation

$$\frac{dP}{dt} = 2P.$$

(b) Find the particular solution satisfying the initial condition $P = 100$ when $t = 0$.

Solution (a) Since $P = Ce^{2t}$ where C is a constant, we find expressions for each side:

$$\text{Left side} = \frac{dP}{dt} = Ce^{2t} \cdot 2 = 2Ce^{2t}$$
$$\text{Right side} = 2P = 2Ce^{2t}.$$

Since the two expressions are equal, $P = Ce^{2t}$ is a solution to the differential equation.

(b) We substitute $P = 100$ and $t = 0$ into the general solution $P = Ce^{2t}$, and solve for C:

$$100 = Ce^{2(0)}$$
$$100 = C \cdot 1$$
$$100 = C.$$

The particular solution for this initial-value problem is $P = 100e^{2t}$.

Example 2 Decide whether or not $y = e^{-2x}$ is a solution of the differential equation $y' - 2y = 0$.

Solution Note that $y' = dy/dx$. Differentiating $y = e^{-2x}$ gives $y' = -2e^{-2x}$. Substituting, we have

$$y' - 2y = -2e^{-2x} - 2e^{-2x} = -4e^{-2x} \neq 0,$$

and so $y = e^{-2x}$ is not a solution to this differential equation.

Example 3 (a) What conditions must be imposed on the constants C and k if $y = Ce^{kt}$ is a solution to the differential equation

$$\frac{dy}{dt} = -0.5y?$$

(b) What additional conditions must be imposed on C and k if $y = Ce^{kt}$ also satisfies the initial condition that $y = 10$ when $t = 0$?

Solution (a) If $y = Ce^{kt}$, then $dy/dt = Cke^{kt}$. Substituting into the equation $dy/dt = -0.5y$ gives

$$Cke^{kt} = -0.5(Ce^{kt}),$$

and therefore, assuming $C \neq 0$, so $Ce^{kt} \neq 0$, we have

$$k = -0.5.$$

So $y = Ce^{-0.5t}$ is a solution to the differential equation. If $C = 0$, then $Ce^{kt} = 0$ is a solution to the differential equation. No conditions are imposed on C.

(b) Since $k = -0.5$, we have $y = Ce^{-0.5t}$. Substituting $y = 10$ when $t = 0$ gives

$$10 = Ce^0$$

so

$$10 = C.$$

So $y = 10e^{-0.5t}$ is a solution to the differential equation together with the initial condition.

Problems for Section 10.2

1. Find the general solution to the differential equation

$$\frac{dy}{dt} = 2t.$$

2. Decide whether or not each of the following is a solution to the differential equation $xy' - 2y = 0$.

(a) $y = x^2$ **(b)** $y = x^3$

3. Check that $y = t^4$ is a solution to the differential equation $t\dfrac{dy}{dt} = 4y$.

In Problems 4–12, use the fact that the derivative gives the slope of a curve to decide which of the graphs (A)–(F) in Figure 10.6 could represent a solution to the differential equation.

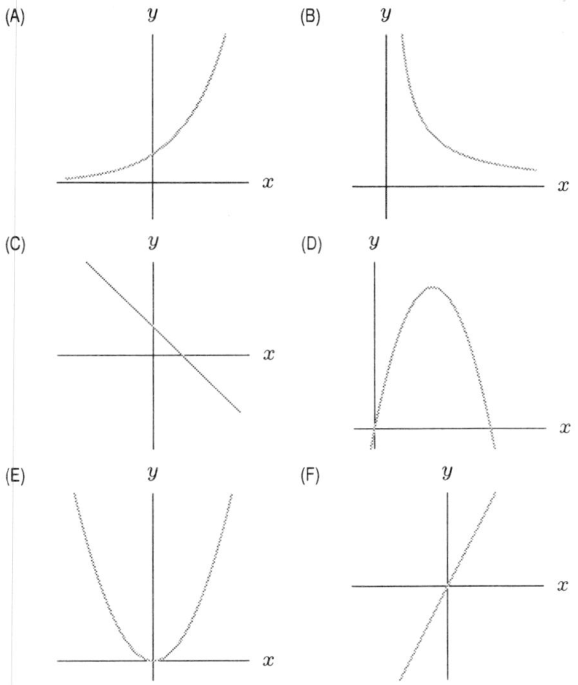

Figure 10.6

4. $\dfrac{dy}{dx} = -1$ **5.** $\dfrac{dy}{dx} = 0.1$ **6.** $\dfrac{dy}{dx} = -y^2$

7. $\dfrac{dy}{dx} = 2x$ **8.** $\dfrac{dy}{dx} = 2$ **9.** $\dfrac{dy}{dx} = y$

10. $\dfrac{dy}{dx} = -\dfrac{1}{x^2}$ **11.** $\dfrac{dy}{dx} = 1 - x$ **12.** $\dfrac{dy}{dx} = 2y$

13. Fill in the missing values in Table 10.2 given that $dy/dt = 0.5t$. Assume the rate of growth, given by dy/dt, is approximately constant over each unit time interval.

Table 10.2

t	0	1	2	3	4
y	8				

14. Fill in the missing values in Table 10.3 given that $dy/dt = 0.5y$. Assume the rate of growth, given by dy/dt, is approximately constant over each unit time interval.

Table 10.3

t	0	1	2	3	4
y	8				

15. Fill in the missing values in Table 10.4 given that $dy/dt = 4 - y$. Assume the rate of growth, given by dy/dt, is approximately constant over each unit time interval.

Table 10.4

t	0	1	2	3	4
y	8				

16. For a certain quantity y, assume that $dy/dt = \sqrt{y}$. Fill in the value of y in Table 10.5. Assume that the rate of growth, dy/dt, is approximately constant over each unit time interval and that the initial value of y is 100.

Table 10.5

t	0	1	2	3	4
y	100				

17. If the initial population of fish is 70 million, use the differential equation $dP/dt = 0.2P - 10$ to estimate the fish population after 1, 2, 3 years.

18. Show that, for any constant P_0, the function $P = P_0e^t$ satisfies the equation
$$\frac{dP}{dt} = P.$$

19. Suppose $Q = Ce^{kt}$ satisfies the differential equation
$$\frac{dQ}{dt} = -0.03Q.$$
What (if anything) does this tell you about the values of C and k?

20. Is there a value of n which makes $y = x^n$ a solution to the equation $13x(dy/dx) = y$? If so, what value?

21. Find the values of k for which $y = x^2 + k$ is a solution to the differential equation $2y - xy' = 10$.

22. Match solutions and differential equations. (Note: Each equation may have more than one solution, or no solution.)

(a) $\frac{dy}{dx} = \frac{y}{x}$	(I) $y = x^3$
(b) $\frac{dy}{dx} = 3\frac{y}{x}$	(II) $y = 3x$
(c) $\frac{dy}{dx} = 3x$	(III) $y = e^{3x}$
(d) $\frac{dy}{dx} = y$	(IV) $y = 3e^x$
(e) $\frac{dy}{dx} = 3y$	(V) $y = x$

10.3 SLOPE FIELDS

In this section, we see how to visualize a differential equation and its solutions. Let's start with the equation
$$\frac{dy}{dx} = y.$$
Any solution to this differential equation has the property that at any point in the plane, the slope of its graph is equal to its y coordinate. (That's what the equation $dy/dx = y$ is telling us!) This means that if the solution goes through the point $(0, 1)$, its slope there is 1; if it goes through a point with $y = 4$ its slope is 4. A solution going through $(0, 2)$ has slope 2 there; at the point where $y = 8$ the slope of this solution is 8. (See Figure 10.7.)

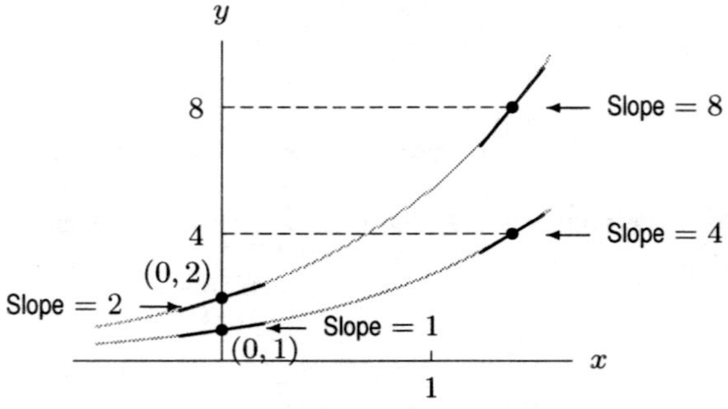

Figure 10.7: Solutions to $\frac{dy}{dx} = y$

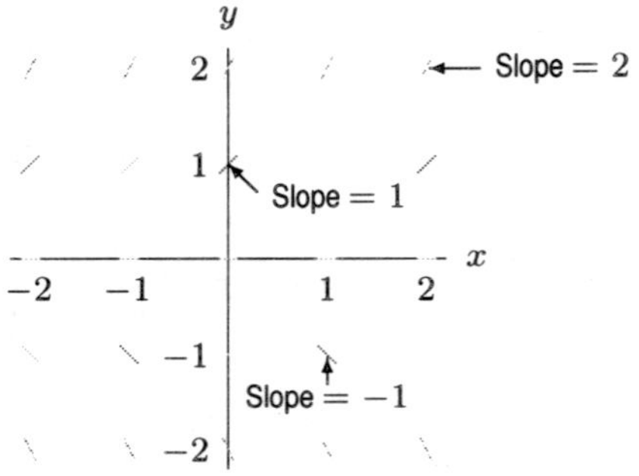

Figure 10.8: Visualizing the slope of y, if $\frac{dy}{dx} = y$

In Figure 10.8 a small line segment is drawn at the marked points showing the slope of the solution curve there. Since $dy/dx = y$, the slope at the point $(1, 2)$ is 2 (the y-coordinate), and so we draw a line segment there with slope 2. We draw a line segment at the point $(0, -1)$ with slope -1, and so on. If we draw many of these line segments, we have the *slope field* for the equation $dy/dx = y$ shown in Figure 10.9. Above the x-axis, the slopes are positive (because y is positive there), and the slopes increase as we move upward (as y increases). Below the x-axis, the slopes are negative, and get more so as we move downward. Notice that on any horizontal line (where y is constant) the slopes are constant. In the slope field you can see the ghost of the solution curve lurking. Start anywhere on the plane and move so that the slope lines are tangent to your path; you will trace out one of the solution curves. Try penciling in some solution curves on Figure 10.9, some above the x-axis and some below. The curves you draw should have the shape of exponential functions. By substituting $y = Ce^x$ into the differential equation, you can check that each curve in the family of exponentials, $y = Ce^x$, is a solution to this differential equation.

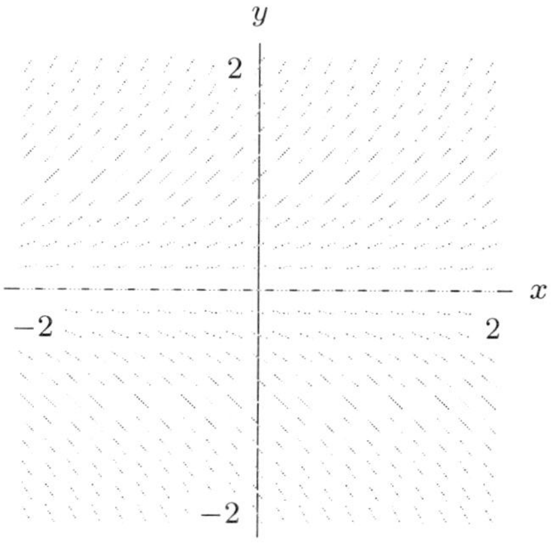

Figure 10.9: Slope field for $\frac{dy}{dx} = y$

In most problems, we are interested in getting the solution curves from the slope field. Think of the slope field as a set of signposts pointing in the direction you should go at each point. Imagine starting anywhere in the plane: look at the slope field at that point and start to move in that direction. After a small step, look at the slope field again, and alter your direction if necessary. Continue to move across the plane in the direction the slope field points, and you'll trace out a solution curve. Notice that the solution curve is not necessarily the graph of a function, and even if it is, we may not have a formula for the function. Geometrically, solving a differential equation means finding the family of solution curves.

Example 1 Figure 10.10 shows the slope field of the differential equation $\dfrac{dy}{dx} = 2x.$

(a) What do you notice about the slope field?

(b) Compare the solution curves in Figure 10.11 with the formula $y = x^2 + C$ for the solutions to this differential equation.

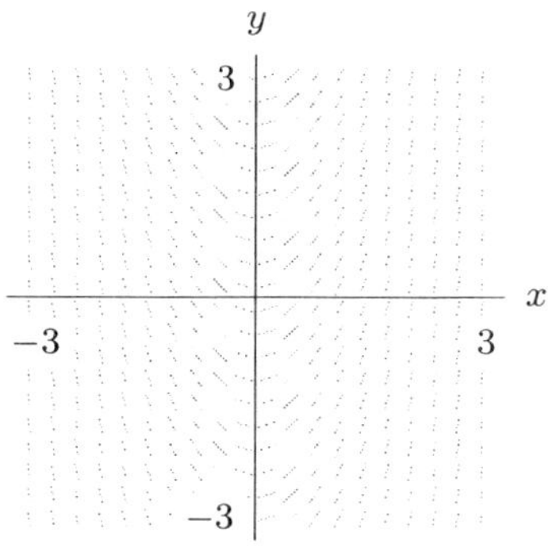

Figure 10.10: Slope field for $\frac{dy}{dx} = 2x$

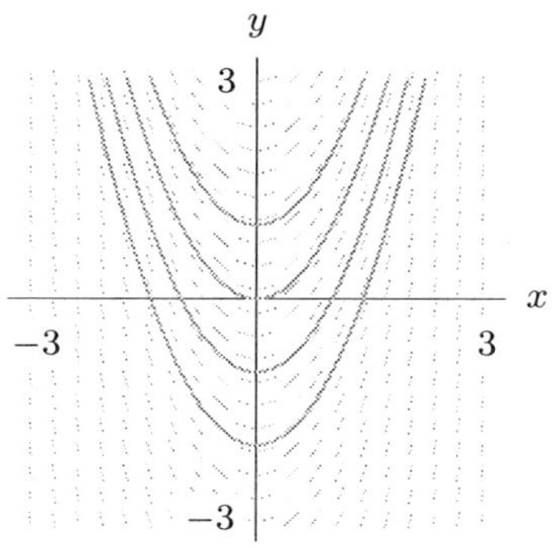

Figure 10.11: Some solutions to $\frac{dy}{dx} = 2x$

Solution

(a) In Figure 10.10, notice that on any vertical line (where x is constant) the slopes are all the same. This is because in this differential equation dy/dx depends on x only. (In the previous example, $dy/dx = y$, the slopes depended on y only.)

(b) The solution curves in Figure 10.11 look like parabolas. It is easy to check by substitution that

$$y = x^2 + C \qquad \text{is a solution to} \qquad \frac{dy}{dx} = 2x,$$

so the parabolas $y = x^2 + C$ are solution curves.

Example 2 Using the slope field, guess the equation of the solution curves of the differential equation

$$\frac{dy}{dx} = -\frac{x}{y}.$$

Solution The slope field is shown in Figure 10.12. Notice that on the y-axis, where x is 0, the slope is 0. On the x-axis, where y is 0, the line segments are vertical and the slope is undefined. At the origin the slope is undefined and there is no line segment.

What do the solution curves of this differential equation look like? The slope field suggests they are circles centered at the origin. We guess that the general solution to this differential equation is

$$x^2 + y^2 = r^2.$$

This solution is derived in the Focus on Theory section at the end of the chapter.

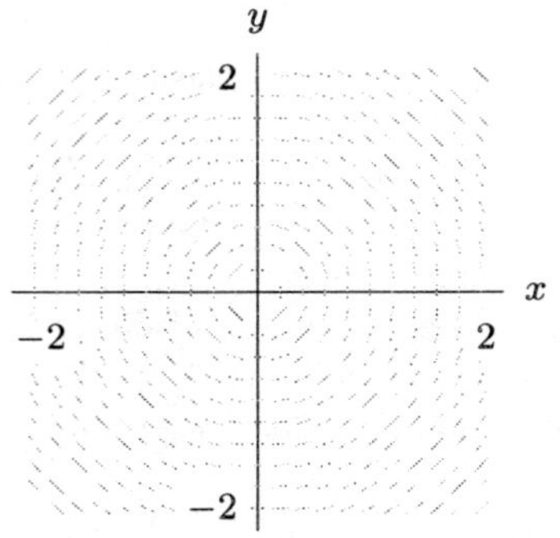

Figure 10.12: Slope field for $\frac{dy}{dx} = -\frac{x}{y}$

The previous example shows that the solutions to differential equations may sometimes be expressed as *implicit functions*. Implicit functions are ones which have not been "solved" for y; in other words, the dependent variable is not expressed as an explicit function of x.

Example 3 The slope fields for $\frac{dy}{dt} = 2 - y$ and $\frac{dy}{dt} = \frac{t}{y}$ are shown in Figure 10.13.

(a) Which slope field corresponds to which differential equation?

(b) Sketch solution curves on each slope field with initial conditions

(i) $y = 1$ when $t = 0$ (ii) $y = 3$ when $t = 0$ (iii) $y = 0$ when $t = 1$

(c) For each solution curve, can you say anything about the long-run behavior of y? In particular, as $t \to \infty$, what happens to the value of y?

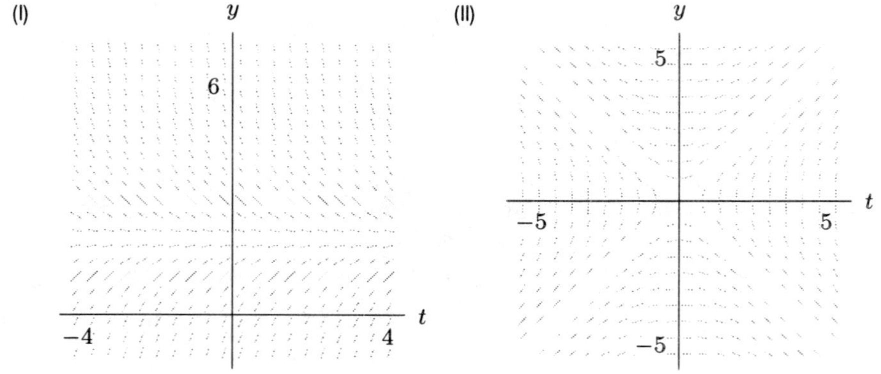

Figure 10.13: Slope fields for $\frac{dy}{dt} = 2 - y$ and $\frac{dy}{dt} = \frac{t}{y}$: Which is which?

Solution

(a) Consider the slopes at different points for the two differential equations. In particular, look at the line $y = 2$ in Figure 10.13. The equation $dy/dt = 2 - y$ has slope 0 all along this line, whereas the line $dy/dt = t/2$ has slope $t/2$. Since slope field (I) looks horizontal at $y = 2$, slope field (I) corresponds to $dy/dt = 2 - y$ and slope field (II) corresponds to $dy/dt = t/y$.

(b) The initial conditions (i) and (ii) give the value of y when t is 0, that is, the y-intercept. To draw the solution curve satisfying the condition (i), draw the solution curve with y-intercept 1. For (ii), draw the solution curve with y-intercept 3. For (iii), the solution goes through the point $(1, 0)$, so draw the solution curve passing through this point. See Figures 10.14 and 10.15.

(c) For $dy/dt = 2 - y$, all solution curves have $y = 2$ as a horizontal asymptote, so $y \to 2$ as $t \to \infty$. For $dy/dt = t/y$ with initial conditions $(0, 1)$ and $(0, 3)$, we see that $y \to \infty$ as $t \to \infty$. The graph has asymptotes which appear to be diagonal lines. In fact, they are $y = t$ and $y = -t$, so $y \to \pm\infty$ as $t \to \infty$.

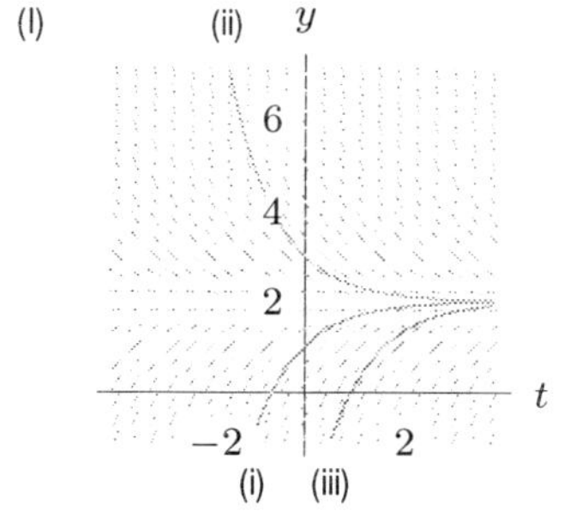

Figure 10.14: Solution curves for $\dfrac{dy}{dt} = 2 - y$

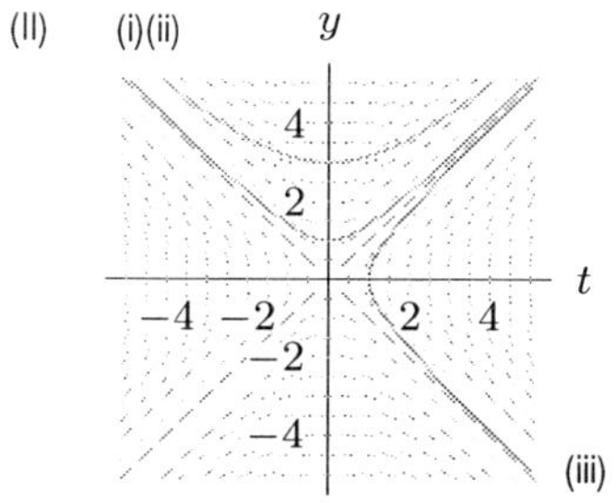

Figure 10.15: Solution curves for $\dfrac{dy}{dt} = \dfrac{t}{y}$

Existence and Uniqueness of Solutions

Since differential equations are used to model many real situations, the question of whether a solution exists and is unique can have great practical importance. If we know how the velocity of a satellite is changing, can we know its velocity for all future time? If we know the initial population of a city, and we know how the population is changing, can we predict the population in the future? Common sense says yes: if we know the initial value of some quantity and we know exactly how it is changing, we should be able to figure out the future value of the quantity.

In the language of differential equations, an initial-value problem (that is, a differential equation and an initial condition) representing a real situation almost always has a unique solution. One way to see this is by looking at the slope field. Imagine starting at the point representing the initial condition. Through that point there will usually be a line segment pointing in the direction the solution curve must go. By following the line segments in the slope field, we trace out the solution curve. Several examples with different starting points are shown in Figure 10.16. In general, at each point there is one line segment and therefore only one direction for the solution curve to go. Thus the solution curve *exists* and is *unique* provided we are given an initial point.

It can be shown that if the slope field is continuous as we move from point to point in the plane, we can be sure that the solution curve exists around every point. Ensuring that each point has only one solution curve through it requires a slightly stronger condition.

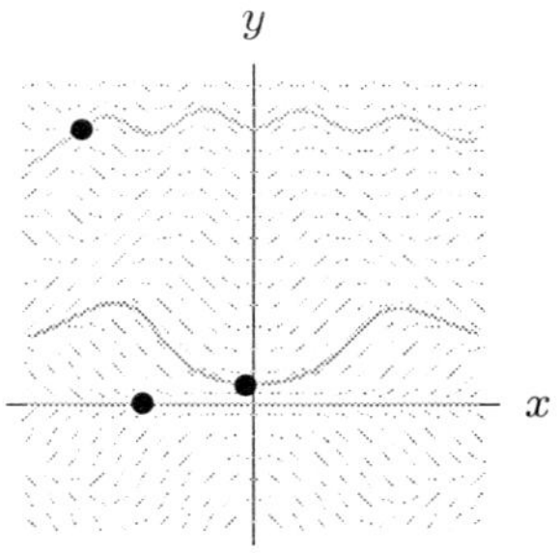

Figure 10.16: There is one and only one solution curve through each point in the plane for this slope field

Problems for Section 10.3

1. Figure 10.17 is a slope field for $dy/dx = y - 10$.
   (a) Draw the solution curve for each of the following initial conditions:
   (i) $y = 8$ when $x = 0$ (ii) $y = 12$ when $x = 0$
   (iii) $y = 10$ when $x = 0$
   (b) Since $dy/dx = y - 10$, when $y = 10$, we have $dy/dx = 10 - 10 = 0$. Explain why this matches your answer to part (iii).

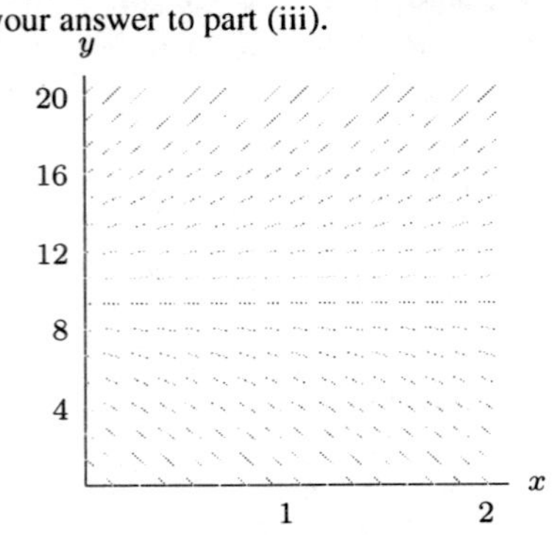

Figure 10.17: Slope field for $dy/dx = y - 10$

2. Sketch three solution curves for each of the slope fields in Figure 10.18.

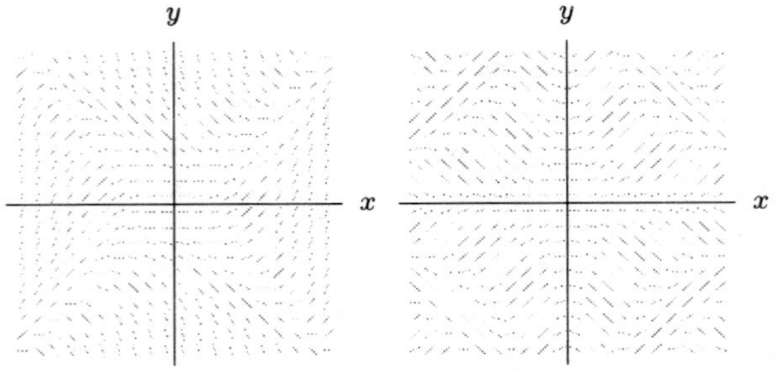

Figure 10.18

3. Figure 10.19 is the slope field for the equation $y' = x+y$.
   (a) Sketch the solutions that pass through the points
   (i) $(0,0)$ (ii) $(-3,1)$ (iii) $(-1,0)$
   (b) From your sketch, guess the equation of the solution passing through $(-1,0)$.
   (c) Check your solution to part (b) by substituting it into the differential equation.

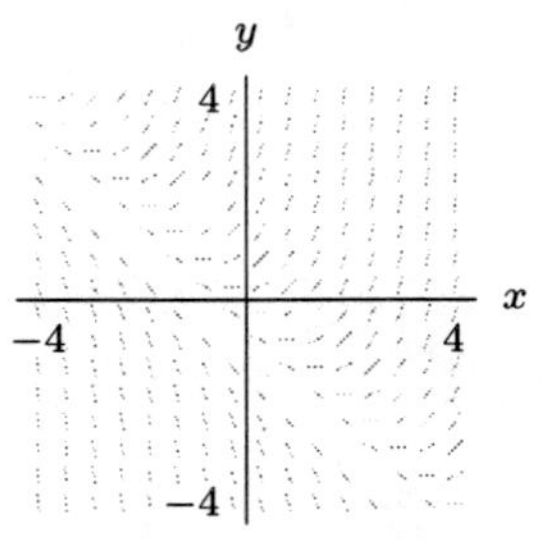

Figure 10.19: Slope field for $y' = x + y$

4. Which one of the following differential equations best fits the slope field shown in Figure 10.20? Explain.

   I. $dP/dt = P - 1$ II. $dP/dt = P(P - 1)$
   III. $dP/dt = 3P(1 - P)$ IV. $dP/dt = 1/3P(1-P)$

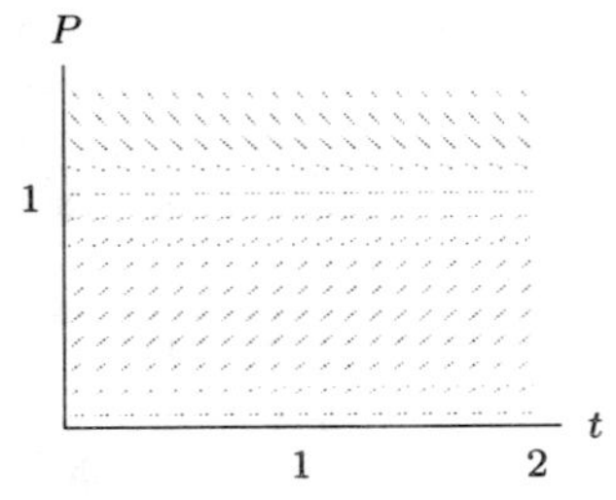

Figure 10.20

5. (a) Consider the slope field for $dy/dx = xy$. What is the slope of the line segment at the point $(2, 1)$? At $(0, 2)$? At $(-1, 1)$? At $(2, -2)$?
   (b) Sketch part of the slope field by drawing line segments with the slopes calculated in part (a).

6. Match the slope fields in Figure 10.21 with their differential equations:

   (a) $y' = -y$ (b) $y' = y$ (c) $y' = x$
   (d) $y' = 1/y$ (e) $y' = y^2$

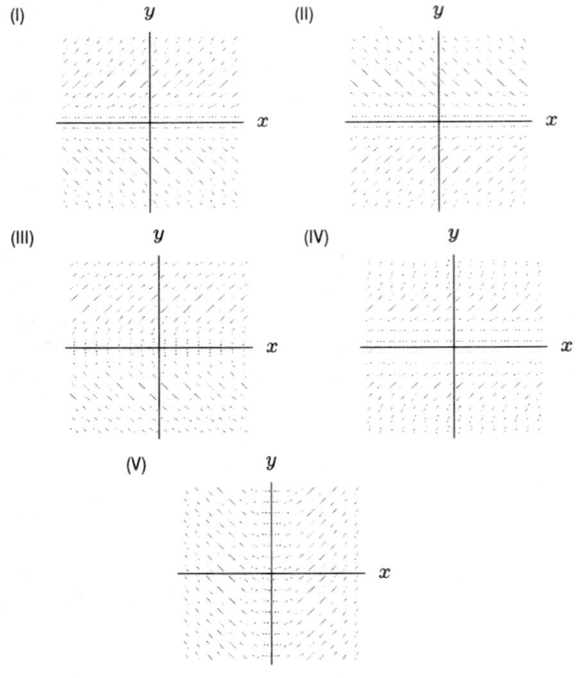

Figure 10.21

7. Match the slope fields in Figure 10.22 with their differential equations:

(a) $y' = 1 + y^2$ (b) $y' = x$ (c) $y' = \sin x$
(d) $y' = y$ (e) $y' = x - y$ (f) $y' = 4 - y$

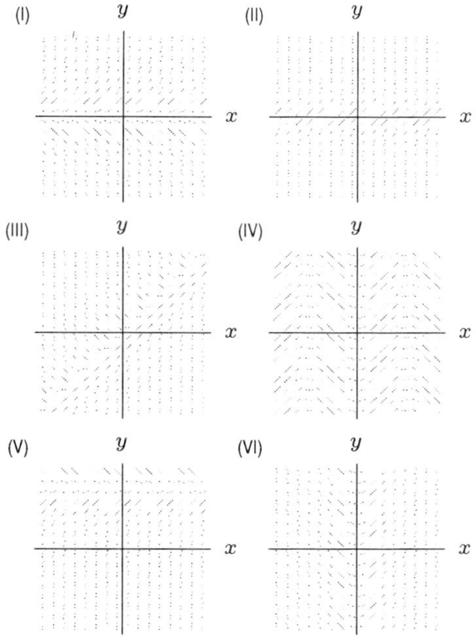

Figure 10.22: Each slope field is graphed for $-5 \le x \le 5$, $-5 \le y \le 5$

For Problems 8–13, consider a solution curve for each of the slope fields in Problem 7. Write one or two sentences describing qualitatively the long-run behavior of y. For example, as x increases, does $y \to \infty$, or does y remain finite? You may get different limiting behavior for different starting points. In each case, your answer should discuss how the limiting behavior depends on the starting point.

8. Slope field (I)

9. Slope field (II)

10. Slope field (III)

11. Slope field (IV)

12. Slope field (V)

13. Slope field (VI)

14. The Gompertz equation, which models growth of animal tumors, is $y' = -ay\ln(y/b)$, where a and b are positive constants. Use Figures 10.23 and 10.24 to write a paragraph describing the similarities and/or differences between solutions to the Gompertz equation with $a = 1$ and $b = 2$ and solutions to the equation $y' = y(2 - y)$.

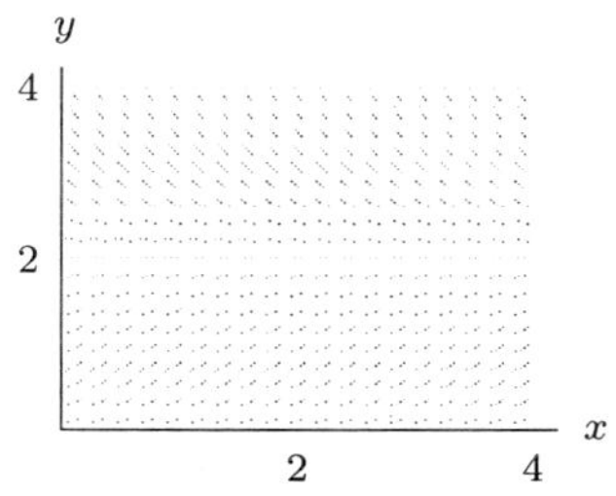

Figure 10.23: Slope field for $y' = -y\ln(y/2)$

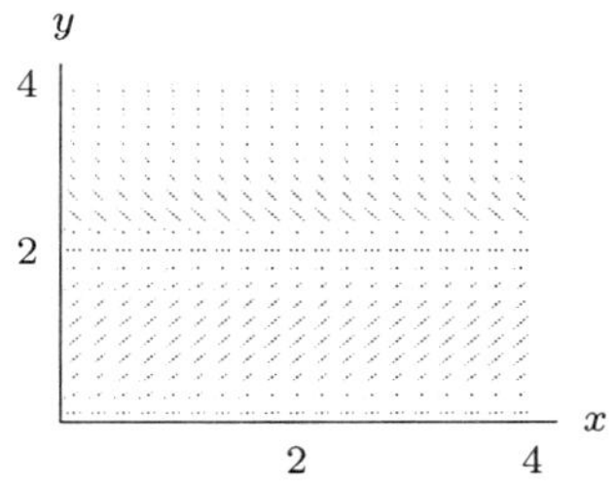

Figure 10.24: Slope field for $y' = y(2 - y)$

10.4 EXPONENTIAL GROWTH AND DECAY

What is a solution to the differential equation

$$\frac{dy}{dt} = y?$$

A solution is a function that is its own derivative. The function $y = e^t$ has this property, so $y = e^t$ is a solution. In fact, any multiple of e^t also has this property. The family of functions $y = Ce^t$ is the general solution to this differential equation. If k is a constant, the differential equation

$$\frac{dy}{dt} = ky,$$

is similar. This differential equation says that the rate of change of y is proportional to y. The constant k is the constant of proportionality. By substituting $y = Ce^{kt}$ into the differential equation, you can check that $y = Ce^{kt}$ is a solution. For another derivation of the solution, see the Focus on Theory section at the end of the chapter. We have the following result:

> The general solution to the differential equation $\frac{dy}{dt} = ky$ is
>
> $$y = Ce^{kt} \qquad \text{for any constant C.}$$
>
> - This is exponential growth for $k > 0$, and exponential decay for $k < 0$.
> - The constant C is the value of y when t is 0.

Graphs of solution curves for some $k > 0$ are in Figure 10.25. For $k < 0$, the graphs are reflected about the y-axis. See Figure 10.26.

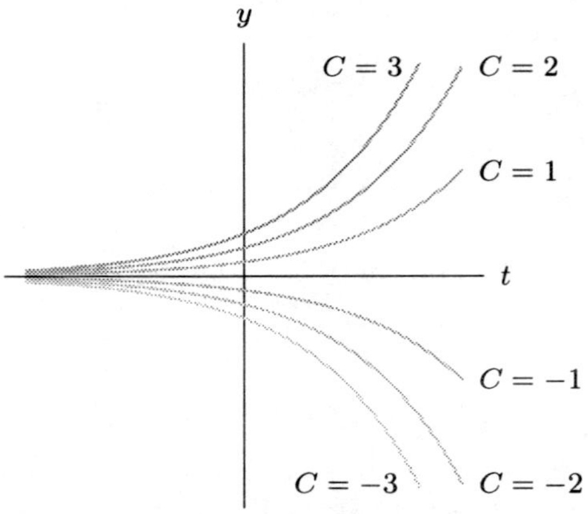

Figure 10.25: Graphs of $y = Ce^{kt}$, which are solutions to $\frac{dy}{dt} = ky$ for some fixed $k > 0$

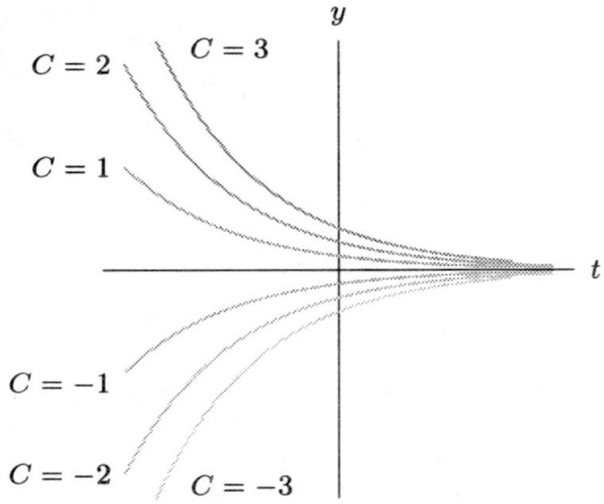

Figure 10.26: Graphs of $y = Ce^{kt}$, which are solutions to $\frac{dy}{dt} = ky$ for some fixed $k < 0$

Example 1 (a) Find the general solution to each of the following differential equations:

(i) $\frac{dy}{dt} = 0.05y$ (ii) $\frac{dP}{dt} = -0.3P$ (iii) $\frac{dw}{dz} = 2z$ (iv) $\frac{dw}{dz} = 2w$

(b) For differential equation (i), find the particular solution satisfying $y = 50$ when $t = 0$.

Solution (a) The differential equations given in (i), (ii), and (iv) are all examples of exponential growth or decay, since each is in the form

$$\text{Derivative} = \text{Constant} \cdot \text{Dependent variable.}$$

Notice that differential equation (iii) is not in this form. Example 1 on page 413 showed that the solution to (iii) is $w = z^2 + C$. The general solutions are

(i) $y = Ce^{0.05t}$ (ii) $P = Ce^{-0.3t}$ (iii) $w = z^2 + C$ (iv) $w = Ce^{2z}$

(b) The general solution to (i) is $y = Ce^{0.05t}$. Substituting $y = 50$ and $t = 0$ gives

$$50 = Ce^{0.05(0)}$$
$$50 = C \cdot 1.$$

So $C = 50$, and the particular solution to this initial-value problem is $y = 50e^{0.05t}$.

Population Growth

Consider the population P of a region where there is no immigration or emigration. The rate at which the population is growing is often proportional to the size of the population. This means larger populations grow faster, as we expect since there are more people to have babies. If the population has a continuous growth rate of 2% per unit time, then we know

$$\text{Rate of growth of population} = 2\% \text{ of Current population},$$

so

$$\frac{dP}{dt} = 0.02P.$$

This equation is of the form $dP/dt = kP$ for $k = 0.02$ and has the general solution $P = Ce^{0.02t}$. If the initial population at time $t = 0$ is P_0, then $P_0 = Ce^{0.02(0)} = C$. So $C = P_0$ and we have

$$P = P_0 e^{0.02t}.$$

Continuously Compounded Interest

In Chapter 1 we introduced continuous compounding as the limiting case in which interest was added more and more often. Here we approach continuous compounding from a different point of view. We imagine interest being accrued at a rate proportional to the balance at that moment. Thus, the larger the balance, the faster interest is earned and the faster the balance grows.

Example 2 A bank account earns interest continuously at a rate of 5% of the current balance per year. Assume that the initial deposit is \$1000 and that no other deposits or withdrawals are made.

(a) Write a differential equation satisfied by the balance in the account.
(b) Solve the differential equation and graph the solution.

Solution

(a) We are looking for B, the balance in the account in dollars, as a function of t, time in years. Interest is being added continuously to the account at a rate of 5% of the balance at that moment, so

$$\text{Rate at which balance is increasing} = 5\% \text{ of Current balance}.$$

Thus, a differential equation that describes the process is

$$\frac{dB}{dt} = 0.05B.$$

Notice that it does not involve the \$1000, the initial condition, because the initial deposit does not affect the process by which interest is earned.

(b) Since $B_0 = 1000$ is the initial value of B, the solution to this differential equation is

$$B = B_0 e^{0.05t} = 1000e^{0.05t}.$$

This function is graphed in Figure 10.27.

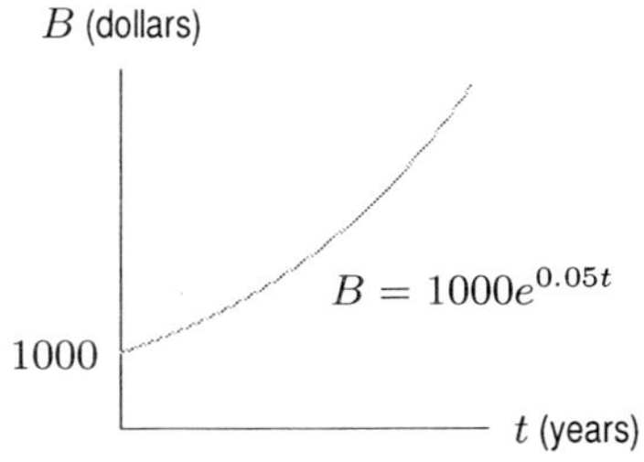

Figure 10.27: Bank balance against time

You may wonder how we can represent an amount of money by a differential equation, since money can only take on discrete values (you can't have fractions of a cent). In fact, the differential equation is only an approximation, but for large amounts of money, it is a pretty good approximation.

Pollution in the Great Lakes

In the 1960s pollution in the Great Lakes became an issue of public concern. We will set up a model for how long it would take the lakes to flush themselves clean, assuming no further pollutants were being dumped in the lake.

Let Q be the total quantity of pollutant in a lake of volume V at time t. Suppose that clean water is flowing into the lake at a constant rate r and that water flows out at the same rate. Assume that the pollutant is evenly spread throughout the lake, and that the clean water coming into the lake immediately mixes with the rest of the water.

How does Q vary with time? First, notice that since pollutants are being taken out of the lake but not added, Q decreases, and the water leaving the lake becomes less polluted, so the rate at which the pollutants leave decreases. This tells us that Q is decreasing and concave up. In addition, the pollutants will never be completely removed from the lake though the quantity remaining will become arbitrarily small. In other words, Q is asymptotic to the t-axis. (See Figure 10.28.)

Setting Up a Differential Equation for the Pollution

To understand how Q changes with time, we write a differential equation for Q. We know that

$$\begin{array}{c}\text{Rate } Q\\ \text{changes}\end{array} = -\left(\begin{array}{c}\text{Rate pollutants}\\ \text{leave in outflow}\end{array}\right)$$

where the negative sign represents the fact that Q is decreasing. At time t, the concentration of pollutants is Q/V and water containing this concentration is leaving at rate r. Thus,

$$\begin{array}{c}\text{Rate pollutants}\\ \text{leave in outflow}\end{array} = \begin{array}{c}\text{Rate of}\\ \text{outflow}\end{array} \times \text{Concentration} = r \cdot \frac{Q}{V}.$$

So the differential equation is

$$\frac{dQ}{dt} = -\frac{r}{V}Q$$

and its general solution is

$$Q = Q_0 e^{-rt/V}.$$

Table 10.6 contains values of r and V for four of the Great Lakes.[1] We use this data to calculate how long it would take for certain fractions of the pollution to be removed from Lake Erie.

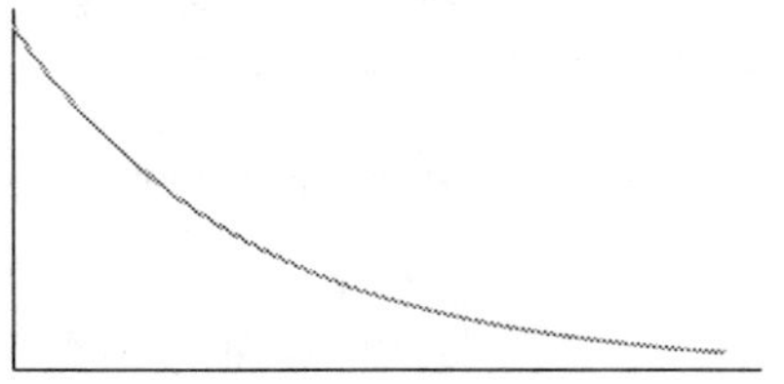

Figure 10.28: Pollutant in lake versus time

Table 10.6 *Volume and outflow in Great Lakes*

	V (km^3)	r (km^3/year)
Superior	12,200	65.2
Michigan	4900	158
Erie	460	175
Ontario	1600	209

Example 3 How long will it take for 90% of the pollution to be removed from Lake Erie? For 99% to be removed?

Solution For Lake Erie, $r/V = 175/460 = 0.38$, so at time t we have

$$Q = Q_0 e^{-0.38t}.$$

When 90% of the pollution has been removed, 10% remains, so $Q = 0.1Q_0$. Substituting gives

$$0.1Q_0 = Q_0 e^{-0.38t}.$$

[1]Data from William E. Boyce and Richard C. DiPrima, *Elementary Differential Equations* (New York: Wiley, 1977).

Canceling Q_0 and solving for t gives

$$t = \frac{-\ln(0.1)}{0.38} \approx 6 \text{ years.}$$

Similarly, when 99% of the pollution has been removed, $Q = 0.01Q_0$, so we solve

$$0.01Q_0 = Q_0e^{-0.38t},$$

giving

$$t = \frac{-\ln(0.01)}{0.38} \approx 12 \text{ years.}$$

The Quantity of a Drug in the Body

As we saw in Section 10.1, the rate at which a drug leaves a patient's body is proportional to the quantity of the drug left in the body. If we let Q represent the quantity of drug left, then

$$\frac{dQ}{dt} = -kQ.$$

The negative sign indicates the quantity of drug in the body is decreasing. The solution to this differential equation is $Q = Q_0e^{-kt}$; the quantity decreases exponentially. The constant k depends on the drug and Q_0 is the amount of drug in the body at time zero. Sometimes physicians convey information about the relative decay rate with a *half life*, which is the time it takes for Q to decrease by a factor of $1/2$.

Example 4 Valproic acid is a drug used to control epilepsy; its half-life in the human body is about 15 hours.

(a) Use the half-life to find the constant k in the differential equation $dQ/dt = -kQ$, where Q represents the quantity of drug in the body t hours after the drug is administered.

(b) At what time will 10% of the original dose remain?

Solution

(a) Since the half-life is 15 hours, we know that the quantity remaining $Q = 0.5Q_0$ when $t = 15$. We substitute into the solution to the differential equation, $Q = Q_0e^{-kt}$, and solve for k:

$$\begin{aligned} Q &= Q_0e^{-kt} \\ 0.5Q_0 &= Q_0e^{-k(15)} \\ 0.5 &= e^{-15k} && \text{(Divide through by } Q_0\text{)} \\ \ln 0.5 &= -15k && \text{(Take the natural logarithm of both sides)} \\ k &= \frac{-\ln 0.5}{15} = 0.0462. && \text{(Solve for } k\text{)} \end{aligned}$$

(b) To find the time when 10% of the original dose remains in the body, we substitute $0.10Q_0$ for the quantity remaining, Q, and solve for the time, t.

$$\begin{aligned} 0.10Q_0 &= Q_0e^{-0.0462t} \\ 0.10 &= e^{-0.0462t} \\ \ln 0.10 &= -0.0462t \\ t &= \frac{\ln 0.10}{-0.0462} = 49.84. \end{aligned}$$

There will be 10% of the drug still in the body at $t = 49.84$, or after about 50 hours.

Problems for Section 10.4

Find solutions to the differential equations in Problems 1–6, subject to the given initial condition.

1. $\dfrac{dw}{dr} = 3w, \quad w = 30 \text{ when } r = 0$
2. $\dfrac{dy}{dx} = -0.14y, \quad y = 5.6 \text{ when } x = 0$
3. $\dfrac{dP}{dt} = 0.02P, \quad P(0) = 20$
4. $\dfrac{dp}{dq} = -0.1p, \quad p = 100 \text{ when } q = 5$
5. $\dfrac{dQ}{dt} = \dfrac{Q}{5}, \quad Q = 50 \text{ when } t = 0$
6. $\dfrac{dy}{dx} + \dfrac{y}{3} = 0, \quad y(0) = 10$
7. Money in a bank account grows continuously at an annual rate of r (when the interest rate is 5%, $r = 0.05$, and so on). Suppose \$1000 is put into the account in 2000.
   - **(a)** Write a differential equation satisfied by M, the amount of money in the account at time t, measured in years since 2000.
   - **(b)** Solve the differential equation.
   - **(c)** Sketch the solution until the year 2030 for interest rates of 5% and 10%.
8. A deposit of \$5000 is made to a bank account paying 1.5% annual interest, compounded continuously.
   - **(a)** Write a differential equation for the balance in the account, B, as a function of time, t, in years.
   - **(b)** Solve the differential equation.
   - **(c)** How much money is in the account in 10 years?
9. A bank account that earns 10% interest compounded continuously has an initial balance of zero. Money is deposited into the account at a continuous rate of \$1000 per year.
   - **(a)** Write a differential equation that describes the rate of change of the balance $B = f(t)$.
   - **(b)** Solve the differential equation.
10. The amount of ozone, Q, in the atmosphere is decreasing at a rate proportional to the amount of ozone present. If time t is measured in years, the constant of proportionality is -0.0025. Write a differential equation for Q as a function of t, and give the general solution for the differential equation. If this rate continues, approximately what percent of the ozone in the atmosphere now will decay in the next 20 years?
11. Using the model in the text and the data in Table 10.6 on page 420, find how long it would take for 90% of the pollution to be removed from Lake Michigan and from Lake Ontario, assuming no new pollutants are added. Explain how you can tell which lake will take longer to be purified just by looking at the data in the table.
12. Use the model in the text and the data in Table 10.6 on page 420 to determine which of the Great Lakes would require the longest time and which would require the shortest time for 80% of the pollution to be removed, assuming no new pollutants are being added. Find the ratio of these two times.
13. The rate at which a drug leaves the bloodstream and passes into the urine is proportional to the quantity of the drug in the blood at that time. If an initial dose of Q_0 is injected directly into the blood, 20% is left in the blood after 3 hours.
   - **(a)** Write and solve a differential equation for the quantity, Q, of the drug in the blood after t hours.
   - **(b)** How much of this drug is in a patient's body after 6 hours if the patient is given 100 mg initially?
14. Oil is pumped continuously from a well at a rate proportional to the amount of oil left in the well. Initially there were 1 million barrels of oil in the well; six years later 500,000 barrels remain.
   - **(a)** At what rate was the amount of oil in the well decreasing when there were 600,000 barrels remaining?
   - **(b)** When will there be 50,000 barrels remaining?
15. In some chemical reactions, the rate at which the amount of a substance changes with time is proportional to the amount present. For example, this is the case as δ-glucono-lactone changes into gluconic acid.
   - **(a)** Write a differential equation satisfied by y, the quantity of δ-glucono-lactone present at time t.
   - **(b)** If 100 grams of δ-glucono-lactone is reduced to 54.9 grams in one hour, how many grams will remain after 10 hours?
16. Hydrocodone bitartrate is used as a cough suppressant. After the drug is fully absorbed, the quantity of drug in the body decreases at a rate proportional to the amount left in the body. The half-life of hydrocodone bitartrate in the body is 3.8 hours and the dose is 10 mg.
   - **(a)** Write a differential equation for the quantity, Q, of hydrocodone bitartrate in the body at time t, in hours since the drug was fully absorbed.
   - **(b)** Solve the differential equation given in part (a).
   - **(c)** Use the half-life to find the constant of proportionality, k.
   - **(d)** How much of the 10-mg dose is still in the body after 12 hours?
17. The amount of land in use for growing crops increases as the world's population increases. Suppose $A(t)$ represents the total number of hectares of land in use in year t. (A hectare is about $2\frac{1}{2}$ acres.)
   - **(a)** Explain why it is plausible that $A(t)$ satisfies the equation $A'(t) = kA(t)$. What assumptions are you

making about the world's population and its relation to the amount of land used?

(b) In 1950 about $1 \cdot 10^9$ hectares of land were in use; in 1980 the figure was $2 \cdot 10^9$. If the total amount of land available for growing crops is thought to be $3.2 \cdot 10^9$ hectares, when does this model predict it is exhausted? (Let $t = 0$ in 1950.)

10.5 APPLICATIONS AND MODELING

In the last section, we considered several situations modeled by the differential equation

$$\frac{dy}{dt} = ky.$$

In this section, we consider situations where the rate of change of y is a linear function of y of the form

$$\frac{dy}{dt} = k(y - A), \qquad \text{where } k \text{ and } A \text{ are constants.}$$

The Quantity of a Drug in the Body

A patient is given the drug warfarin, an anticoagulant, intravenously at the rate of 0.5 mg/hour. Warfarin is metabolized and leaves the body at the rate of about 2% per hour. A differential equation for the quantity, Q (in mg), of warfarin in the body after t hours is given by

$$\begin{aligned} \text{Rate of change} &= \text{Rate in} - \text{Rate out} \\ \frac{dQ}{dt} &= 0.5 - 0.02Q. \end{aligned}$$

What does this tell us about the quantity of warfarin in the body for different initial values of Q?

If Q is small, then $0.02Q$ is also small and the rate the drug is excreted is less than the rate at which the drug is entering the body. Since the rate in is greater than the rate out, the rate of change is positive and the quantity of drug in the body is increasing. If Q is large enough that $0.02Q$ is greater than 0.5, then $0.5 - 0.02Q$ is negative, so dQ/dt is negative and the quantity is decreasing.

For small Q, the quantity will increase until the rate in equals the rate out. For large Q, the quantity will decrease until the rate in equals the rate out. What is the value of Q at which the rate in exactly matches the rate out? We have

$$\begin{aligned} \text{Rate in} &= \text{Rate out} \\ 0.5 &= 0.02Q \\ Q &= 25. \end{aligned}$$

If the amount of warfarin in the body is initially 25 mg, then the amount being excreted exactly matches the amount being added. The quantity of drug Q will stay constant at 25 mg. Notice also that when $Q = 25$, the derivative dQ/dt is zero, since

$$\frac{dQ}{dt} = 0.5 - 0.02(25) = 0.5 - 0.5 = 0.$$

If the initial quantity is 25, then the solution is the horizontal line $Q = 25$. This solution is called an *equilibrium solution*.

The slope field for this differential equation is shown in Figure 10.29, with solution curves drawn for $Q_0 = 20$, $Q_0 = 25$, and $Q_0 = 30$. In each case, we see that the quantity of drug in the body is approaching the equilibrium solution of 25 mg. The solution curve with $Q_0 = 30$ should remind you of an exponential decay function. It is, in fact, an exponential decay function that has been shifted up 25 units.

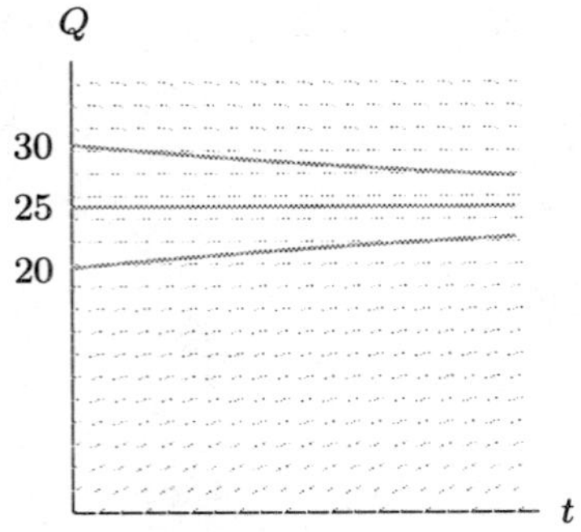

Figure 10.29: Slope field for $dQ/dt = 0.5 - 0.02Q$

Solving the Differential Equation $dy/dt = k(y - A)$

The drug concentration in the previous example satisfies a differential equation of the form

$$\frac{dy}{dt} = k(y - A).$$

Let us find the general solution to this equation. Since A is a constant, $dA/dt = 0$ so that we have

$$\frac{d}{dt}(y - A) = \frac{dy}{dt} - \frac{dA}{dt} = \frac{dy}{dt} - 0 = k(y - A).$$

Thus $y - A$ satisfies an exponential differential equation, so $y - A$ must be of the form

$$y - A = Ce^{kt}.$$

For an alternative derivation of the solution, see the Focus on Theory section at the end of the chapte.

> The general solution to the differential equation
>
> $$\frac{dy}{dt} = k(y - A)$$
>
> is
>
> $$y = A + Ce^{kt}, \qquad \text{for any constant } C.$$

Warning: Notice that, for differential equations of this form, the arbitrary constant C is *not* the initial value of the variable, but rather the initial value of $y - A$.

Example 1 Give the solution to each of the following differential equations:

(a) $\dfrac{dy}{dt} = 0.02(y - 50)$

(b) $\dfrac{dP}{dt} = 5(P - 10), \quad P = 8$ when $t = 0$

(c) $\dfrac{dy}{dt} = 3y - 300$

(d) $\dfrac{dW}{dt} = 500 - 0.1W$

Solution (a) The general solution is $y = 50 + Ce^{0.02t}$.

(b) The general solution is $P = 10 + Ce^{5t}$. Use the initial condition to solve for C:

$$8 = 10 + C(e^0)$$
$$8 = 10 + C$$

So $C = -2$, and the particular solution is $P = 10 - 2e^{5t}$.

(c) First rewrite the right-hand side of the equation in the form $k(y - A)$ by factoring out a 3:

$$\frac{dy}{dt} = 3(y - 100).$$

The general solution to this differential equation is $y = 100 + Ce^{3t}$.

(d) We begin by factoring out the coefficient of W:

$$\frac{dW}{dt} = 500 - 0.1W = -0.1\left(W - \frac{500}{0.1}\right) = -0.1(W - 5000).$$

The general solution to this differential equation is $W = 5000 + Ce^{-0.1t}$.

Example 2 At the start of this section, we gave the following differential equation for the quantity of warfarin in the body:

$$\frac{dQ}{dt} = 0.5 - 0.02Q.$$

Write the general solution to this differential equation. Find particular solutions for $Q_0 = 20$, $Q_0 = 25$, and $Q_0 = 30$.

Solution We first rewrite the differential equation in the form $dQ/dt = k(Q - A)$ by factoring out -0.02:

$$\frac{dQ}{dt} = 0.5 - 0.02Q = -0.02(Q - 25).$$

The general solution to this differential equation is

$$Q = 25 + Ce^{-0.02t}.$$

To find the particular solution when $Q_0 = 20$, we use the initial condition to solve for C:

$$20 = 25 + C(e^0)$$
$$C = -5.$$

The particular solution when $Q_0 = 20$ is $Q = 25 - 5e^{-0.02t}$.

When $Q_0 = 25$, we have $C = 0$ and the particular solution is the horizontal line $Q = 25$. When $Q_0 = 30$, we have $C = 5$ and the particular solution is $Q = 25 + 5e^{-0.02t}$. These three solutions are the three we saw earlier in Figure 10.29.

Example 3 A company's revenue is earned at a continuous annual rate of 5% of its net worth. At the same time, the company's payroll obligations are paid out at a constant rate of 200 million dollars a year. We saw in Section 10.1 that the differential equation governing the net worth, W (in millions of dollars), of this company in year t is given by

$$\text{Rate of change of } W = \text{Rate in} - \text{Rate out}$$
$$\frac{dW}{dt} = 0.05W - 200.$$

(a) Solve the differential equation, assuming an initial net worth of W_0 million dollars.

(b) Sketch the solution for $W_0 = 3000$, 4000 and 5000. For which of these values of W_0 does the company go bankrupt? In which year?

Solution (a) Factor out 0.05 to get

$$\frac{dW}{dt} = 0.05(W - 4000).$$

The general solution is

$$W = 4000 + Ce^{0.05t}.$$

To find C we use the initial condition that $W = W_0$ when $t = 0$.

$$W_0 = 4000 + Ce^0$$
$$W_0 - 4000 = C$$

Substituting this value for C into $W = 4000 + Ce^{0.05t}$ gives

$$W = 4000 + (W_0 - 4000)e^{0.05t}.$$

(b) If $W_0 = 4000$, then $W = 4000$, the equilibrium solution.
If $W_0 = 5000$, then $W = 4000 + 1000e^{0.05t}$.
If $W_0 = 3000$, then $W = 4000 - 1000e^{0.05t}$. The graphs of these functions are in Figure 10.30. Notice that if the net worth starts with W_0 near, but not equal to, \$4000 million, then W moves further away. We see that if $W_0 = 3000$, the value of W goes to 0, and the company goes bankrupt. Solving $W = 0$ gives $t \approx 27.7$, so the company goes bankrupt in its twenty-eighth year.

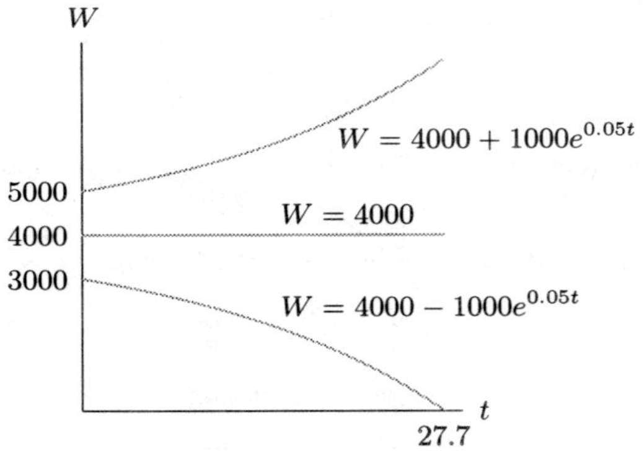

Figure 10.30: Solutions to $\frac{dW}{dt} = 0.05W - 200$

Equilibrium Solutions

Figure 10.29 shows the quantity of warfarin in the body for several different initial quantities. All these curves are solutions to the differential equation

$$\frac{dQ}{dt} = 0.5 - 0.02Q = -0.02(Q - 25),$$

and all the solutions have the form

$$Q = 25 + Ce^{-0.02t}$$

for some C. Notice that $Q \to 25$ as $t \to \infty$ for all solutions because $e^{-0.02t} \to 0$ as $t \to \infty$. In other words, in the long run, the quantity approaches the *equilibrium solution* of $Q = 25$ no matter what the initial quantity.

Notice that the equilibrium solution can be found directly from the differential equation by solving $dQ/dt = 0$:

$$\frac{dQ}{dt} = -0.02(Q - 25) = 0,$$

giving $Q = 25$. Because Q always gets closer and closer to the equilibrium value of 25 as $t \to \infty$, we call $Q = 25$ a *stable* equilibrium for Q.

A different situation is shown in Figure 10.30 with the solutions to the differential equation $dW/dt = 0.05W - 200$. We find the equilibrium by looking at the solution curves or by setting $dW/dt = 0$:

$$\frac{dW}{dt} = 0.05W - 200 = 0.05(W - 4000) = 0,$$

giving $W = 4000$ as the equilibrium solution. This equilibrium solution is called *unstable* because if W starts near, but not equal to, 4000, the net worth W moves further away from 4000 as $t \to \infty$.

- An **equilibrium solution** is constant for all values of the independent variable. The graph is a horizontal line. Equilibrium solutions can be identified by setting the derivative of the function to zero.
- An equilibrium solution is **stable** if a small change in the initial conditions gives a solution which tends toward the equilibrium as the independent variable tends to positive infinity.
- An equilibrium solution is **unstable** if a small change in the initial conditions gives a solution curve which veers away from the equilibrium as the independent variable tends to positive infinity.

In general, a differential equation may have more than one equilibrium solution or no equilibrium solution.

Example 4 Find the equilibrium solution for each of the following differential equations. Determine whether the equilibrium solution is stable or unstable.

(a) $\dfrac{dH}{dt} = -2(H-20)$ (b) $\dfrac{dB}{dt} = 2(B-10)$

Solution (a) To find equilibrium solutions, we set $dH/dt = 0$:

$$\frac{dH}{dt} = -2(H-20) = 0,$$

giving $H = 20$ as the equilibrium solution. The general solution to this differential equation is $H = 20 + Ce^{-2t}$. The solution curves for $H_0 = 10$, $H_0 = 20$, and $H_0 = 30$ are shown in Figure 10.31. We see that the equilibrium solution is stable.

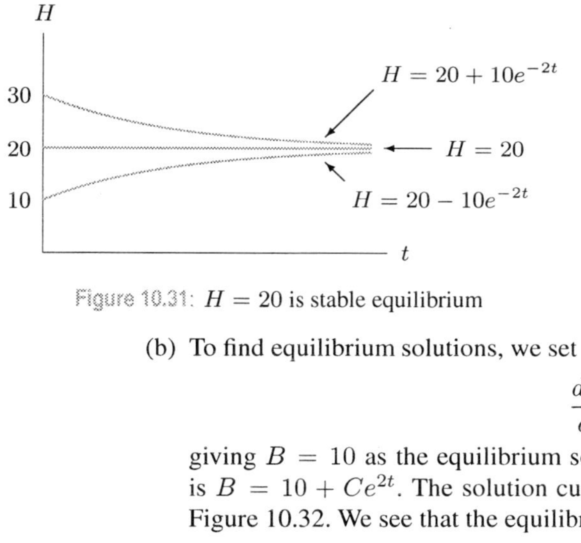

Figure 10.31: $H = 20$ is stable equilibrium

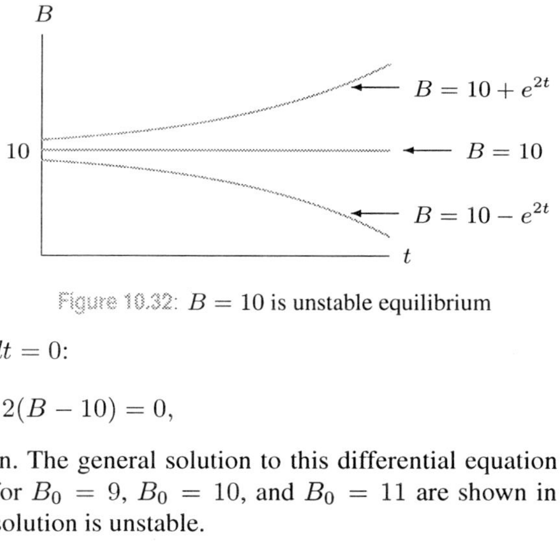

Figure 10.32: $B = 10$ is unstable equilibrium

(b) To find equilibrium solutions, we set $dB/dt = 0$:

$$\frac{dB}{dt} = 2(B-10) = 0,$$

giving $B = 10$ as the equilibrium solution. The general solution to this differential equation is $B = 10 + Ce^{2t}$. The solution curves for $B_0 = 9$, $B_0 = 10$, and $B_0 = 11$ are shown in Figure 10.32. We see that the equilibrium solution is unstable.

Newton's Law of Heating and Cooling

Newton proposed that the temperature of a hot object decreases at a rate proportional to the difference between its temperature and that of its surroundings. Similarly, a cold object heats up at a rate proportional to the temperature difference between the object and its surroundings.

For example, a hot cup of coffee standing on a table cools at a rate proportional to the temperature difference between the coffee and the surrounding air. As the coffee cools, the rate at which it cools decreases because the temperature difference between the coffee and the air decreases. In the long run, the rate of cooling tends to zero and the temperature of the coffee approaches room temperature. Figure 10.33 shows the temperature of two cups of coffee against time, one starting at a higher temperature than the other, but both tending toward room temperature in the long run.

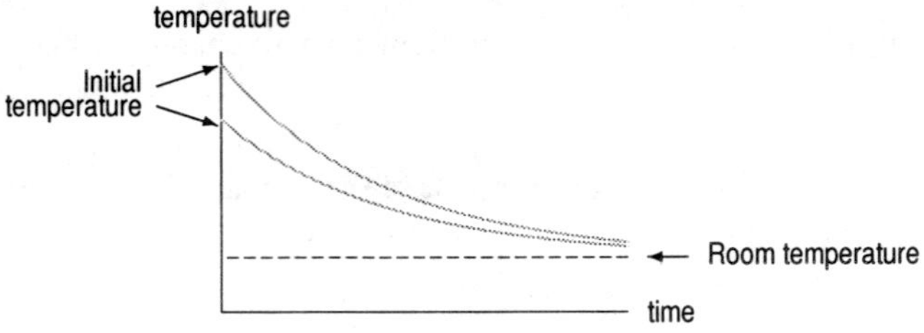

Figure 10.33: Temperature of coffee versus time

Let H be the temperature at time t of a cup of coffee in a 70°F room. Newton's Law says that the rate of change of H is proportional to the temperature difference between the coffee and the room:

$$\text{Rate of change of temperature} = \text{Constant} \cdot \text{Temperature difference.}$$

The rate of change of temperature is dH/dt. The temperature difference between the coffee and the room is $(H - 70)$, so

$$\frac{dH}{dt} = \text{Constant} \cdot (H - 70).$$

What about the sign of the constant? If the coffee starts out hotter than 70° (that is, $H - 70 > 0$), then the temperature of the coffee decreases (i.e., $dH/dt < 0$) and so the constant must be negative:

$$\frac{dH}{dt} = -k(H - 70) \qquad k > 0.$$

What can we learn from this differential equation? Suppose we take $k = 1$. The slope field for this differential equation in Figure 10.34 shows several solution curves. Notice that, as we expect, the temperature of the coffee is approaching the temperature of the room. The general solution to this differential equation is

$$H = 70 + Ce^{-t},$$

where C is an arbitrary constant.

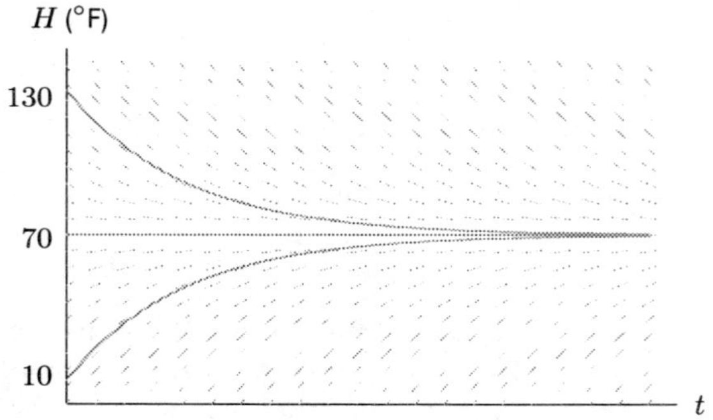

Figure 10.34: Slope field for $\frac{dH}{dt} = -(H - 70)$

Example 5 The body of a murder victim is found at noon in a room with a constant temperature of 20°C. At noon the temperature of the body is 35°C; two hours later the temperature of the body is 33°C.

(a) Find the temperature, H, of the body as a function of t, the time in hours since it was found.
(b) Graph H against t. What happens to the temperature in the long run?
(c) At the time of the murder, the victim's body had the normal body temperature, 37°C. When did the murder occur?

Solution

(a) Newton's Law of Cooling says that

$$\text{Rate of change of temperature} = \text{Constant} \cdot \text{Temperature difference.}$$

Since the temperature difference is $H - 20$, we have for some constant k

$$\frac{dH}{dt} = -k(H - 20).$$

The general solution is

$$H = 20 + Ce^{-kt}.$$

To determine C, we use the fact that $H = 35$ at $t = 0$:

$$35 = 20 + C \cdot e^0$$
$$35 = 20 + C$$

So $C = 15$ and we have

$$H = 20 + 15e^{-kt}.$$

To find k, we use the fact that $H = 33$ when $t = 2$:

$$33 = 20 + 15e^{-k(2)}.$$

We isolate the exponential and solve for k:

$$13 = 15e^{-2k}$$
$$\frac{13}{15} = e^{-2k}$$
$$\ln\left(\frac{13}{15}\right) = -2k$$
$$k = -\frac{\ln(13/15)}{2} = 0.072.$$

Therefore, the temperature, H, of the body as a function of time, t, is given by

$$H = 20 + 15e^{-0.072t}.$$

(b) The graph of $H = 20 + 15e^{-0.072t}$ has a vertical intercept of $H = 35$, the initial temperature. The temperature decays exponentially with a horizontal asymptote of $H = 20$. (See Figure 10.35.) "In the long run" means as $t \to \infty$. The graph shows that $H \to 20$ as $t \to \infty$.

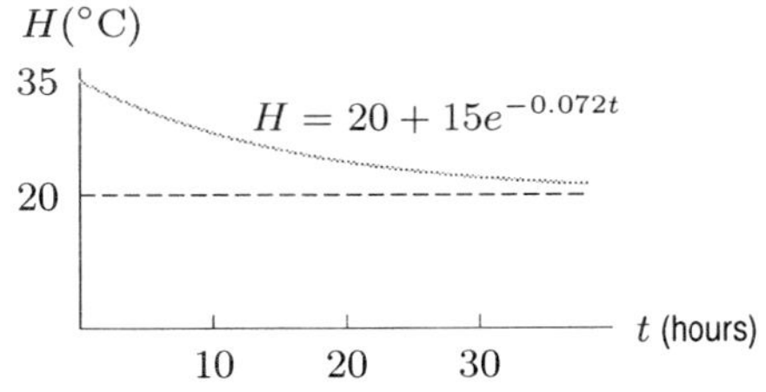

Figure 10.35: Temperature of dead body

(c) We want to know when the temperature was 37°C. We substitute $H = 37$ and solve for t:

$$37 = 20 + 15e^{-0.072t}$$
$$\frac{17}{15} = e^{-0.072t}.$$

Taking natural logs on both sides gives

$$\ln\left(\frac{17}{15}\right) = -0.072t$$

so

$$t = -\frac{\ln(17/15)}{0.072} = -1.74 \text{ hours.}$$

The murder occurred about 1.74 hours before noon, that is, about 10:15 am.

Problems for Section 10.5

Find particular solutions in Problems 1–8.

1. $\frac{dH}{dt} = 3(H - 75), \quad H = 0$ when $t = 0$
2. $\frac{dy}{dt} = 0.5(y - 200), \quad y = 50$ when $t = 0$
3. $\frac{dP}{dt} = P + 4, \quad P = 100$ when $t = 0$
4. $\frac{dB}{dt} = 4B - 100, \quad B = 20$ when $t = 0$
5. $\frac{dQ}{dt} = 0.3Q - 120, \quad Q = 50$ when $t = 0$
6. $\frac{dm}{dt} = 0.1m + 200, \quad m(0) = 1000$
7. $\frac{dB}{dt} + 2B = 50, \quad B(1) = 100$
8. $\frac{dB}{dt} + 0.1B - 10 = 0 \quad B(2) = 3$
9. Check that $y = A + Ce^{kt}$ is a solution to the differential equation
$$\frac{dy}{dt} = k(y - A).$$
10. A bank account earns 7% annual interest compounded continuously. You deposit \$10,000 in the account, and withdraw money continuously from the account at a rate of \$1000 per year.
    - **(a)** Write a differential equation for the balance, B, in the account after t years.
    - **(b)** What is the equilibrium solution to the differential equation? (This is the amount that must be deposited now for the balance to stay the same over the years.)
    - **(c)** Find the solution to the differential equation.
    - **(d)** How much is in the account after 5 years?
    - **(e)** Graph the solution. What happens to the balance in the long run?
11. A company earns 2% per month on its assets, paid continuously, and its expenses are paid out continuously at a rate of \$80,000 per month.
    - **(a)** Write a differential equation for the value, V, of the company as a function of time, t, in months.
    - **(b)** What is the equilibrium solution for the differential equation? What is the significance of this value for the company?
    - **(c)** Solve the differential equation found in part (a).
    - **(d)** If the company has assets worth \$3 million at time $t = 0$, what are its assets worth one year later?
12. Money in an account earns interest at a continuous rate of 8% per year, and payments are made continuously out of the account at the rate of \$5000 a year. The account initially contains \$50,000. Write a differential equation for the amount of money in the account, B, in t years. Solve the differential equation. Does the account ever run out of money? If so, when?
13. A bank account earns 5% annual interest, compounded continuously. Money is deposited in a continuous cash flow at a rate of \$1200 per year into the account.
    - **(a)** Write a differential equation that describes the rate at which the balance $B = f(t)$ is changing.
    - **(b)** Solve the differential equation given an initial balance $B_0 = 0$.
    - **(c)** Find the balance after 5 years.
14. A patient is given the drug theophylline intravenously at a rate of 43.2 mg/hour to relieve acute asthma. The rate at which the drug leaves the patient's body is proportional to the quantity there, with proportionality constant 0.082 if time, t, is in hours. The patient's body contains none of the drug initially.
    - **(a)** Describe in words how you expect the quantity of theophylline in the patient to vary with time.
    - **(b)** Write a differential equation satisfied by the quantity of theophylline in the body, $Q(t)$.
    - **(c)** Solve the differential equation and graph the solution. What happens to the quantity in the long run?
15. One theory on the speed an employee learns a new task claims that the more the employee already knows, the more slowly he or she learns. Suppose that the rate at which a person learns is equal to the percentage of the task not yet learned. If y is the percentage learned by time t, the percentage not yet learned by that time is $100 - y$, so we can model this situation with the differential equation
$$\frac{dy}{dt} = 100 - y.$$
    - **(a)** Find the general solution to this differential equation.
    - **(b)** Sketch several solutions.
    - **(c)** Find the particular solution if the employee starts learning at time $t = 0$ (so $y = 0$ when $t = 0$).
16. A chain smoker smokes five cigarettes every hour. From each cigarette, 0.4 mg of nicotine is absorbed into the person's bloodstream. Nicotine leaves the body at a rate proportional to the amount present, with constant of proportionality -0.346 if t is in hours.
    - **(a)** Write a differential equation for the level of nicotine in the body, N, in mg, as a function of time, t, in hours.
    - **(b)** Solve the differential equation from part (a). Initially there is no nicotine in the blood.
    - **(c)** The person wakes up at 7 am and begins smoking. How much nicotine is in the blood when the person goes to sleep at 11 pm (16 hours later)?

17. As you know, when a course ends, students start to forget the material they have learned. One model (called the Ebbinghaus model) assumes that the rate at which a student forgets material is proportional to the difference between the material currently remembered and some positive constant, a.

(a) Let $y = f(t)$ be the fraction of the original material remembered t weeks after the course has ended. Set up a differential equation for y. Your equation will contain two constants; the constant a is less than y for all t.
(b) Solve the differential equation.
(c) Describe the practical meaning (in terms of the amount remembered) of the constants in the solution $y = f(t)$.

18. (a) What are the equilibrium solutions for the differential equation

$$\frac{dy}{dt} = 0.2(y-3)(y+2)?$$

(b) Use a graphing calculator or computer to sketch a slope field for this differential equation. Use the slope field to determine whether each equilibrium solution is stable or unstable.

19. (a) Find the equilibrium solution of the equation

$$\frac{dy}{dt} = 0.5y - 250.$$

(b) Find the general solution of this equation.
(c) Graph several solutions with different initial values.
(d) Is the equilibrium solution stable or unstable?

20. Figure 10.36 gives the slope field for a differential equation. Estimate all equilibrium solutions and indicate whether each is stable or unstable.

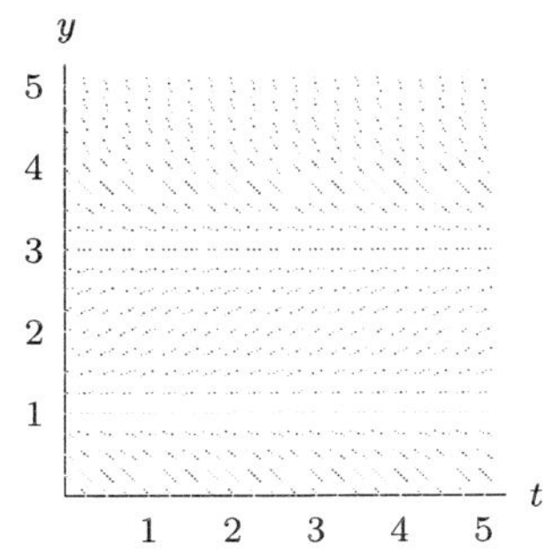

Figure 10.36

21. A yam is put in a 200°C oven and heats up according to the differential equation

$$\frac{dH}{dt} = -k(H-200), \quad \text{for } k \text{ a positive constant.}$$

(a) If the yam is at 20°C when it is put in the oven, solve the differential equation.
(b) Find k using the fact that after 30 minutes the temperature of the yam is 120°C.

22. At 1:00 pm one winter afternoon, there is a power failure at your house in Wisconsin, and your heat does not work without electricity. When the power goes out, it is 68°F in your house. At 10:00 pm, it is 57°F in the house, and you notice that it is 10°F outside.

(a) Assuming that the temperature, T, in your home obeys Newton's Law of Cooling, write the differential equation satisfied by T.
(b) Solve the differential equation to estimate the temperature in the house when you get up at 7:00 am the next morning. Should you worry about your water pipes freezing?
(c) What assumption did you make in part (a) about the temperature outside? Given this (probably incorrect) assumption, would you revise your estimate up or down? Why?

23. A detective finds a murder victim at 9 am. The temperature of the body is measured at 90.3°F. One hour later, the temperature of the body is 89.0°F. The temperature of the room has been maintained at a constant 68°F.

(a) Assuming the temperature, T, of the body obeys Newton's Law of Cooling, write a differential equation for T.
(b) Solve the differential equation to estimate the time the murder occurred.

24. A drug is administered intravenously at a constant rate of r mg/hour and is excreted at a rate proportional to the quantity present, with constant of proportionality $\alpha > 0$.

(a) Solve a differential equation for the quantity, Q, in milligrams, of the drug in the body at time t hours. Assume there is no drug in the body initially. Your answer will contain r and α. Graph Q against t. What is Q_∞, the limiting long-run value of Q?
(b) What effect does doubling r have on Q_∞? What effect does doubling r have on the time to reach half the limiting value, $\frac{1}{2}Q_\infty$?
(c) What effect does doubling α have on Q_∞? On the time to reach $\frac{1}{2}Q_\infty$?

25. Some people write the solution of the initial value problem

$$\frac{dy}{dt} = k(y-A) \qquad y = y_0 \text{ at } t = 0$$

in the form

$$\frac{y-A}{y_0-A} = e^{kt}.$$

Show that this formula gives the correct solution for y, assuming $y_0 \neq A$.

Chapter Eleven

GEOMETRIC SERIES

Contents

11.1 GEOMETRIC SERIES

Repeated Drug Dosage

Malaria is a parasitic infection transmitted by mosquito bites, mainly in tropical areas of the world. The disease has existed since ancient times, and currently there are hundreds of millions of cases each year, with millions of deaths. In the 1640s, the Jesuits in Peru introduced the bark of the cinchona tree to the West as the first treatment for malaria. The drug quinine is the active ingredient in the bark, and it is still used today.

Suppose a person is given a 50-mg dose of quinine at the same time every day for the prevention of malaria. After the first dose, the person has 50 mg of quinine in the body. What about after the second dose? Each day, the person's body metabolizes some of the quinine so that, after one day, 23% of the original amount remains. After the second dose, the amount of quinine in the body is the amount from the second dose (50 mg) plus the remnants of the first dose (that is, $50 \cdot 0.23 = 11.5$ mg) for a total of 61.5 mg.

Let Q_n represent the quantity, in mg, of quinine in the body right after the nth dose. Then

$$\begin{aligned} Q_1 &= \text{First dose} = 50. \\ Q_2 &= \text{Second dose + Remnants of first dose} = 50 + 50(0.23) = 61.5. \\ Q_3 &= \text{Third dose + Remnants of previous doses} = 50 + 61.5(0.23) = 64.145. \end{aligned}$$

Notice that we can multiply out the expression for Q_3 to show the contributions of the first and second dose separately:

$$\begin{aligned} Q_3 &= 50 + 61.5(0.23) = 50 + (50 + 50(0.23))(0.23) \\ Q_3 &= 50 + 50(0.23) + 50(0.23)^2, \end{aligned}$$

so we have

$$Q_3 = \text{Third dose + Remnants of second dose + Remnants of first dose.}$$

The multiplied-out form of Q_3 enables us to guess formulas for later values of Q_n:

$$\begin{aligned} Q_4 &= 50 + 50(0.23) + 50(0.23)^2 + 50(0.23)^3 = 64.753. \\ Q_5 &= 50 + 50(0.23) + 50(0.23)^2 + 50(0.23)^3 + 50(0.23)^4 = 64.893. \\ Q_6 &= 50 + 50(0.23) + 50(0.23)^2 + 50(0.23)^3 + 50(0.23)^4 + 50(0.23)^5 = 64.925. \\ &\vdots \\ Q_{10} &= 50 + 50(0.23) + 50(0.23)^2 + \cdots + 50(0.23)^8 + 50(0.23)^9 = 64.935. \end{aligned}$$

The values of Q_6 and Q_{10} suggest that the quantity is stabilizing at around 64.9 mg. See Figure 11.1.

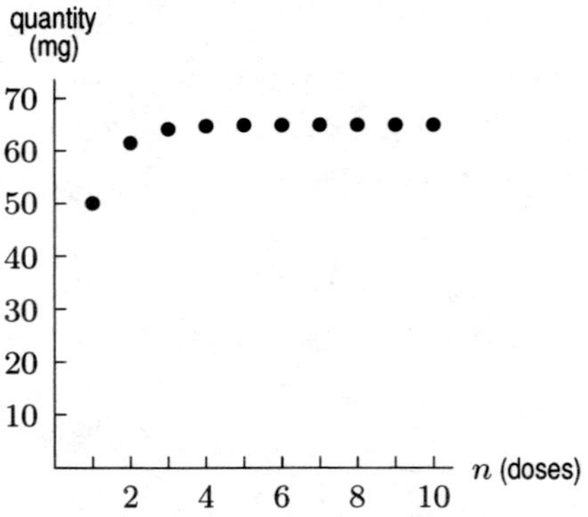

Figure 11.1: Quantity of quinine levels off

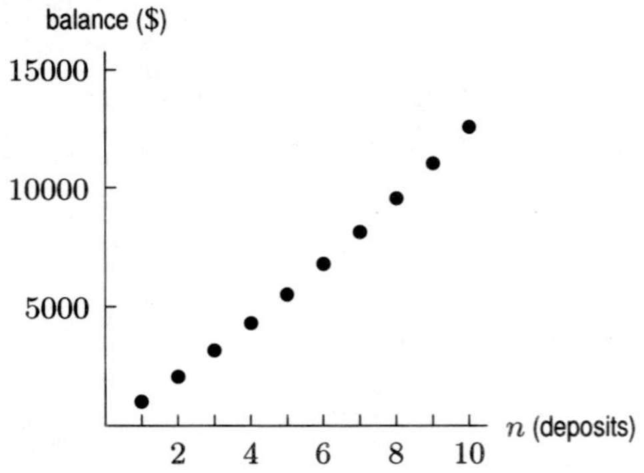

Figure 11.2: Bank balance grows without bound

Repeated Deposits into a Savings Account

People who save money often do so by putting a fixed amount aside regularly. Suppose \$1000 is deposited every year in a savings account earning 5% interest a year, compounded annually. Let B_n represent the balance, in dollars, in the account right after the nth deposit. Then

$$\begin{aligned} B_1 &= \text{First deposit } = 1000. \\ B_2 &= \text{Second deposit + Amount from first deposit } = 1000 + 1000(1.05) = 2050. \\ B_3 &= \text{Third deposit + Amount from previous deposits } = 1000 + 2050(1.05) = 3152.5. \end{aligned}$$

As before, we multiply out the expression for B_3 to show the contributions of the first and second deposits separately:

$$\begin{aligned} B_3 &= 1000 + 2050(1.05) = 1000 + (1000 + 1000(1.05))(1.05) \\ B_3 &= 1000 + 1000(1.05) + 1000(1.05)^2 \\ B_3 &= \text{Third deposit + Amount from second deposit + Amount from first deposit.} \end{aligned}$$

The multiplied-out formula for B_3 enables us to guess formulas for B_6 and B_{10}. Evaluating gives

$$B_6 = 1000 + 1000(1.05) + 1000(1.05)^2 + 1000(1.05)^3 + 1000(1.05)^4 + 1000(1.05)^5 = 6801.91.$$

$$B_{10} = 1000 + 1000(1.05) + 1000(1.05)^2 + \cdots + 1000(1.05)^8 + 1000(1.05)^9 = 12{,}577.89.$$

Notice that the balance is growing without bound. See Figure 11.2.

Finite Geometric Series

In the two previous examples, we encountered sums of the form $a+ar+ar^2+ar^3+\cdots+ar^8+ar^9$. Such a sum is called a *finite geometric series*. A geometric series is a sum in which each term is a constant multiple of the preceding one. The first term is a, and the constant multiplier, or common ratio, of successive terms is r.

> A **finite geometric series** with n terms has (for n a positive integer) the form
>
> $$a + ar + ar^2 + ar^3 + \cdots + ar^{n-2} + ar^{n-1}.$$

Sum of a Finite Geometric Series

In the quinine example, suppose we want to find Q_{40}, the quantity in the body after 40 doses:

$$Q_{40} = 50 + 50(0.23) + 50(0.23)^2 + \cdots + 50(0.23)^{38} + 50(0.23)^{39}.$$

To calculate Q_{40}, we could add these 40 terms. Fortunately, there is a better way.

We write S_n for the sum of the first n terms of the series, that is, up to the term ar^{n-1}:

$$S_n = a + ar + ar^2 + ar^3 + \cdots + ar^{n-2} + ar^{n-1}.$$

Multiplying both sides by r gives

$$rS_n = ar + ar^2 + ar^3 + ar^4 + \cdots + ar^{n-1} + ar^n.$$

Now subtract rS_n from S_n, so all the terms except two on the right cancel, giving

$$S_n - rS_n = a - ar^n,$$

so

$$(1-r)S_n = a(1-r^n).$$

Provided $r \neq 1$, we can solve for S_n. The result is called a *closed form* for S_n.

The **sum of a finite geometric series** is given by

$$S_n = a + ar + ar^2 + ar^3 + \cdots + ar^{n-1} = \frac{a(1 - r^n)}{1 - r}, \quad \text{provided } r \neq 1.$$

Note that the value of n in the formula is the number of terms in the sum S_n.

Example 1 In the quinine example, calculate and interpret Q_{40} and Q_{100}.

Solution We saw earlier that

$$Q_{40} = 50 + 50(0.23) + 50(0.23)^2 + \cdots + 50(0.23)^{39}.$$

This is a finite geometric series with $a = 50$ and $r = 0.23$. Using the formula for the sum with $n = 40$, we have

$$Q_{40} = \frac{50(1 - (0.23)^{40})}{1 - 0.23} = 64.935.$$

The amount of quinine in the body right after the 40th dose is 64.935 mg.

Similarly, using $n = 100$, we have

$$Q_{100} = \frac{50(1 - (0.23)^{100})}{1 - 0.23} = 64.935.$$

Right after the 100th dose, the amount of quinine in the body is still 64.935 mg. To three decimal places, the amount appears to have stabilized.

Example 2 In the bank deposit example, calculate and interpret B_{40} and B_{100}.

Solution We have

$$B_{40} = 1000 + 1000(1.05) + 1000(1.05)^2 + \cdots + 1000(1.05)^{39}.$$

This is a finite geometric series with $a = 1000$ and $r = 1.05$. The formula for the sum with $n = 40$ gives

$$B_{40} = \frac{1000(1 - (1.05)^{40})}{1 - 1.05} = 120{,}799.77.$$

The balance in the account right after the 40th deposit is \$120,799.77.

Similarly, using $n = 100$, we have

$$B_{100} = \frac{1000(1 - (1.05)^{100})}{1 - 1.05} = 2{,}610{,}025.16.$$

Right after the 100th deposit, the balance in the account is \$2,610,025.16. Compound interest has increased the \$100,000 investment to over \$2 million.

Infinite Geometric Series

Suppose a finite geometric series has n terms in it. What happens as $n \to \infty$? We get an infinite geometric series that goes on forever.

An **infinite geometric series** has the form

$$a + ar + ar^2 + ar^3 + \cdots + ar^{n-1} + ar^n + \cdots.$$

The "$\cdots$" at the end of the series tells us that the series is going on forever—it is infinite.

Sum of an Infinite Geometric Series

Given an infinite geometric series

$$a + ar + ar^2 + ar^3 + \cdots,$$

we call the sum of the first n terms a *partial sum*, written S_n. To calculate S_n, we use the formula

$$S_n = \frac{a(1 - r^n)}{1 - r}.$$

What happens to S_n as $n \to \infty$? It depends on the value of r. If $|r| < 1$, that is, $-1 < r < 1$, then $r^n \to 0$ as $n \to \infty$, so as $n \to \infty$,

$$S_n = \frac{a(1 - r^n)}{1 - r} \to \frac{a(1 - 0)}{1 - r} = \frac{a}{1 - r}.$$

Thus, provided $|r| < 1$, as $n \to \infty$ the partial sums S_n approach a limit of $a/(1 - r)$. When this happens, we define the sum of the infinite geometric series to be that limit and say the series *converges* to $a/(1 - r)$.

> For $|r| < 1$, the **sum of the infinite geometric series** is given by
>
> $$S = a + ar + ar^2 + ar^3 + \cdots + ar^{n-1} + ar^n + \cdots = \frac{a}{1 - r}.$$

If, on the other hand, $|r| > 1$, then r^n and the partial sums have no limit as $n \to \infty$ (if $a \neq 0$). In this case, we say the series *diverges*. If $r > 1$, the terms in the series become larger and larger in magnitude, and the partial sums diverge to $+\infty$ if $a > 0$, or to $-\infty$ if $a < 0$. If $r < -1$, the terms become larger in magnitude, the partial sums oscillate as $n \to \infty$, and the series diverges.

What happens if $r = 1$? The series is

$$a + a + a + a + \cdots,$$

so if $a \neq 0$, the partial sums grow without bound, and the series does not converge. If $r = -1$, the series is

$$a - a + a - a + a - \cdots,$$

and, if $a \neq 0$, the partial sums oscillate between a and 0, and the series does not converge.

Example 3 For each of the following infinite series, find the first three partial sums and the sum (if it exists):

(a) $10 + 10(0.75) + 10(0.75)^2 + \cdots$

(b) $250 + 250(1.2) + 250(1.2)^2 + \cdots$

Solution (a) This is an infinite geometric series with $a = 10$ and $r = 0.75$. The first three partial sums are:

$$S_1 = 10.$$
$$S_2 = 10 + 10(0.75) = 10 + 7.5 = 17.5.$$
$$S_3 = 10 + 10(0.75) + 10(0.75)^2 = 10 + 7.5 + 5.625 = 23.125.$$

Since $|r| < 1$, the series converges and the sum is

$$S = \frac{a}{1 - r} = \frac{10}{1 - 0.75} = 40.$$

If we find partial sums for larger and larger n, they get closer and closer to 40. (See Problem 15.)

(b) This is an infinite geometric series with $a = 250$ and $r = 1.2$. The first three partial sums are:

$$S_1 = 250.$$
$$S_2 = 250 + 250(1.2) = 250 + 300 = 550.$$
$$S_3 = 250 + 250(1.2) + 250(1.2)^2 = 250 + 300 + 360 = 910.$$

Since $r > 1$, the series diverges, and the partial sums grow without bound. (See Problem 16.)

Example 4 Suppose 50-mg doses of quinine are taken daily forever. Find the long-run quantities of quinine in the body right after and right before a dose is given.

Solution Since quinine is given forever, from Example 1 we know that the long-run quantity of quinine, right after a dose, is given by

$$Q = 50 + 50(0.23) + 50(0.23)^2 + \cdots.$$

This is an infinite geometric series with $a = 50$ and $r = 0.23$. Since $-1 < r < 1$, the series converges to a finite sum given by

$$Q = \frac{a}{1-r} = \frac{50}{1-0.23} = 64.935.$$

The long-run quantity of quinine in the body right after a dose is 64.935 mg. The quantity levels off to this value in Figure 11.1.

What is the long-run quantity of quinine right before a dose? Since a dose is 50 mg, the quantity of quinine in the body right before a dose is $64.935 - 50 = 14.935$ mg. Thus, in the long run, the quinine level oscillates between 15 mg and 65 mg.

Example 5 Suppose that \$1000 a year is deposited forever into the bank account in Example 2. Does the balance in the account stabilize at a fixed amount? Explain.

Solution Since the deposits are made forever, the account balance right after a deposit is represented by

$$B = 1000 + 1000(1.05) + 1000(1.05)^2 + \cdots.$$

This is an infinite geometric series with $a = 1000$ and $r = 1.05$. Since r is larger than 1, the series diverges. This makes sense, since if you keep depositing \$1000 in an account, your balance grows without bound, even if you don't earn interest. This matches Figure 11.2 and Example 2.

Problems for Section 11.1

1. Find the sum of the following series in two ways: by adding terms and by using the geometric series formula.

$$3 + 3\cdot 2 + 3\cdot 2^2$$

2. Find the sum of the following series in two ways: by adding terms and by using the geometric series formula.

$$50 + 50(0.9) + 50(0.9)^2 + 50(0.9)^3$$

In Problems 3–14, find the sum, if it exists.

3. $5 + 5\cdot 3 + 5\cdot 3^2 + \cdots + 5\cdot 3^{12}$
4. $20 + 20(1.45) + 20(1.45)^2 + \cdots + 20(1.45)^{14}$
5. $100 + 100(0.85) + 100(0.85)^2 + \cdots + 100(0.85)^{10}$
6. $1000 + 1000(1.05) + 1000(1.05)^2 + \cdots$
7. $75 + 75(0.22) + 75(0.22)^2 + \cdots$
8. $500(0.4) + 500(0.4)^2 + 500(0.4)^3 + \cdots$
9. $31500 + 6300 + 1260 + 252 + \cdots$
10. $3 + \frac{3}{2} + \frac{3}{4} + \frac{3}{8} + \cdots + \frac{3}{2^{10}}$
11. $65 + \frac{65}{1.02} + \frac{65}{(1.02)^2} + \cdots + \frac{65}{(1.02)^{18}}$
12. $1000 + 1500 + 2250 + 3375 + 5062.5 + \cdots$
13. $200 + 100 + 50 + 25 + 12.5 + \cdots$
14. $-2 + 1 - \frac{1}{2} + \frac{1}{4} - \frac{1}{8} + \frac{1}{16} - \cdots$
15. In Example 3(a), we found partial sums of the geometric series with $a = 10$ and $r = 0.75$ and showed that the sum of this series is 40. Find the partial sums S_n for $n = 5, 10, 15, 20$. As n gets larger, do the partial sums appear to be approaching 40?
16. In Example 3(b), we found partial sums for the geometric series with $a = 250$ and $r = 1.2$. Find the partial sums S_n for $n = 5, 10, 15, 20$. As n gets larger, do the partial sums appear to grow without bound, as expected if $r > 1$?

17. Every month, \$500 is deposited into an account earning 0.5% interest a month, compounded monthly.

(a) How much is in the account right after the 6th deposit? Right before the 6th deposit?
(b) How much is in the account right after the 12th deposit? Right before the 12th deposit?

18. Each year, a family deposits \$5000 into an account paying 8.12% interest per year, compounded annually. How much is in the account right after the 20th deposit?

19. Each morning, a patient receives a 25 mg injection of an anti-inflammatory drug, and 40% of the drug remains in the body after 24 hours. Find the quantity in the body:

(a) Right after the 3rd injection.
(b) Right after the 6th injection.
(c) In the long run, right after an injection.

20. A smoker inhales 0.4 mg of nicotine from a cigarette. After one hour, 71% of the nicotine remains in the body. If a person smokes one cigarette every hour beginning at 7 am, how much nicotine is in the body right after the 11 pm cigarette?

21. In Example 4, we saw that if 50 mg of quinine is given every 24 hours, the long-run quantity of quinine in the body is about 65 mg right after a dose and about 15 mg right before a dose. The concentration of quinine in the body is measured in milligrams of quinine per kilogram of body weight. To be effective, the average concentration of quinine in the body must be at least 0.4 mg/kg. Concentrations above 3.0 mg/kg are not safe.

(a) Estimate the average quantity of quinine in the body in the long run by averaging the long-run quantities of quinine in the body right after a dose and right before a dose.
(b) Find the average concentration for a person weighing 70 kilograms. Is this treatment safe and effective for such a person?
(c) For what range of weights would this treatment produce a long-run average concentration that is

(i) Too low? (ii) Unsafe?

22. Figure 11.3 shows the quantity of the drug atenolol in the blood as a function of time, with the first dose at time $t = 0$. Atenolol is taken in 50 mg doses once a day to lower blood pressure.

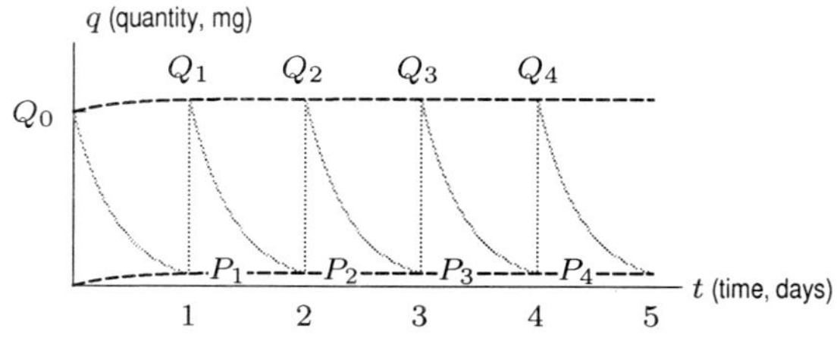

Figure 11.3

(a) If the half-life of atenolol in the blood is 6.3 hours, what percentage of the atenolol present at the start of a 24-hour period is still there at the end?
(b) Find expressions for the quantities $Q_0, Q_1, Q_2, Q_3, \ldots$, and Q_n shown in Figure 11.3. Write the expression for Q_n in closed form.
(c) Find expressions for the quantities $P_1, P_2, P_3, \ldots$, and P_n shown in Figure 11.3. Write the expression for P_n in closed form.

11.2 APPLICATIONS TO BUSINESS AND ECONOMICS

Annuities

An *annuity* is a sequence of equal payments made at regular intervals. We can use the sum of a geometric series to calculate the total value of an annuity.

Example 1 An annuity pays \$5000 every year into an account that earns 7% interest per year, compounded annually. What is the balance in the account right after the 10th deposit?

Solution The 10th deposit contributes \$5000 to the balance. The previous deposit has earned interest for a year, so it contributes \$5000(1.07). The deposit the year before that has earned interest for two years, so it contributes $\$5000(1.07)^2$. Continuing, we see that

$$\text{Balance after } 10^{\text{th}} \text{ deposit} = 5000 + 5000(1.07) + 5000(1.07)^2 + \cdots + 5000(1.07)^9.$$

This sum is a finite geometric series with $a = 5000$ and $r = 1.07$. We use the formula for the sum with $n = 10$:

$$\text{Balance after } 10^{\text{th}} \text{ deposit} = \frac{a(1-r^n)}{1-r} = \frac{5000(1-(1.07)^{10})}{1-1.07} = 69{,}082.24 \text{ dollars.}$$

The balance in the account right after the 10th deposit is \$69,082.24.

Present Value of an Annuity

The *present value of an annuity* is the amount of money that must be deposited today to make a series of fixed payments in the future. How can we compute this present value? We begin by considering a single payment. Suppose a payment of \$1000 is to be made three years in the future from an account paying interest at a rate of 8% per year, compounded annually. The present value is the amount P such that

$$P(1.08)^3 = 1000,$$

so we have

$$\text{Present value} = P = 1000(1.08)^{-3}.$$

To find the present value of four payments of \$1000 at the same 8% interest, one made now, one in one year's time, one in two years, and one in three years, we add their present values:

$$\text{Present value of four payments} = 1000 + 1000(1.08)^{-1} + 1000(1.08)^{-2} + 1000(1.08)^{-3}.$$

This pattern allows us to find the present value of any annuity, as in the following example.

Example 2 An account earns 8% interest per year, compounded annually. Twenty payments of \$10,000 each, made once a year starting now, are to be made out of the account. How much must be deposited in the account now to cover these payments? In other words, what is the present value of this annuity?

Solution The present value of the payment to be made immediately is \$10,000. The present value of next year's payment is $\$10{,}000(1.08)^{-1}$. Since the 20th payment is made 19 years in the future, the present value of the 20th payment is $\$10{,}000(1.08)^{-19}$. The present value, P, of the entire annuity, in dollars, is the sum

$$P = 10{,}000 + 10{,}000(1.08)^{-1} + 10{,}000(1.08)^{-2} + \cdots + 10{,}000(1.08)^{-19}.$$

Rewriting $(1.08)^{-2} = \left((1.08)^{-1}\right)^2$ and $(1.08)^{-3} = \left((1.08)^{-1}\right)^3$, and so on, shows that P is the sum of the finite geometric series

$$P = 10{,}000 + 10{,}000(1.08)^{-1} + 10{,}000((1.08)^{-1})^2 + \cdots + 10{,}000((1.08)^{-1})^{19}.$$

We use the formula for the sum with $a = 10{,}000$ and $r = (1.08)^{-1}$ and $n = 20$, giving

$$\text{Present value} = P = \frac{a(1-r^n)}{1-r} = \frac{10{,}000(1-((1.08)^{-1})^{20})}{1-(1.08)^{-1}} = 106{,}035.99.$$

Thus, \$106,035.99 must be deposited now to cover the payments for this annuity. Notice that the annuity pays out a total of $20 \cdot \$10{,}000 = \$200{,}000$, so the present value is considerably less than the amount eventually paid out.

Example 3 The annuity in Example 2 now makes annual payments of \$10,000 in perpetuity (that is, forever), rather than just twenty times. What is the present value of this annuity?

Solution Since the payments are made forever, the present value is given by the infinite sum:

$$\text{Present value} = 10{,}000 + 10{,}000(1.08)^{-1} + 10{,}000((1.08)^{-1})^2 + 10{,}000((1.08)^{-1})^3 + \cdots.$$

This is an infinite geometric series with $a = 10{,}000$ and $r = (1.08)^{-1} = 0.925926$. Since $-1 < r < 1$, this series converges to a finite sum. We have

$$\text{Present value} = \frac{a}{1-r} = \frac{10{,}000}{1-(1.08)^{-1}} = 135{,}000.$$

The present value of this annuity in perpetuity is \$135,000. Notice that the amount needed to make annual payments forever is only about \$29,000 more than the amount needed to make 20 annual payments. This is the power of compound interest.

The Multiplier Effect

A government decides to give a tax rebate to stimulate the economy. What is the total effect of the rebate on spending? Each individual who receives a tax rebate spends some proportion of the additional income, and this money becomes additional income for a second individual, who spends a proportion of it, providing additional income for a third individual, and so on. The total effect of the rebate on the economy is much larger than the size of the rebate itself; this is called the *multiplier effect*.

Example 4 A government gives tax rebates totaling 3 billion dollars. Everyone who receives money spends 75% of it and saves the other 25%. Find the total additional spending resulting from this tax rebate.

Solution The additional spending refers to all the additional money spent by consumers as a result of this tax rebate. The recipients of the 3 billion dollars spend 75% of what they receive, that is $3(0.75) = 2.25$ billion dollars. The recipients of this 2.25 billion dollars spend 75% of that, or $2.25(0.75) = 1.6875$ billion dollars. The recipients of this money spend 75% of this amount, and so on. Thus,

$$\text{Total additional spending} = 2.25 + 2.25(0.75) + 2.25(0.75)^2 + 2.25(0.75)^3 + \cdots \text{ billion dollars.}$$

(Notice that the initial amount spent was 2.25, not 3 billion dollars.) This is an infinite geometric series with $a = 2.25$ and $r = 0.75$. Since $-1 < r < 1$, this series converges to a finite sum:

$$\text{Total additional spending} = \frac{a}{1-r} = \frac{2.25}{1-0.75} = 9 \text{ billion dollars.}$$

Thus, a 3 billion dollar tax rebate generates 9 billion dollars of additional spending.

Market Stabilization

Each year, a manufacturer produces a fixed number of units of a product and each year a fixed percentage of these units (regardless of age) fail or go out of use. The total number of units in use in the long run is called the *market stabilization point*.

Example 5 The US Mint produces about 7.4 billion pennies a year and about 5% of them are removed from circulation each year.[1] Approximately how many pennies are in circulation?

Solution We can estimate the number of pennies in circulation using geometric series if we assume (for example) that the yearly production takes place at the start of the year and that the 5% are removed at the end of the year. In any year, 7.4 billion pennies are produced, and 7.4(0.95) billion pennies remain from the previous year's production (since 5% went out of circulation). There are $7.4(0.95)^2$ billion pennies remaining from those produced two years before, and so on. If N is the number of pennies in circulation, in billions, then

$$N = 7.4 + 7.4(0.95) + 7.4(0.95)^2 + \cdots.$$

This is an infinite geometric series with $a = 7.4$ and $r = 0.95$. Since $-1 < r < 1$, the series converges and its sum is given by

$$N = \frac{a}{1-r} = \frac{7.4}{1-0.95} = 148.$$

Thus, the market has stabilized with about 148 billion pennies in circulation today. Of the 148 billion pennies in circulation, 5% or $(148)(0.05) = 7.4$ billion are removed each year, which exactly equals the number produced each year.

[1]From www.usmint.gov.

Problems for Section 11.2

1. A yearly deposit of $1000 is made into a bank account that pays 8.5% interest per year, compounded annually. What is the balance in the account right after the 20th deposit? How much of the balance comes from the annual deposits and how much comes from interest?

2. Annual deposits of $2000 are made into an account paying 6% interest per year, compounded continuously. What is the balance in the account right after and right before the 5th deposit?

3. An annuity earning 0.5% per month, compounded monthly, is to make 36 monthly payments of $1000 each, starting now. What is the present value of this annuity?

4. An annuity makes annual payments of $50,000, starting now, from an account paying 7.2% interest per year, compounded annually. Find the present value of the annuity if it makes

   (a) Ten payments **(b)** Payments in perpetuity

5. Twenty annual payments of $5000 each, with the first payment one year from now, are to be made from an account earning 10% per year, compounded annually. How much must be deposited now to cover the payments?

6. What is the present value of an annuity that pays $20,000 each year, forever, starting today, from an account that pays 1% interest per year, compounded annually?

7. A deposit of $100,000 is made into an account paying 8% interest per year, compounded annually. Annual payments of $10,000 each, starting right after the deposit, are made out of the account. How many payments can be made before the account runs out of money?

8. An employer pays you 1 penny the first day you work and doubles your wages each day after that. Find your total earnings after working 7 days a week for

   (a) One week **(b)** Two weeks
   (c) Three weeks **(d)** Four weeks

9. An employee accepts a job with a starting salary of $30,000 and a cost-of-living increase of 4% every year for the next 10 years. What is the employee's salary right before the start of the 11th year? What are her total earnings during the first 10 years?

10. Find the market stabilization point for a product if 10,000 new units of the product are manufactured at the start of each year and 25% of the total number of units in use fail at the end of each year.

11. Since 2007, the Bureau of Engraving and Printing has produced about 11 million new $1 bills a day; worn bills are removed by Federal Reserve banks. There are about 9.2 billion dollar bills currently in circulation. Assuming that a fixed percentage of $1 bills are removed from circulation each day, use a geometric series to estimate this percentage.[2]

Problems 12–14 are about *bonds*, which are issued by a government to raise money. An individual who buys a $1000 bond gives the government $1000 and in return receives a fixed sum of money, called the *coupon*, every six months or every year for the life of the bond. At the time of the last coupon, the individual also gets back the $1000, or *principal*.

12. What is the present value of a $1000 bond which pays $50 a year for 10 years, starting one year from now? Assume the interest rate is 6% per year, compounded annually.

13. What is the present value of a $1000 bond which pays $50 a year for 10 years, starting one year from now? Assume the interest rate is 4% per year, compounded annually.

14. **(a)** What is the present value of a $1000 bond which pays $50 a year for 10 years, starting one year from now? Assume the interest rate is 5% per year, compounded annually.
    (b) Since $50 is 5% of $1000, this bond is called a 5% bond. What does your answer to part (a) tell you about the relationship between the principal and the present value of this bond if the interest rate is 5%?
    (c) If the interest rate is more than 5% per year, compounded annually, which is larger: the principal or the present value of the bond? Why is the bond then described as *trading at a discount*?
    (d) If the interest rate is less than 5% per year, compounded annually, why is the bond described as *trading at a premium*?

15. This problem illustrates how banks create credit and can thereby lend out more money than has been deposited. Suppose that initially $100 is deposited in a bank. Experience has shown bankers that on average only 8% of the money deposited is withdrawn by the owner at any time. Consequently, bankers feel free to lend out 92% of their deposits. Thus $92 of the original $100 is loaned out to other customers (to start a business, for example). This $92 becomes someone else's income and, sooner or later, is redeposited in the bank. Thus 92% of $92, or $92(0.92) = $84.64, is loaned out again and eventually redeposited. Of the $84.64, the bank again loans out 92%, and so on.

    (a) Find the total amount of money deposited in the bank as a result of these transactions.
    (b) The total amount of money deposited divided by the original deposit is called the *credit multiplier*. Calculate the credit multiplier for this example and explain what this number tells us.

[2] www.dallasnews.com/sharedcontent/dws/bus/stories/DN-obsoletedollarbill_22bus.State.Edition1.c8a2b4.html.

16. To stimulate the economy in 2002, the government gave a tax rebate totaling 40 billion dollars. Find the total additional spending resulting from this tax rebate if everyone who receives money spends

(a) 80% of it **(b)** 90% of it

17. To stimulate the economy in 2008, the government gave a tax rebate totaling 100 billion dollars. Find the total additional spending resulting from this tax rebate if everyone who receives money spends

(a) 80% of it **(b)** 90% of it

18. In April 2009, economists predicted that each dollar in tax cuts would generate \$3 in economic growth. What spending rate was assumed in arriving at this estimated multiplier effect?

19. A person who deposits money in a bank account starts a long process described by the reserve-deposit ratio, r. For every dollar deposited, the bank keeps r dollars and lends $1 - r$ dollars to someone else, who deposits the loan in a bank account. The same fraction of the second deposit is loaned out, to be deposited in turn, and so on. If the initial deposit is N dollars, find the total value of the bank accounts generated by this deposit:

(a) After the second deposit
(b) After the third deposit
(c) If the process continues forever

11.3 APPLICATIONS TO THE NATURAL SCIENCES

Steady-State Drug Levels

A patient is given a drug at regular intervals. In Section 11.1, we saw that in the long run, the quantity of drug in the body varies between a maximum level right after a dose and a minimum level right before a dose. At this *steady state*, the quantity eliminated between doses is equal to one dose. (See Problem 19.)

Example 1 A child with an ear infection is given a 200-mg ampicillin tablet once every 4 hours. About 12% of the drug in the body at the start of a four-hour period is still there at the end of that period. What is the quantity of ampicillin in the body

(a) Right after taking the 3rd tablet? (b) Right after taking the 6th tablet?

(c) At the steady state, right after and right before taking a tablet?

Solution Let Q_n be the quantity of ampicillin, in mg, in the body right after taking the nth tablet. Then

$$Q_n = 200 + 200(0.12) + 200(0.12)^2 + \cdots + 200(0.12)^{n-1}.$$

(a) Using the formula for the sum of a finite geometric series with $a = 200$, $r = 0.12$, and $n = 3$, we have

$$Q_3 = \frac{a(1-r^n)}{1-r} = \frac{200(1-(0.12)^3)}{1-0.12} = 226.88 \text{ mg}.$$

(b) Using $n = 6$, we have

$$Q_6 = \frac{a(1-r^n)}{1-r} = \frac{200(1-(0.12)^6)}{1-0.12} = 227.2720 \text{ mg}.$$

(c) At the steady state, the quantity Q in mg, right after a tablet is taken is the sum of the infinite geometric series

$$Q = 200 + 200(0.12) + 200(0.12)^2 + 200(0.12)^3 + \cdots.$$

Since $r = 0.12$ and $-1 < r < 1$, the series converges to a finite sum. At the steady state

$$\text{Quantity right after a dose } = Q = \frac{a}{1-r} = \frac{200}{1-0.12} = 227.2727 \text{ mg}.$$

The quantity of ampicillin in the body right before a dose is exactly 200 mg less than the quantity right after the dose, so at the steady state

$$\begin{aligned}\text{Quantity before dose } &= \text{Quantity after dose} - \text{Size of one dose}\\ &= 227.2727 - 200\\ &= 27.2727 \text{ mg}.\end{aligned}$$

Problem 4 shows how this relationship can be used to calculate Q, the long-run quantity of ampicillin right after a dose.

Example 2 Valproic acid, a drug used to control epilepsy, has a half-life of 15 hours in the body. If D mg of valproic acid is taken every 12 hours, at the steady state, what quantity of the drug is in the body right after taking a tablet?

Solution After a dose D of valproic acid, the quantity, Q, in the body decays exponentially, so $Q = Db^t$, where t is time in hours. The half-life is 15 hours, so

$$\begin{aligned} 0.5D &= Db^{15} \\ 0.5 &= b^{15} \\ b &= (0.5)^{1/15}. \end{aligned}$$

Right before the next dose at $t = 12$, we have

$$\text{Fraction of drug remaining after 12 hours } = b^{12} = ((0.5)^{1/15})^{12} = (0.5)^{12/15} = (0.5)^{0.8}.$$

Since D mg of valproic acid is given every 12 hours, in the long run

$$\text{Quantity right after dose } = D + D(0.5)^{0.8} + D\left((0.5)^{0.8}\right)^2 + \cdots.$$

This is an infinite geometric series with $r = (0.5)^{0.8} = 0.57435$. Since $-1 < r < 1$, the series converges and its sum is

$$\text{Quantity right after dose } = \frac{D}{1-(0.5)^{0.8}} = 2.35D.$$

Thus, in the long run, the quantity of the drug right after taking a dose is about 2.35 times the dose.

Accumulation of Toxins in the Body

Toxins found in pesticides can get into the food chain and accumulate in people's bodies through the food they eat. We use a geometric series to calculate the total accumulation of toxin in the body.

Example 3 Every day, a person consumes 5 micrograms (μg) of a toxin, which leaves the body at a continuous rate of 2% per day. In the long run, how much toxin is in the body at the end of the day?

Solution Since the toxin leaves the body at a continuous rate of 2% every day, the 5 μg amount consumed one day earlier has decayed to $5e^{-0.02}$, the 5 μg consumed two days earlier has decayed to $5e^{-0.02(2)} = 5(e^{-0.02})^2$, and so on. We assume that the 5 μg consumed during the day has not yet decayed. In the long run, at the end of each day we have, in micrograms,

$$\text{Total accumulation of toxin } = 5 + 5(e^{-0.02}) + 5(e^{-0.02})^2 + 5(e^{-0.02})^3 + \cdots.$$

This is an infinite geometric series with $a = 5$ and $r = e^{-0.02} = 0.9802$. Since $-1 < r < 1$, the series converges and its sum is

$$\text{Total accumulation of toxin } = \frac{a}{1-r} = \frac{5}{1-e^{-0.02}} = 252.5 \text{ micrograms.}$$

In the long run, there are 252.5 micrograms of toxin in the body at the end of the day.

Depletion of Natural Resources

Geometric series can be used to estimate how long a natural resource (such as oil) will last, assuming that usage levels increase at a constant percentage rate.

Example 4 At the end of 2008, world oil reserves were about 1950 billion barrels.[3] During 2008, about 29.3 billion barrels of oil were consumed.[4] Over the past decade, oil consumption has been increasing at about 1% per year.[5] Assuming yearly oil consumption increases at this rate in the future, how long will the reserves last?

[3] www.theoildrum.com/node/5395.
[4] *CIA World Factbook 2008*.
[5] www.eia.doe.gov/emeu/international/oilcomsumption.html.

Solution Under these assumptions, the oil used in 2009, in billions of barrels, is predicted to be $29.3(1.01)$. In 2010, we predict $29.3(1.01)^2$ billion barrels to be used, and $29.3(1.01)^3$ the next year, and so on. Thus, starting with 2009, the total quantity of oil used in n years, Q_n, is

$$Q_n = 29.3(1.01) + 29.3(1.01)^2 + 29.3(1.01)^3 + \cdots + 29.3(1.01)^n \text{ billions of barrels.}$$

This is a finite geometric series with n terms where $a = 29.3(1.01)$ and $r = 1.01$. (Notice that the last term can be written $29.3(1.01)(1.01)^{n-1}$.) The sum is

$$Q_n = \frac{a(1-r^n)}{1-r} = 29.3(1.01)\left(\frac{1-(1.01)^n}{1-1.01}\right) = 2959((1.01)^n - 1) \text{ billions of barrels.}$$

We want to find the value of n for which Q_n reaches the total reserves, 1950. Solving gives

$$\begin{aligned} 2959((1.01)^n - 1) &= 1950 \\ (1.01)^n - 1 &= \frac{1950}{2959} = 0.6590 \\ (1.01)^n &= 1.6590. \end{aligned}$$

Taking logarithms and using $\ln(A^p) = p\ln A$, we have

$$\begin{aligned} n\ln(1.01) &= \ln(1.6590) \\ n &= \frac{\ln(1.6590)}{\ln(1.01)} = 50.9 \text{ years.} \end{aligned}$$

Thus, if present consumption patterns are maintained, the world's oil supply will be exhausted in about 51 years. However, if consumption patterns change, the length of time until the reserves run out can be very different. Problem 2 concerns predictions using a 2.5% yearly increase and a 5.5% yearly decrease, the maximum and minimum figures for the past decade.

Geometric Series and Differential Equations

In this section, we used geometric series to model drug levels; in Chapter 10, we used differential equations to model drug levels. How do we know whether to use a geometric series or a differential equation? The answer depends on whether the drug is given in *discrete* doses (such as a tablet each morning) or *continuously* (such as intravenously).

Example 5 A patient receives 25 mg of a drug each day, and the drug is metabolized and eliminated at a continuous rate of 10% per day. Find the quantity of drug in the patient's body in the long run:

(a) Using a geometric series, assuming the 25-mg dose of the drug is administered in a single injection each morning. (Find the quantity both before and after an injection is given.)
(b) Using a differential equation, assuming the 25-mg dose of the drug is administered continuously throughout the day, using an intravenous drip.

Solution (a) A 25-mg injection is given each day. Since the drug is metabolized at a continuous rate of 10% per day, the quantity remaining a day later is $25e^{-0.1}$ mg. The quantity remaining two days later is $25(e^{-0.1})^2$ mg. In the long run,

$$\text{Quantity after injection} = 25 + 25(e^{-0.1}) + 25(e^{-0.1})^2 + 25(e^{-0.1})^3 + \cdots \text{ mg.}$$

We use the formula for the sum of an infinite geometric series with $a = 25$ and $r = e^{-0.1}$:

$$\text{Quantity after injection} = \frac{a}{1-r} = \frac{25}{1-e^{-0.1}} = 262.7 \text{ mg.}$$

Since each injection is 25 mg, in the long run

$$\text{Quantity before injection} = 262.7 - 25 = 237.7 \text{ mg.}$$

(b) The drug is entering the body at the continuous rate of 25 mg per day and leaving at the continuous rate of 0.1 times the current level in the body. Thus, if Q is the quantity of drug in the body after t days, then

$$\frac{dQ}{dt} = \text{Rate in} - \text{Rate out} = 25 - 0.1Q.$$

In Chapter 10, we saw that Q tends toward the equilibrium solution which occurs when $dQ/dt = 0$. Then

$$\frac{dQ}{dt} = 25 - 0.1Q = 0$$
$$Q = 250 \text{ mg}.$$

When the drug is given continuously, the quantity of the drug in the body levels off at 250 mg.

Problems for Section 11.3

1. In 2008, world oil consumption was 29.3 billion barrels,[6] a decrease of 5.5% from 2007. Assuming that consumption continues to decrease at the same percentage rate, make a table showing yearly consumption between 2008 and 2017, inclusive. Find the total quantity of oil consumed during this decade.

2. As in Example 4, assume that oil reserves at the end of 2008 were 1950 billion barrels and that consumption in 2008 was 29.3 billion barrels. What happens in the long run if oil consumption

(a) Decreases by 5.5% per year
(b) Increases by 2.5% per year

3. Every morning, a patient receives a 50-mg injection of a drug. At the end of a 24-hour period, 60% of the drug remains in the body. What quantity of drug is in the body

(a) Right after the 3rd injection?
(b) Right after the 7th injection?
(c) Right after an injection, at the steady state?

4. This problem gives another way of finding, Q, the long-run ampicillin quantity right after a dose in Example 1. (See page 463.) In the four hours after a dose, 200 mg of ampicillin must be excreted, as the next 200-mg tablet elevates the ampicillin quantity back to Q mg. Use this information to solve for Q.

5. A dose of 120 mg is taken by a patient at the same time every day. In one day, 30% of the drug is excreted.

(a) At the steady state, find the quantity of drug in the body right after a dose.
(b) Check that at the steady state, the quantity excreted in one day is equal to the dose.

6. At the same time every day, a patient takes 50 mg of the antidepressant fluoxetine, whose half-life is 3 days.

(a) What fraction of the dose remains in the body after a 24-hour period?
(b) What is the quantity of fluoxetine in the body right after taking the 7th dose?
(c) In the long run, what is the quantity of fluoxetine in the body right after a dose?

7. A person with chronic pain takes a 30 mg tablet of morphine every 4 hours. The half-life of morphine is 2 hours.

(a) How much morphine is in the body right after and right before taking the 6th tablet?
(b) At the steady state, find the quantity of morphine in the body right after and right before taking a tablet.

8. **(a)** An allergy drug with a half-life of 18 weeks is given in 100-mg doses once a week. At the steady state, find the quantity of the drug in the body right after a dose.
(b) The drug does not become effective until the quantity in the body right after a dose reaches 2000 mg. How many weeks after the first dose does the drug become effective?

9. A cigarette puts 1.2 mg of nicotine into the body. Nicotine leaves the body at a continuous rate of 34.65% per hour, but more than 60 mg can be lethal. If a person smokes a cigarette with each of the following frequencies, find the long-run quantity of nicotine in the body right after a cigarette. Does the nicotine reach the lethal level?

(a) Every hour
(b) Every half hour
(c) Every 15 minutes
(d) Every 6 minutes
(e) Every 3 minutes

10. Each day at lunch a person consumes 8 micrograms of a toxin found in a pesticide; the toxin is metabolized at a continuous rate of 0.5% per day. In the long run, how much of this toxin accumulates in the person's body? Give the quantities right after and right before lunch.

11. At the end of 2008, the total reserve of a mineral was 350,000 m^3. In the year 2009, about 5000 m^3 was used. Each year, consumption of the mineral is expected to increase by 8%. Under these assumptions, in how many years will all reserves of the mineral be depleted?

[6] *Statistical Review of World Energy 2008*, www.bp.com.

12. We use 1500 kg of a mineral this year and consumption of the mineral is increasing annually by 4%. The total reserves of the mineral are estimated to be 120,000 kg. Approximately when will the reserves run out?

13. At the end of 2007, natural gas reserves were 180 trillion m^3; during 2007, about 3 trillion m^3 of natural gas were consumed.[7] Estimate how long natural gas reserves will last if consumption increases at 2% per year.

14. Over the past decade, natural gas consumption has been increasing at between 0% and 5% a year. Using the data from Problem 13, estimate how long the natural gas reserves will last assuming the rate of increase is

(a) 0% (b) 5%

Problems 15–17 concern how long reserves of the mineral in Problem 11 last if usage patterns change. For example, as reserves get lower, substitutes may be developed.

15. How long will the reserves last if the annual increase in usage is 4%?

16. How long will the reserves last if the annual usage stays constant at 5000 m^3 per year?

17. How long will the reserves last if the usage decreases each year by 4%?

18. (a) A dose D of a drug is administered at intervals equal to the half-life. (That is, the second dose is given when half the first dose remains.) At the steady state, find the quantity of drug in the body right after a dose.
(b) If the quantity of a drug in the body after a dose is 300 mg at the steady state and if the interval between doses equals the half-life, what is the dose?

19. A dose, D, of a drug is taken at regular time intervals, and a fraction r remains after one time interval. Show that at the steady state, the quantity of the drug excreted between doses equals the dose.

20. Cephalexin is an antibiotic with a half-life in the body of 0.9 hours, taken in tablets of 250 mg every six hours.

(a) What percentage of the cephalexin in the body at the start of a six-hour period is still there at the end (assuming no tablets are taken during that time)?
(b) Write an expression for Q_1, Q_2, Q_3, Q_4, where Q_n mg is the amount of cephalexin in the body right after the nth tablet is taken.
(c) Express Q_3, Q_4 in closed form and evaluate them.
(d) Write an expression for Q_n and put it in closed form.
(e) If the patient keeps taking the tablets, use your answer to part (d) to find the quantity of cephalexin in the body in the long run, right after taking a tablet.

CHAPTER SUMMARY

- **Geometric series**
  Finite and infinite.
- **Sums of geometric series**
  Partial sums, convergence of infinite series.
- **Applications to business and economics**
  Annuities, present value, multiplier effect, market stabilization.
- **Applications to life sciences**
  Repeated drug doses, accumulation of toxins, depletion of natural resources.

REVIEW PROBLEMS FOR CHAPTER ELEVEN

In Problems 1–8, find the sum, if it exists.

1. $2 + 2^2 + 2^3 + \cdots + 2^{10}$
2. $20 + 20(1.4) + 20(1.4)^2 + \cdots + 20(1.4)^8$
3. $1000 + 1000(1.08) + 1000(1.08)^2 + 1000(1.08)^3 + \cdots$
4. $500 + 500(0.6) + 500(0.6)^2 + \cdots + 500(0.6)^{15}$
5. $30 + 30(0.85) + 30(0.85)^2 + 30(0.85)^3 + \cdots$
6. $25 + 25(0.2) + 25(0.2)^2 + 25(0.2)^3 + \cdots$
7. $1 + \frac{1}{2} + \frac{1}{2^2} + \cdots + \frac{1}{2^8}$
8. $1 + \frac{1}{3} + \frac{1}{3^2} + \frac{1}{3^3} + \cdots$
9. Around January 1, 1993, Barbra Streisand signed a contract with Sony Corporation for \$2 million a year for 10 years. Suppose the first payment was made on the day of signing and that all other payments were made on the first day of the year. Suppose also that all payments were made into a bank account earning 4% a year, compounded annually.

(a) How much money was in the account
  (i) On the night of December 31, 1999?
  (ii) On the day the last payment was made?
(b) What was the present value of the contract on the day it was signed?

10. A drug is given in daily doses of 100 mg. After 24 hours, 82% of the previous day's dose remains in the body. What is the long-run quantity of drug in the body, right after and right before a dose is given?

[7] www.worldenergyoutlook.org/docs/weo2008/weo2008_es_english.pdf

11. A used car costs \$15,000. The repair contract costs \$500 at the end of the first year and increases by 20% at the end of each subsequent year. Find the total cost of owning the car for ten years. Include the payment at the end of the tenth year.

12. A donor sets up an endowment to fund an annual scholarship of \$10,000. The endowment earns 6% interest per year, compounded annually. Find the amount that must be deposited now if the endowment is to fund one award each year, with one award now and continuing

(a) Until twenty awards have been made
(b) Forever

13. This problem shows how to estimate the cumulative effect of a tax cut on a country's economy. Suppose the government proposes a tax cut totaling \$100 million. We assume that all the people who have extra money spend 80% of it and save 20%. Thus, of the extra income generated by the tax cut, \$100(0.8) million = \$80 million is spent and becomes extra income to someone else. These people also spend 80% of their additional income, or \$80(0.8) million, and so on. Calculate the total additional spending created by such a tax cut.

14. To stimulate the economy, the government gives a tax rebate totaling 5 billion dollars. Find the total additional spending resulting from this tax rebate if everyone who receives money spends

(a) 80% of it (b) 90% of it

15. A government gives a tax rebate of N dollars to stimulate the economy. Everyone who receives money spends a fixed fraction, k, of the money received, with $0 < k < 1$.

(a) Find a formula (in terms of N and k) for the total additional spending resulting from the tax rebate.
(b) If $k = 0.85$, what is the total additional spending as a multiple of the size, N, of the tax rebate?

16. Before World War I, the British government issued what are called *consols*, which pay the owner or his heirs a fixed amount of money every year forever. (Cartoonists of the time described aristocrats living off such payments as "pickled in consols.") What should a person expect to pay for consols which pay £10 a year forever? Assume the first payment is one year from the date of purchase and that interest remains 4% per year, compounded annually. (£ denotes pounds, the British unit of currency.)

17. A repeating decimal can always be expressed as a fraction. This problem shows how writing a repeating decimal as a geometric series enables you to find the fraction.

(a) Write the repeating decimal 0.232323... as a geometric series using the fact that $0.232323\ldots = 0.23 + 0.0023 + 0.000023 + \cdots$.
(b) Use the formula for the sum of a geometric series to show that $0.232323\ldots = 23/99$.

18. A ball is dropped from a height of 10 feet and bounces. Each bounce is $\frac{3}{4}$ of the height of the bounce before. Thus, after the ball hits the floor for the first time, the ball rises to a height of $10(\frac{3}{4}) = 7.5$ feet, and after it hits the floor for the second time, it rises to a height of $7.5(\frac{3}{4}) = 10(\frac{3}{4})^2 = 5.625$ feet.

(a) Find an expression for the height to which the ball rises after it hits the floor for the nth time.
(b) Find an expression for the total vertical distance the ball has traveled when it hits the floor for the first, second, third, and fourth times.
(c) Find an expression for the total vertical distance the ball has traveled when it hits the floor for the nth time. Express your answer in closed form.

19. One way of valuing a company is to calculate the present value of all its future earnings. A farm expects to sell \$1000 worth of Christmas trees once a year forever, with the first sale in the immediate future. What is the present value of this Christmas tree business? The interest rate is 4% per year, compounded continuously.

20. Every year, a company sells 1000 units of a product while 20% of the total number in use fail. Assume sales are at the start of the year and failures are at the end of the year.

(a) Find the market stabilization point for this product.
(b) If the stabilization point is approached very slowly, the number of units in use may not get close to this value because market conditions change first. Make a table for S_n, the number of units in use right after the nth annual sale, for $n = 5, 10, 15, 20$, to see how rapidly this market approaches the stabilization point.

CHECK YOUR UNDERSTANDING

In Problems 1–30, indicate whether the statement is true or false.

1. The sum $1 + 2 + 4 + 8 + 16 + 32$ is a finite geometric series with 6 terms.

2. The sum $1 + 2 + 4 + 8 + 16 + 32 + \cdots + 2^{10}$ is a finite geometric series with 10 terms.

3. The sum $1 + 3 + 5 + 7 + 9 + 11$ is a finite geometric series with 6 terms.

4. The sum $1 - 2 + 4 - 8 + 16 - 32$ is a finite geometric series with 6 terms.

5. The sum of $1 + (1/3) + (1/9) + \cdots + (1/3)^{10}$ is $(1 - (1/3)^{11})/(1 - (1/3))$.

6. The sum of $3 \cdot 1 + 3 \cdot 2 + 3 \cdot 4 + \cdots + 3(2^{20})$ is $3(1 - 2^{20})/(1 - 2)$.

7. The sum of the infinite geometric series $3 + 3^2 + 3^3 + \cdots + 3^n + \cdots$ is $3/(1 - 3)$.

8. The sum of the infinite geometric series $(1/2)+(1/2)^2+(1/2)^3+\cdots+(1/2)^n+\cdots$ is $1/(1-(1/2))$.
9. The sum of the infinite geometric series $(1/3)+(1/3)^2+(1/3)^3+\cdots+(1/3)^n+\cdots$ is $1/2$.
10. The sum of the infinite geometric series $5(1/2)+5(1/2)^2+5(1/2)^3+\cdots+5(1/2)^n+\cdots$ is 5.
11. An annuity is a sequence of equal payments or deposits made at regular intervals.
12. The present value of an annuity is the amount of money deposited today to make a series of fixed payments in the future.
13. The present value of an annuity that makes payments of \$6000 per year for 10 years is less than \$60,000, assuming an annual interest rate of 3%.
14. The present value of a series of constant annual payments made forever is infinite.
15. If annual deposits of \$3000 are made at the beginning of each year into an account paying 2% interest per year, compounded annually, then the account balance after the fourth deposit is $3000(1.02)^4+3000(1.02)^3+3000(1.02)^2+3000(1.02)$ dollars.
16. If annual deposits of \$3000 are made at the beginning of each year into an account paying 2% interest per year, compounded annually, then the account balance after the tenth deposit is $3000(1-(1.02)^{10})/(1-1.02)$ dollars.
17. A deposit of \$735,000 into an account earning 5% compounded annually is sufficient to generate annual payments of \$35,000 in perpetuity, starting now.
18. The present value of a payment 5 years in the future of \$2000, assuming annual interest rate of 3%, is $2000(1.03)^5$ dollars.
19. The present value of 3 annual payments of \$600, starting now, assuming annual interest rate of 4%, is $600+600(1.04)^{-1}+600(1.04)^{-2}$ dollars.
20. The present value of 10 annual payments of \$600, starting now, assuming annual interest rate of 4%, is $600(1-(1.04)^{-10})/(1-(1.04)^{-1})$ dollars.
21. If a person consumes a 25-mg tablet of a drug four times throughout the day, at the end of the day the person will have 100 mg of the drug in the body.
22. If the quantity of a drug in the body has reached the steady-state level for a person taking a drug at regular intervals, then the quantity of drug in the person's body stays constant.
23. A person consuming 3 micrograms of a toxin with breakfast each day, of which 5% leaves the body each day, has 3/(1-0.05) micrograms of toxin in the body right after breakfast, in the long run.
24. If a country used 10 billion barrels of oil last year, and oil consumption increases at 2% annually, then over the next 5 years, the total oil consumption will be $10(1.02)(1-(1.02)^5)/(1-1.02)$ billion barrels.
25. In a person taking a constant dose D mg of medication everyday, the steady-state level is the same before and after each daily dose.
26. In a person taking a constant dose D mg of medication every day, at the steady-state level the quantity of the drug eliminated daily is equal to the daily dose D.
27. If a drug is administered intravenously, then a geometric series is a better model than a differential equation for the quantity of drug in the bloodstream.
28. If a certain drug has a half-life of three hours in the body, then nine hours after a dosage of 160 mg, the quantity in the body drops to 20 mg.
29. If a 50-mg injection of a drug is given once every day, and the drug is metabolized at a continuous rate of 5% per day, then the long-term steady-state quantity in the body after injection is $50/(1-e^{0.05})$.
30. If a drug is administered intravenously at a rate of 50 mg per day, and the drug is metabolized at a continuous rate of 5% per day, then the long-term steady-state quantity in the body is 1000 mg.

PROJECTS FOR CHAPTER ELEVEN

1. Do You Have Any Common Ancestors?

In this project, we estimate the number of ancestors you have and determine whether you have any common ancestors. (A common ancestor is one who appears on two sides of your family tree. For example, if your great-grandmother on your mother's mother's side is also your grandmother on your father's side, then she would be a common ancestor.)

(a) In general, each person has 2 biological parents, 4 biological grandparents, 8 biological great-grandparents, and so on. Write a formula for the number of ancestors you have, going back n generations.

(b) How long is one generation? Estimate the age of typical parents when a baby is born. This is the length of time for a generation. How many generations are included if we go back 100 years? 500 years? 1000 years? 2000 years?

(c) Use your answers to parts (a) and (b) to estimate the number of ancestors you have if we go back 100 years, 500 years, 1000 years, or 2000 years.

(d) In parts (a) and (c), we counted every ancestor separately, so we assumed that you have no common ancestors. Use the fact that the population of the world was about 6 billion people in 1999 and was about 200 million people in the year 1 AD to determine whether this is a reasonable assumption. Explain your reasoning.

2. Harrod-Hicks Model of an Expanding National Economy

In an expanding national economy, the Harrod-Hicks model relates the national income in one year to the national income in the preceding year. If $f(n)$ is the national income in year n, then the model predicts, for some constants k and h with $k > 1$ and $h > 0$, that

$$f(n+1) = kf(n) - h.$$

(a) Let $C = f(0)$. Write $f(1)$, $f(2)$, and $f(3)$ in terms of k, h, and C.

(b) Show that

$$f(1) = kC - h,$$

$$f(2) = k^2C - (1+k)h,$$

$$f(3) = k^3C - (1+k+k^2)h.$$

Use these formulas to guess a formula for $f(n)$.

(c) Use the formula for the sum of a finite geometric series to rewrite the formula for $f(n)$ in closed form.

3. Probability of Winning in Sports

In certain sports, winning a game requires a lead of two points. That is, if the score is tied you have to score two points in a row to win.

(a) For some sports (e.g. tennis), a point is scored every play. Suppose your probability of scoring the next point is always p. Then, your opponent's probability of scoring the next point is always $1 - p$.

(i) What is the probability that you win the next two points?

(ii) What is the probability that you and your opponent split the next two points, that is, that neither of you wins both points?

(iii) What is the probability that you split the next two points but you win the two after that?

(iv) What is the probability that you either win the next two points or split the next two and then win the next two after that?

(v) Give a formula for your probability w of winning a tied game.

(vi) Compute your probability of winning a tied game when $p = 0.5$; when $p = 0.6$; when $p = 0.7$; when $p = 0.4$. Comment on your answers.

(b) In other sports (e.g. volleyball prior to 1999), you can score a point only if it is your turn, with turns alternating until a point is scored. Suppose your probability of scoring a point when it is your turn is p, and your opponent's probability of scoring a point when it is her turn is q.

(i) Find a formula for the probability S that you are the first to score the next point, assuming it is currently your turn.

(ii) Suppose that if you score a point, the next turn is yours. Using your answers to part (a) and your formula for S, compute the probability of winning a tied game (if you need two points in a row to win).

- Assume $p = 0.5$ and $q = 0.5$ and it is your turn.
- Assume $p = 0.6$ and $q = 0.5$ and it is your turn.

Section 1.1

1 (a) (IV)
(b) (II)
(c) (III)

3 Argentina produced 9 million metric tons of wheat in 2002

5
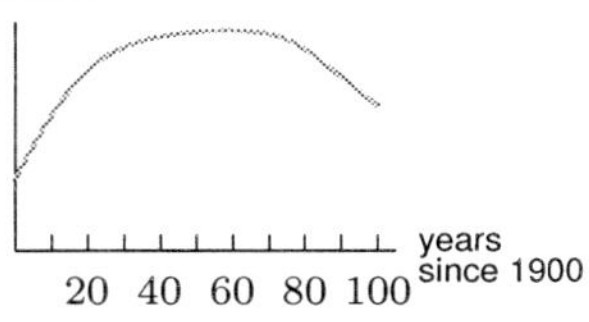

7 $f(5) = 13$

9 $f(5) = 3$

11 $f(5) = 4.1$

13 (b) CFC consumption in 1987
(c) Year CFC consumption is zero

17
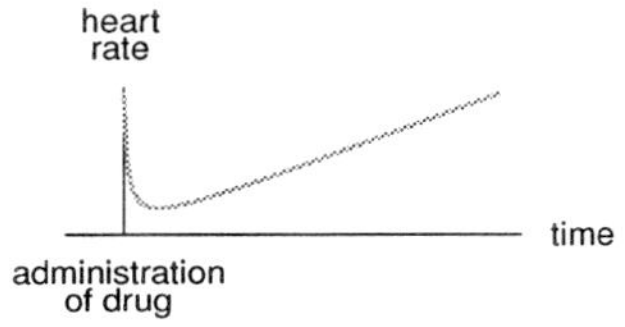

19
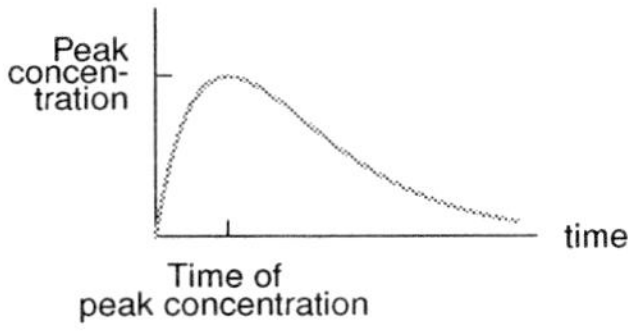

21
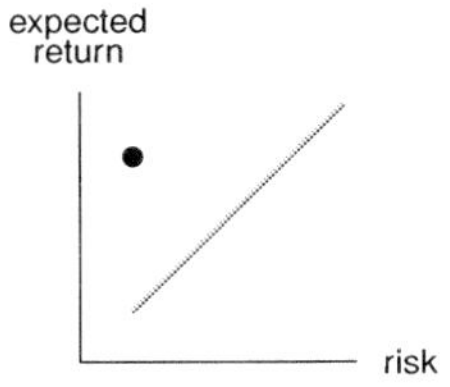

23 (a) (III)
(b) Potato's temperature before put in oven

25
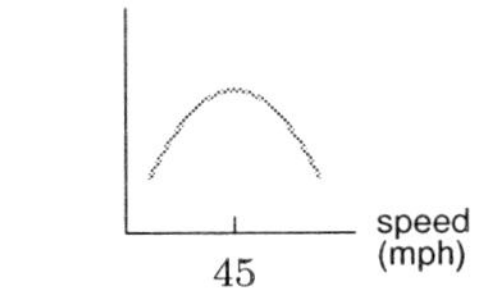

Section 1.2

1 Slope: $-12/7$
Vertical intercept: $2/7$

3 Slope: 2
Vertical intercept: $-2/3$

5 $y = (1/2)x + 2$

7 $y = (1/2)x + 2$

9 (a) l_1
(b) l_3
(c) l_2
(d) l_4

11 (a) $P = 30{,}700 + 850t$
(b) 39,200 people
(c) In 2016

13 (a) $C_1 = 40 + 0.15m$
$C_2 = 50 + 0.10m$
(b)
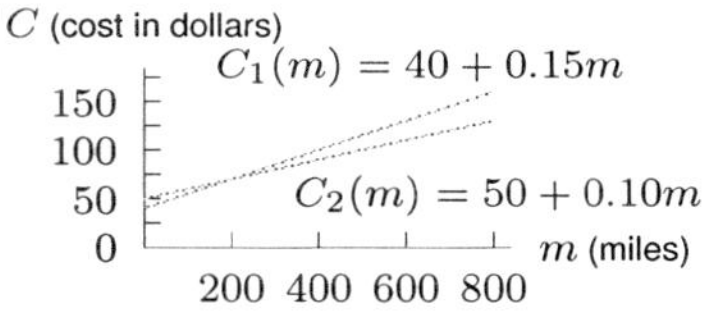

(c) For distances less than 200 miles, C_1 is cheaper.
For distances more than 200 miles, C_2 is cheaper.

15 (a) 1.8 billion dollars/year
(b) 19.1 billion dollars
(c) 28.1 billion dollars
(d) 2011

17 (a) Linear
(b) Linear
(c) Not linear

19 (a) $q = -(1/3)p + 8$
(b) $p = -3q + 24$

21 (a) $P = 11.3 + 0.4t$
(b) 13.7%
(c) 1.4%

23 (b) $P = 100 - 0.5d$
(c) -0.5%/ft
(d) 100%; 200 ft

25 (a) $C = 3.68 + 0.12w$
(b) 0.12 \$/gal
(c) \$3.68

27 (a) $\Delta w/\Delta h$ constant
(b) $w = 5h - 174$; 5 lbs/in
(c) $h = 0.2w + 34.8$; 0.2 in/lb

29 (c)

31 (a) 60, 40 years
(b) (ii)
(c) 6.375 beats/minute more under new formula

33 No

Section 1.3

1 Concave down

3 Concave up

5 Increases by 12.5%

7 Decreases by 6%

9 Decreasing
Concave up

11
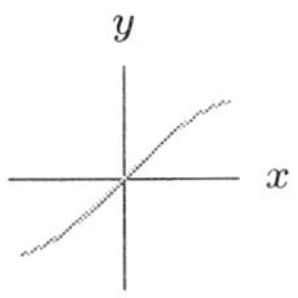

13 -3

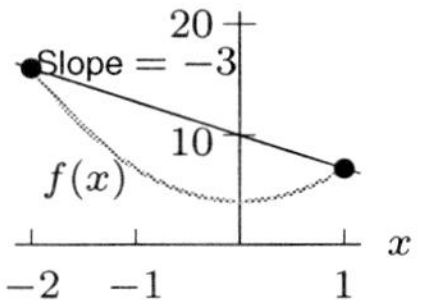

15 (a) 90 million bicycles
(b) 1.8 million bicycles per year

17 (a) 115,000 people/year
(b) 0.07, 0.08, 0.41, 0.06
(c) 115,000 people/year

19 (a) \$18,280 million
(b) \$3656 million per year

21 1 meter/sec

23 $72/7 = 10.286$ cm/sec

25 (a) Negative
(b) Positive
(c) Negative
(d) Negative
(e) Positive

27 1490 thousand people/year
912.9 thousand people/year
1879 thousand people/year

29 (a) $-\$35$ billion dollars
(b) $-\$7$ billion dollars per year
(c) Yes; 2006–2007, 2007–2008

31 (a) Negative
(b) -0.087 mg/hour

33 15.468, 57.654, 135.899, 146.353, 158.549 people/min

35 (a) -11 cm/sec
(b) -5.5 (cm/sec)/kg

37 (a) Concave up; no
(b) 2.6 m/sec

39 Decreasing, concave down

41 (a)
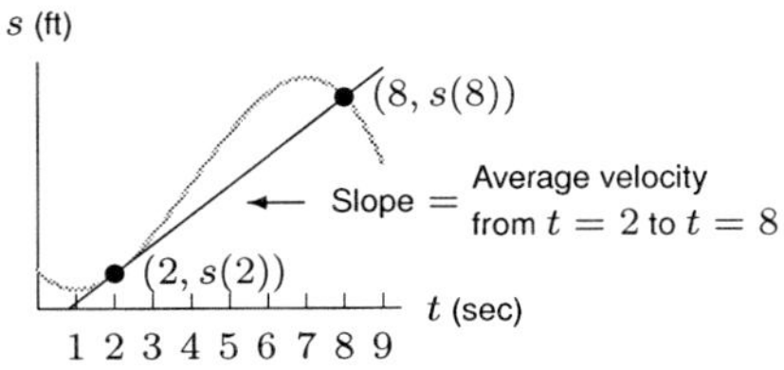

(b) Between $t = 3$ and $t = 6$
(c) Negative

43 The change in 1800-1810

45 The increase from \$100,000 to \$500,000

47 Decrease 12.8%

49

Year	2005	2006	2007	2008
Inflation	4.0%	2.1%	4.3%	0.0%

Section 1.4

1 (a) When more than roughly 335 items are produced and sold
(b) About \$650

3 (a) \$75; \$7.50 per unit
(b) \$150

5 (a) Price \$12, sell 60
(b) Decreasing

7 Vertical intercept: $p = 4$ dollars
Horizontal intercept: $q = 6$ units

9 (a) $C(q) = 5000 + 30q$
$R(q) = 50q$
(b) \$30/unit, \$50/unit
(c)

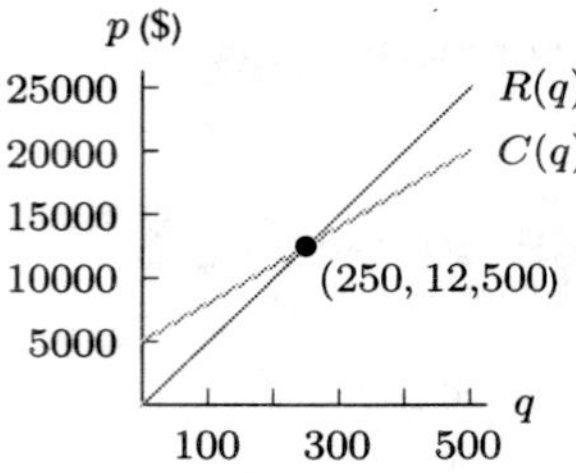

(d) 250 chairs and \$12,500

11 (a) \$4000
(b) \$2
(c) \$10
(d)

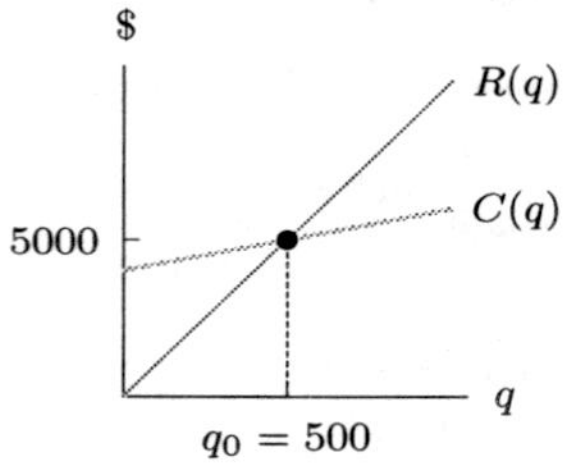

(e) 500

13 (a) When there are are more than 1000 customers
(b)

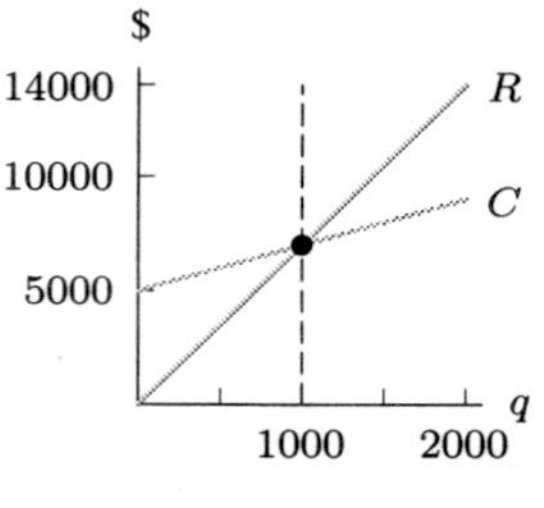

15 (a) $C(q) = 650,000 + 20q$
$R(q) = 70q$
$\pi(q) = 50q - 650,000$
(b) \$20/pair, \$70/pair, \$50/pair
(c) More than 13,000 pairs

17 (a) Between 20 and 60 units
(b) About 40 units

19 (a) $V(t) = -1500t + 15,000$
(b) $V(3) = \$10,500$

21 (a) $V(t) = -2000t + 50,000$
(b)

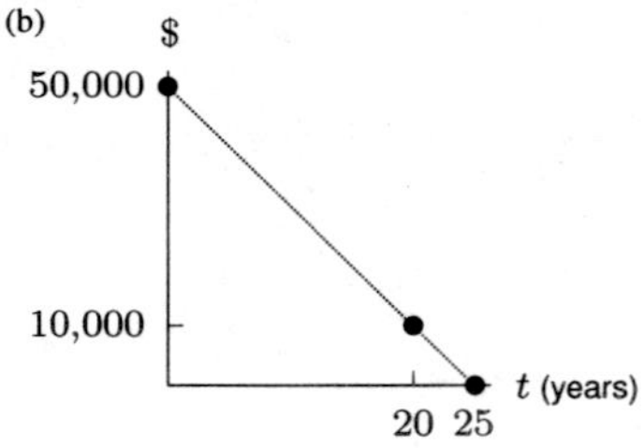

(c) (0 years, \$50,000) and (25 years, \$0)

23 (a) $p = \$10$; $q = 3000$
(b) Suppliers produce 3500 units; Consumers buy 2500
(c) Suppliers produce 2500 units; Consumers buy 3500

25 (a) $C = 5q + 7000$
$R = 12q$
(b) $q = 1520$, $\pi(12) = \$3640$
(c) $C = 17,000 - 200p$
$R = 2000p - 40p^2$
$\pi(p) = -40p^2 + 2200p - 17,000$
(d) At \$27.50 per shirt the profit is \$13,250

27 (a) $q = 820 - 20p$
(b) $p = 41 - 0.05q$

29

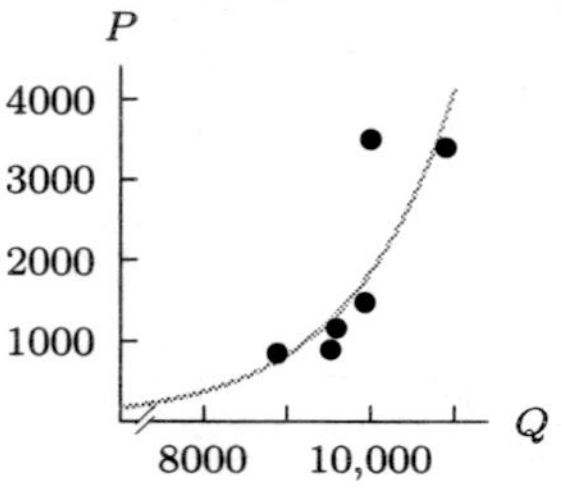

31 (a) $40b + 10s = 1000$
(b)

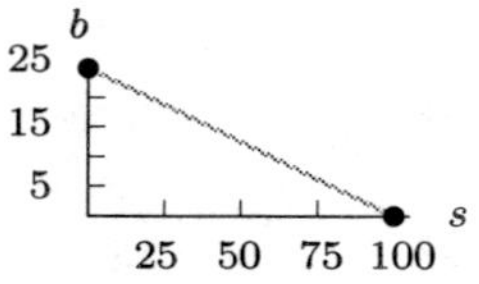

(c) The intercepts are $(0, 25)$ and $(100, 0)$

33 (a)

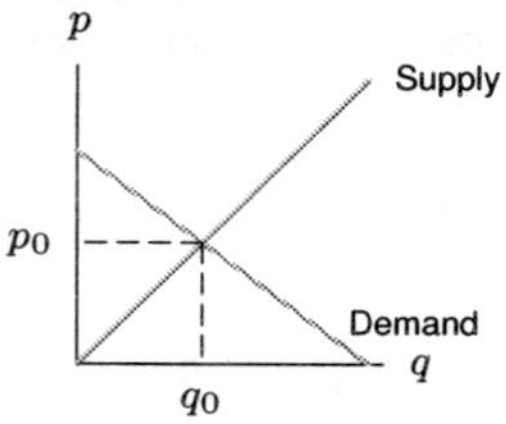

(b) Equilibrium price will increase; equilibrium quantity will decrease
(c) Equilibrium price and quantity will decrease

35 $q = 4p - 28$

37 (a) $p = 100$, $q = 500$
(b) $p = 102$, $q = 460$
(c) Consumer pays \$2
Producer pays \$4
(d) \$2760

39 (a) Demand: $q = 100 - 2p$
Supply: $q = 2.85p - 50$
(b) New equilibrium price $p \approx \$30.93$
New equilibrium quantity $q \approx 38.14$ units
(c) Consumer pays \$0.93
Producer pays \$0.62
Total \$1.55
(d) \$59.12

Section 1.5

1 (a) (i), 12%
(b) (ii), 1000
(c) Yes, (iv)

3 (a) II
(b) I
(c) III
(d) V

5 (a) $G = 310(1.03)^t$
(b) $G = 310 + 8t$

7 (a) $Q = 30 - 2t$

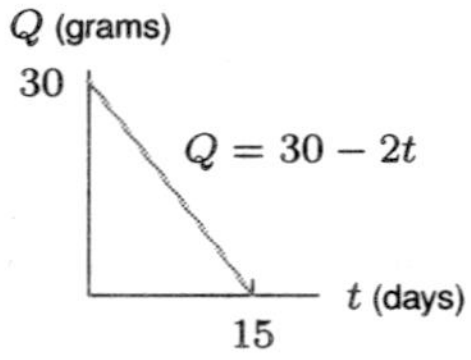

(b) $Q = 30(0.88)^t$

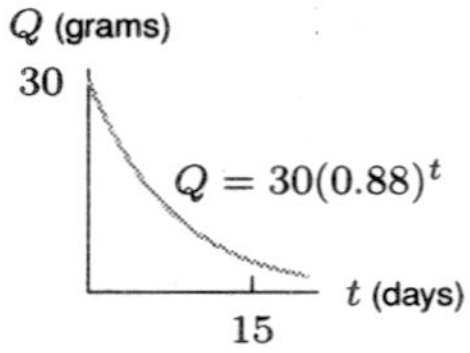

9 (a) $A = 50(0.94)^t$
(b) 11.33 mg
(c)

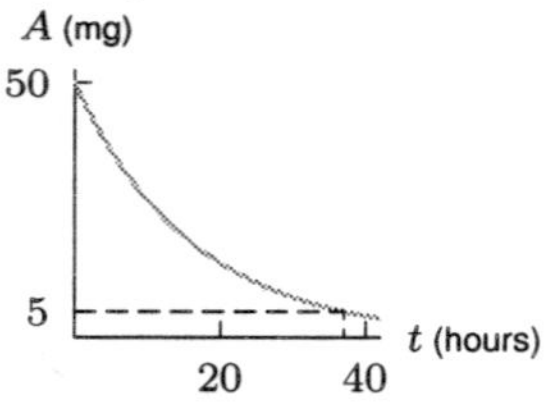

(d) About 37 hours

11 $\text{CPI} = 211(1.028)^t$

13 (a)

x	0	1	2	3
e^x	1	2.72	7.39	20.09

(b)

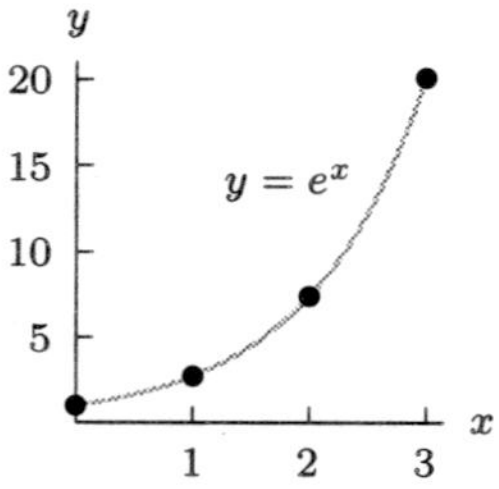

(c)

x	0	1	2	3
e^{-x}	1	0.37	0.14	0.05

(d)

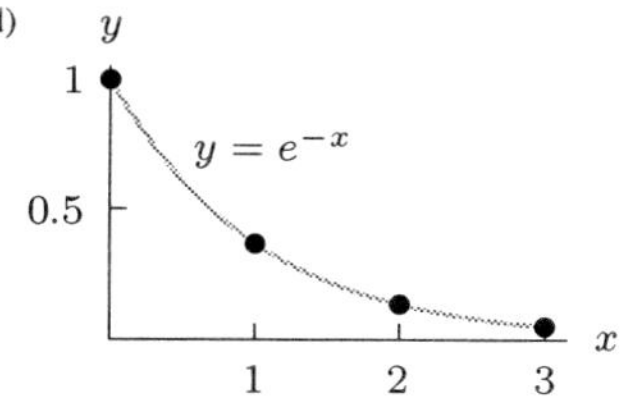

15 $f(x) = 4.30(1.4)^x$

17 $y = 500(1.59)^t$

19 About 6.97 billion; close

21 22.6% per year

23 (a) $W = 40{,}300 + 16{,}178t$
(b) $W = 40{,}300(1.25)^t = 40{,}300e^{0.22t}$
(c)

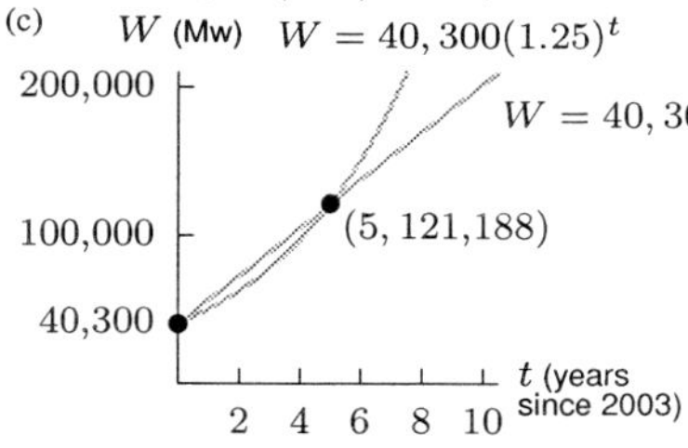

25 (a) Neither
(b) Exponential: $s(t) = 30.12(0.6)^t$
(c) Linear: $g(u) = -1.5u + 27$

27 Min. wage grew 4.69% per year

29 (a) 125%
(b) 9 times

31 $d = 670(1.096)^{h/1000}$

Section 1.6

1 $t = (\ln 7)/(\ln 5) \approx 1.209$

3 $t = (\ln 2)/(\ln 1.02) \approx 35.003$

5 $t = (\ln 5)/(\ln 3) \approx 1.465$

7 $t = \ln 10 \approx 2.3026$

9 $t = (\ln 100)/3 \approx 1.535$

11 $t = 30.54$

13 $t = (\ln B - \ln P)/r$

15 $t = \ln 8 - \ln 5 \approx 0.47$

17 5; 7%

19 15; −6% (continuous)

21 (a) D
(b) C
(c) B

23 $P = 15(1.2840)^t$; growth

25 $P = P_0(1.2214)^t$; growth

27 $P = 15e^{0.4055t}$

29 $P = 174e^{-0.1054t}$

31 $P = 6.4e^{0.01252t}$

33 (a) \$5 million; \$3.704 million dollars
(b) 4.108 years

35 (a) 12%
(b) $P = 25e^{-0.128t}$, 12.8%

37 (a) $P = 5.4(1.034)^t$
(b) $P = 5.4e^{0.0334t}$
(c) Annual = 3.4%
Continuous = 3.3%

39 9.53%

41 2009

Section 1.7

1 (a) $W = 18000e^{0.27t}$
(b) About 2010

3 \$14,918.25

5 About 11.6 years

7 About 10.24 years

9 (a) \$5068.93
(b) \$4878.84

11 (a) $A = 100e^{-0.17t}$
(b) $t \approx 4$ hours

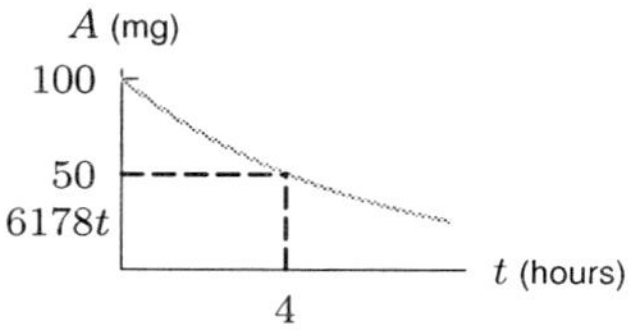

(c) $t = 4.077$ hours

13 A: continuous
B: annual
\$20

15 (a) 47.6%
(b) 23.7%

17 8.45%

19 (a) $P(t) = (0.975)^t$
(b)

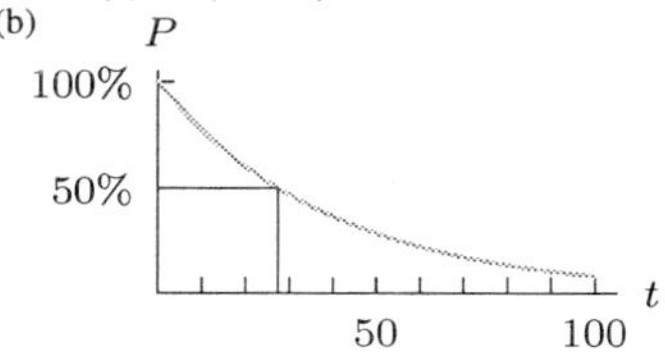

(c) About 27 years
(d) About 8%

21 (a) 4 years
(b) 4 years

23 About 173 hours

25 96.34 years

27 0.0345; 191 million tons

29 (a) (i) About 4.07 billion dollars
(ii) About 7.98 trillion dollars
(b) 1803

31 It is a fake

33 \$35,365.34

35 \$14,522.98

37 (a) Option 1
(b) \$2102.54, \$2051.27, \$2000
(c) \$2000, \$1951.23, \$1902.46

39 (a) Option 1
(b) Option 1: \$10.929 million;
Option 2: \$10.530 million

41 Loan

43 No

Section 1.8

1 (a) $h^2 + 6h + 11$
(b) 11
(c) $h^2 + 6h$

3 (a) 4
(b) 2
(c) $(x + 1)^2$
(d) $x^2 + 1$
(e) $t^2(t + 1)$

5 (a) e
(b) e^2
(c) e^{x^2}
(d) e^{2x}
(e) $e^t t^2$

7 (a) 9
(b) 20
(c) 25
(d) 11

9 (a) $10x^2 + 3$
(b) $20x^2 + 60x + 45$
(c) $4x + 9$

11 (a) 3
(b) 4
(c) 11
(d) 8
(e) 12

13 (a) $y = 2^u$, $u = 3x - 1$
(b) $P = \sqrt{u}$, $u = 5t^2 + 10$
(c) $w = 2 \ln u$, $u = 3r + 4$

15 $2z + 1$

17 $2zh - h^2$

19 0.4

21 −0.9

23 $2(y - 1)^3 - (y - 1)^2$

25 18

27 Can't be done

29 (a)

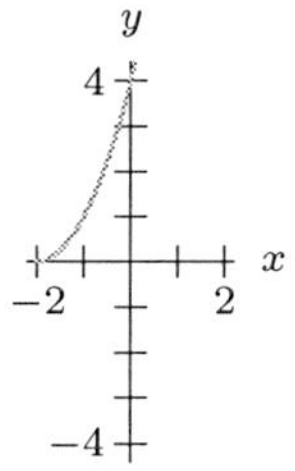

(b)

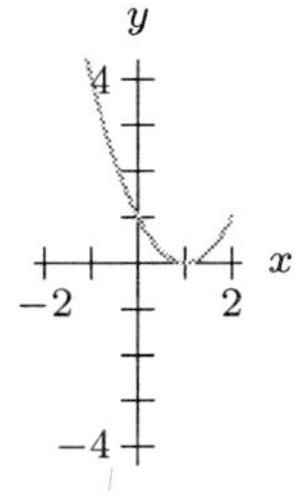

(c)

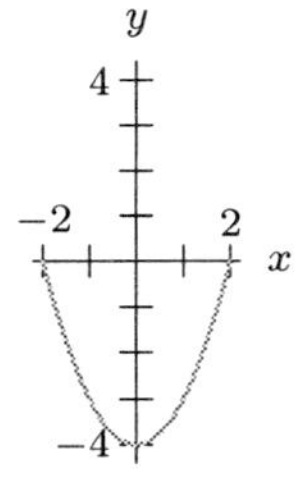

(d)

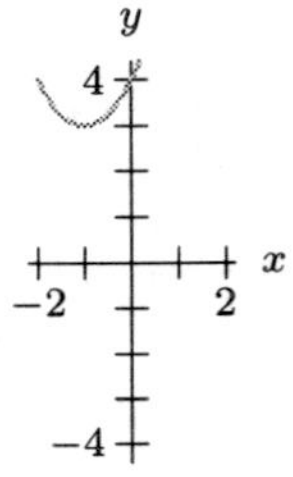

(e)

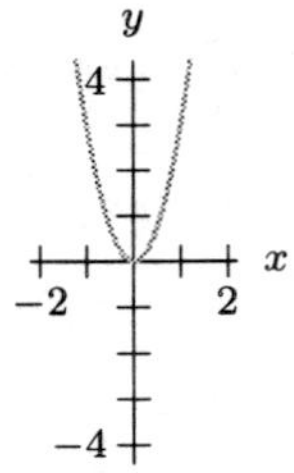

(f)

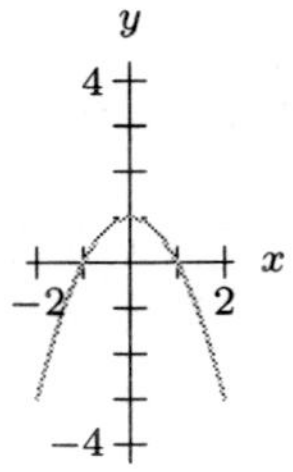

31 (a)

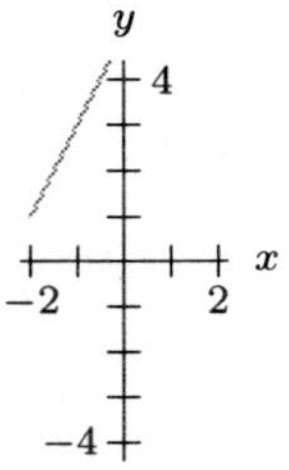

(b)

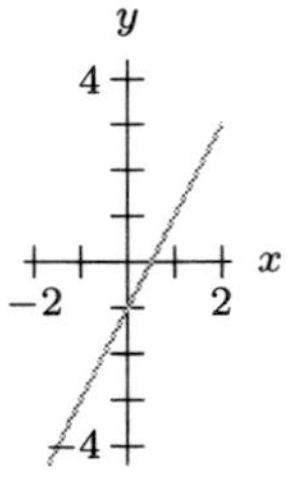

(c)

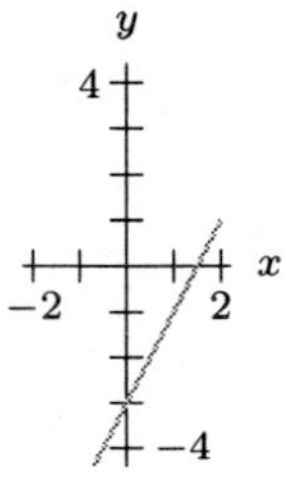

(d)

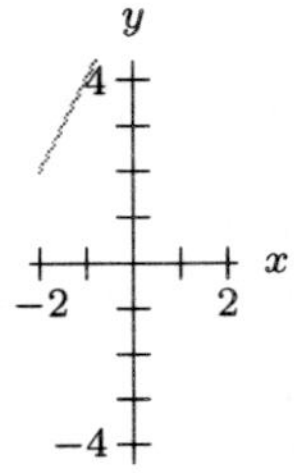

(e)

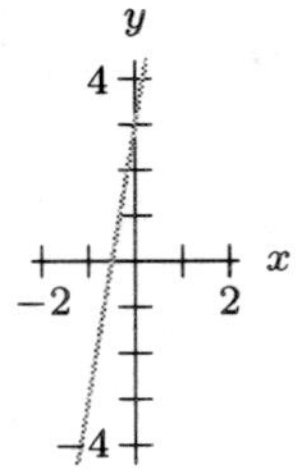

(f)

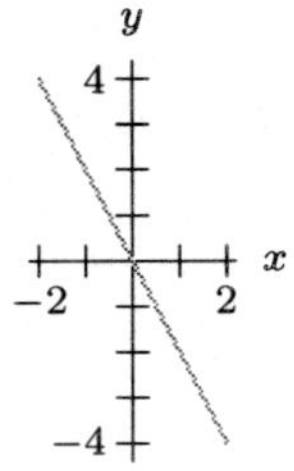

33

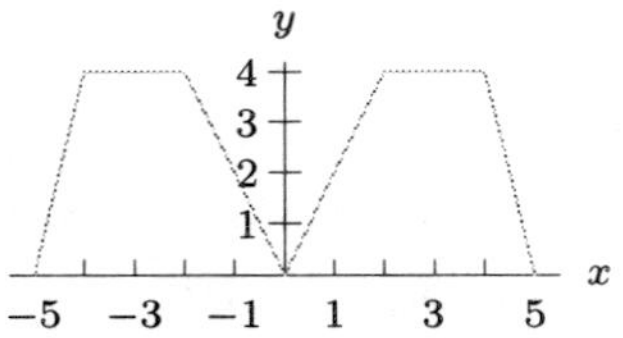

35

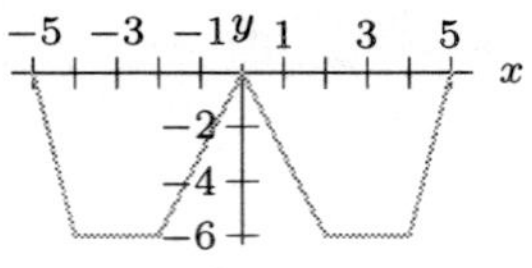

37

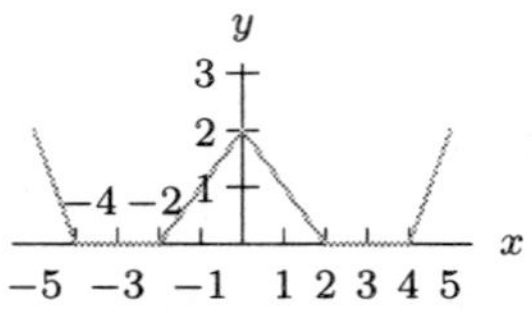

39 **(a)**

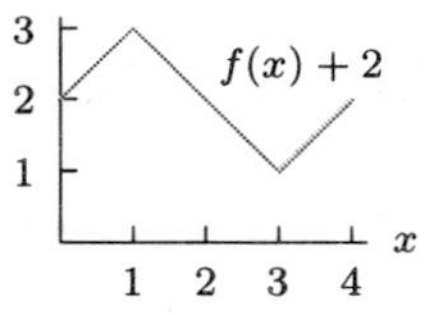

(b)

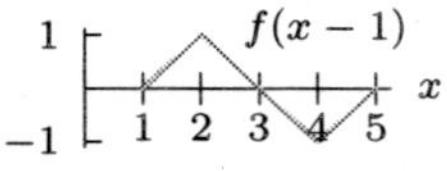

(c)

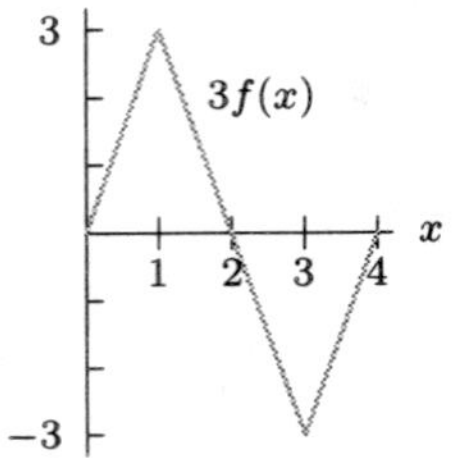

(d)

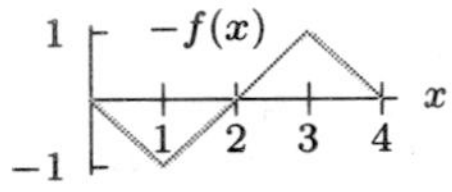

41 **(a)**

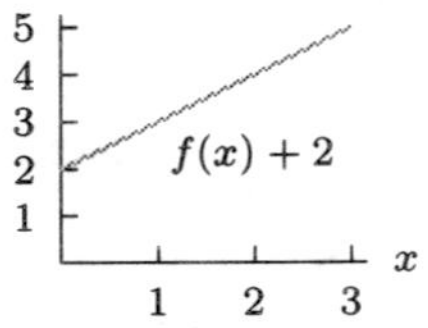

(b)

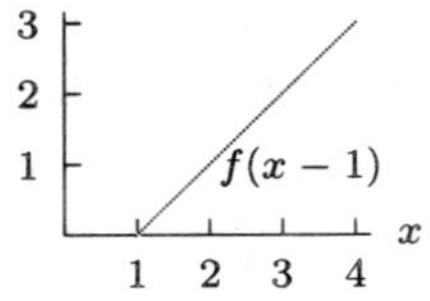

(c)

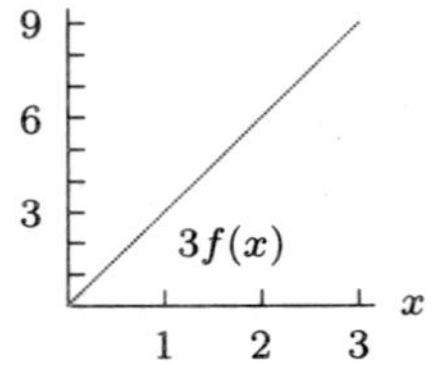

(d)

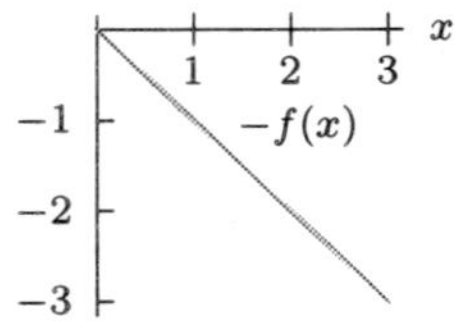

43 (a) $y = 2x^2 + 1$

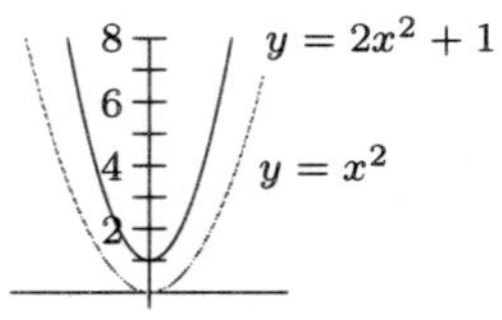

(b) $y = 2(x^2 + 1)$
(c) No

Section 1.9

1 $y = 5x^{1/2}$

3 Not a power function.

5 $y = 9x^{10}$

7 Not a power function

9 $y = 8x^{-1}$

11 Not a power function

13 $S = kh^2$

15 $v = d/t$

17 $N = kA^{1/4}$, with $k > 0$,
Increasing, concave down

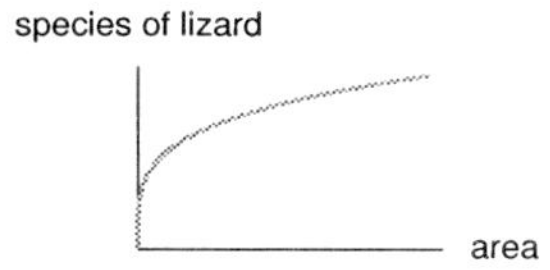

19 (a) $y = (x-2)^3 + 1$
(b) $y = -(x+3)^2 - 2$

21 Yes; $k \approx 0.0087$

23 $N = k/L^2$; small

25 (a) $T = kB^{1/4}$
(b) $k = 17.4$
(c) 50.3 seconds

27 (a) $N = kP^{0.77}$
(b) A has 5.888 times more than B
(c) Town

29 (a) 0.5125
(b) 0.3162
(c) 201,583 dynes/cm^2

31 (a) $C = 115{,}000 - 700p$
$R = 3000p - 20p^2$
(b)

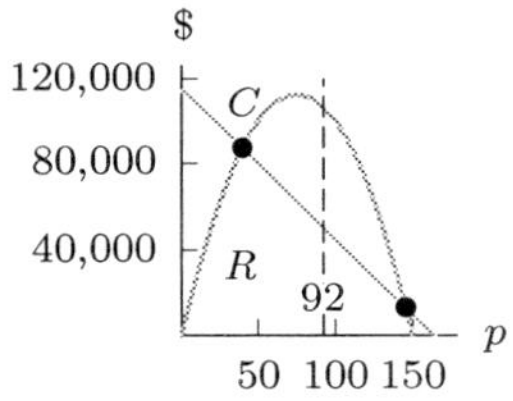

(d) When it charges between \$40 and \$145
(e) About \$92

Section 1.10

1 (b) Max at 7 pm; min at 9 am
(c) Period = one day; amplitude $\approx 2°$C

3

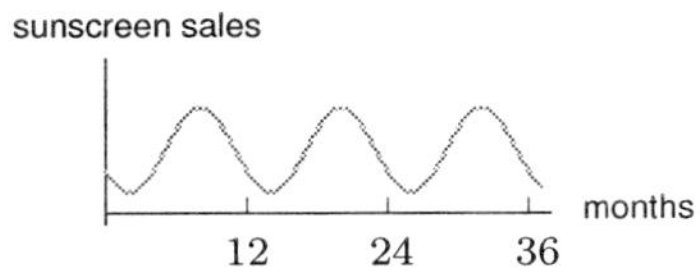

5 Amplitude = 3; Period = π

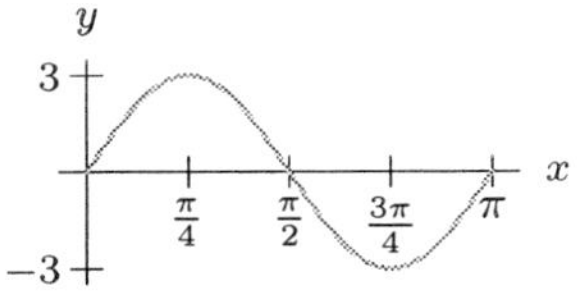

7 Amplitude = 4; Period = π

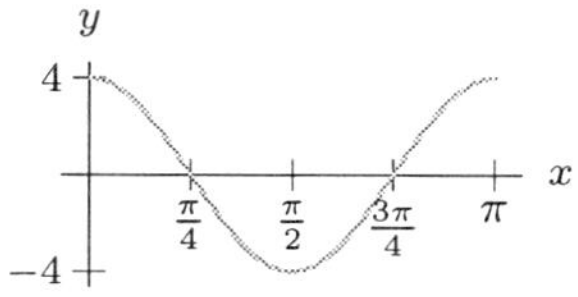

9 Amplitude = 1; Period = π

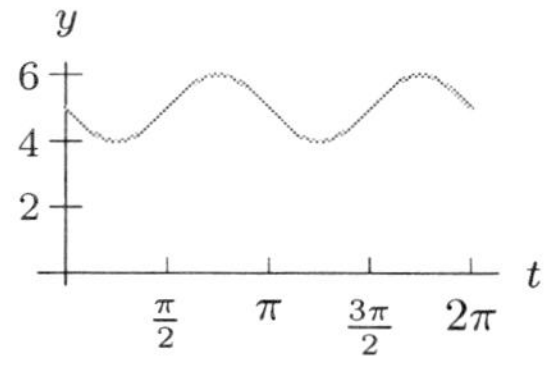

11 (a) 5
(b) 8
(c) $f(x) = 5\cos((\pi/4)x)$

13 $H = -16\cos((\pi/6)t) + 10$

15 $0.35\cos(2\pi t/5.4) + 4$

17 $y = 7\sin(\pi t/5)$

19 $f(x) = 5\cos(x/3)$

21 $f(x) = -4\sin(2x)$

23 $f(x) = -8\cos(x/10)$

25 $f(x) = 5\sin((\pi/3)x)$

27 $f(x) = 3 + 3\sin((\pi/4)x)$

29 Depth $= 7 + 1.5\sin(\pi t/3)$

31 (a) Period is 12; amplitude is 4
(b) $g(34) = 11$; $g(60) = 14$

Chapter 1 Review

1 5 kilometers, 23 minutes

3

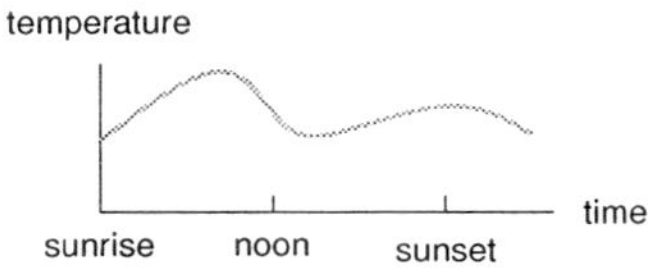

5 (b) 200 bushels
(c) 80 lbs
(d) $0 \leq Y \leq 550$
(e) Decreasing
(f) Concave down

7

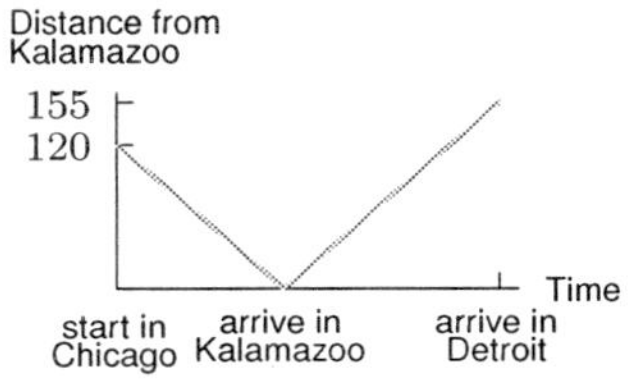

9 $y = 8/3 - x/3$

11 $x = -1$

13 $y = 14x - 45$

15 (a) \$0.025/cubic foot
(b) $c = 15 + 0.025w$
c = cost of water
w = cubic feet of water
(c) 3400 cubic feet

17 $(\ln 3)/2 = 0.549$ mm/sec

19 0 mm/sec

21 0 mm/sec

23 (a)

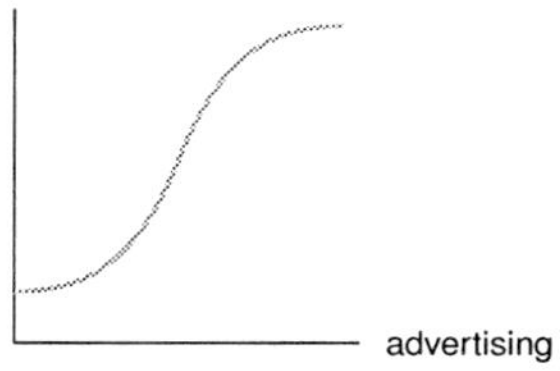

(b)

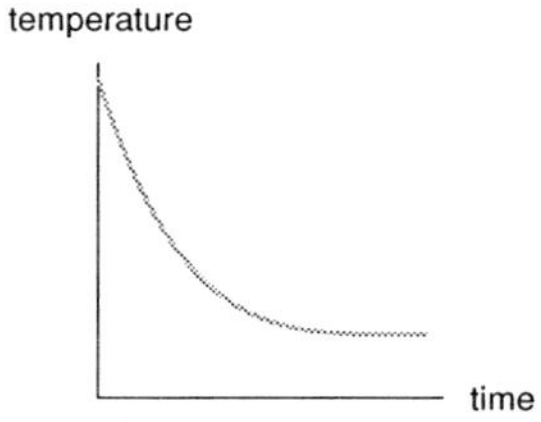

25 (a)

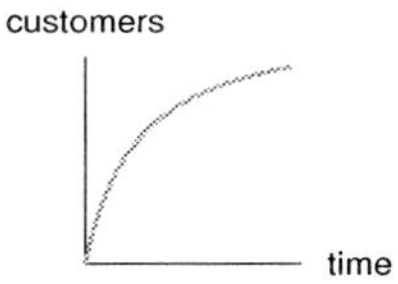

(b) Concave down

27 (a) (i) Positive
(ii) Positive
(iii) Negative
(iv) Positive
(b) (i) $0 \leq t \leq 5$
(ii) $0 \leq t \leq 20$
(c) 25 m^3/week

29 (a) \$1000; \$15
(b) It costs the company \$2500 to produce 100 items

31 $y = -(950/7)x + 950$

33 (a) $p = \$250$; $q = 750$
(b) Suppliers produce 875 units;
Consumers buy 625
(c) Suppliers produce 625 units;
Consumers would buy 875

35 $y = (-3/7)x + 3$

37 $y = 3e^{0.2197t}$
or $y = 3(1.2457)^t$

39 $z = 1 - \cos\theta$

41 $f(x)$ is neither,
$g(x) = 30.8 - 3.2x$ is linear,
$h(x) = 15{,}000(0.6)^x$ is exponential

43 $(\log(2/5))/(\log 1.04) = -23.4$

45 $\ln(0.4)/3 = -0.305$

47 (a) 15%
(b) $P = 10(1.162)^t$
(c) 16.2%

(d) Graphs are the same since functions are equal

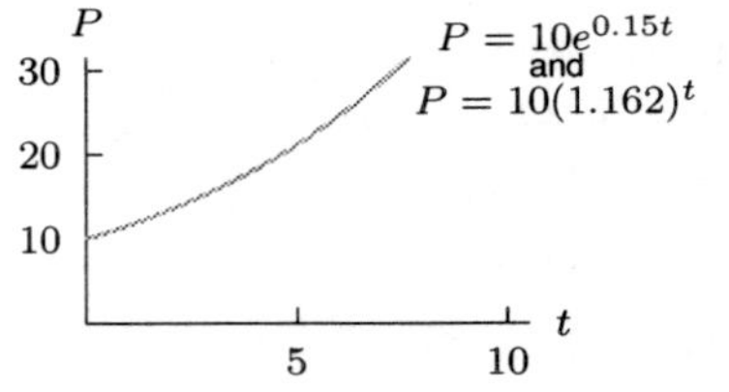

49 (a) 15,678.7 years
(b) 5728.489, or about 5730 years

51 7.925 hours

53 Yes

55 (a) $\ln(2x+3)$
(b) $2\ln x + 3$
(c) $4x+9$

57

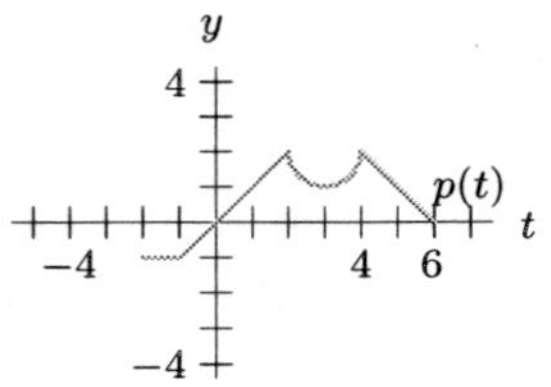

59

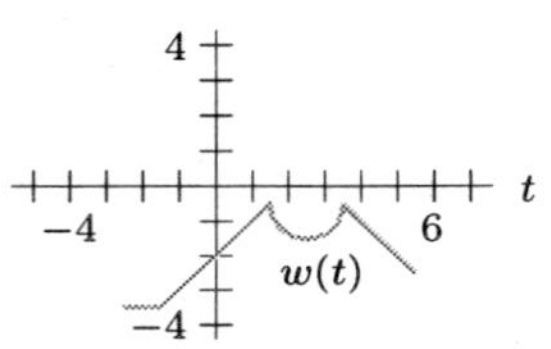

61

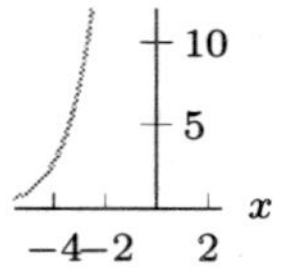

63 (a) A, B positive; C negative
(b) $A+B$
(c) A

65 $y = (x-2)^2 - 5$

67 (a) $P = k/R$
(b) 300,000; 150,000; 100,000
(c) 6 million; 3 million, 2 million
(d) k is population of largest city

69 8π; 3

71 (a) Period = 12 months; Amplitude = 4,500 cases
(b) About 2000 cases and 2000 cases

73 $f(x) = 2\sin(x/4) + 2$

Ch. 1 Understanding

1 False

3 True

5 True

7 True

9 False

11 False

13 True

15 False

17 False

19 False

21 False

23 False

25 True

27 True

29 True

31 True

33 False

35 Possible answer:
$$f(x) = \begin{cases} 1 & x \le 2 \\ x & x > 2 \end{cases}$$

37 Possible answer $f(x) = 1/(x+7\pi)$

39 Possible answer:
$f(x) = (x-1)/(x-2)$

41 $f(x) = x, g(x) = -2x$

43 $f(x) = e^x, g(x) = e^{-2x}$

45 False; $f(x) = x/(x^2+1)$

47 False; $y = x+1$ at points $(1,2)$ and $(2,3)$

49 False; $y = 4x+1$ starting at $x = 1$

51 False

53 True; $f(x) = 0$

55 True

57 True

59 False

61 True

63 False

65 False

67 (a) Follows
(b) Does not follow (although true)
(c) Follows
(d) Does not follow

Section 2.1

1 (a) Positive
(b) Negative

3 (a) (i) 6.3 m/sec
(ii) 6.03 m/sec
(iii) 6.003 m/sec
(b) 6 m/sec

5 (a) 8 ft/sec
(b) 6 ft/sec

7 (a) 63 cubic millimeters
(b) 10.5 cubic millimeters/month
(c) 44.4 cubic millimeters/month

9 $f'(2) \approx 40.268$

11 (a) Positive at C and G
Negative at A and E. Zero at B, D, and F
(b) Largest at G
Most negative at A

13 $f'(d) = 0$, $f'(b) = 0.5$, $f'(c) = 2$, $f'(a) = -0.5$, $f'(e) = -2$

15 (a) The slopes of the two tangent lines at $x = a$ are equal for all a
(b) A vertical shift does not change the slope

17 $P'(0) = 10$

19 (a) $g(2) = 5$
(b) $g'(2) = -0.4$

21 (a) $f(4)$
(b) $f(2) - f(1)$
(c) $(f(2) - f(1))/(2-1)$
(d) $f'(1)$

23 300 million people;
2.867 million people/yr

Section 2.2

1 1.0, 0.3, −0.5, −1

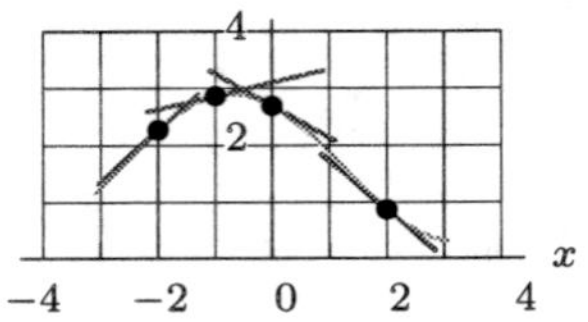

3

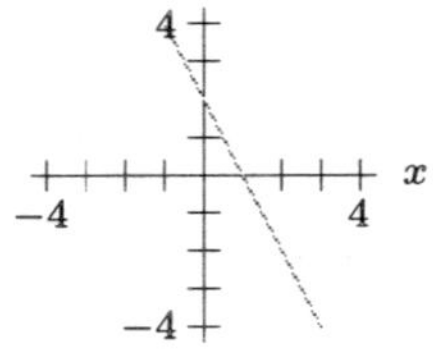

5

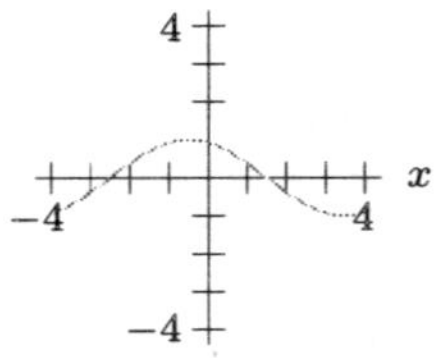

7

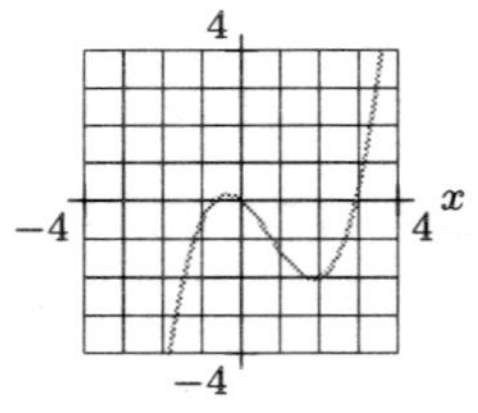

9 −6, −3, −1.8, −1.2, −1.2

11

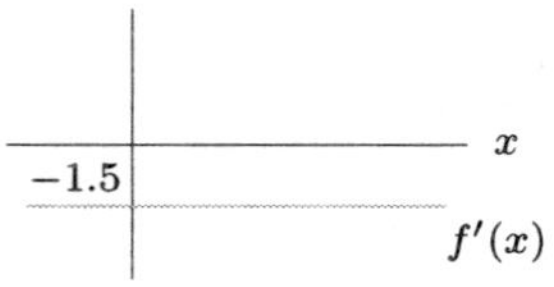

13

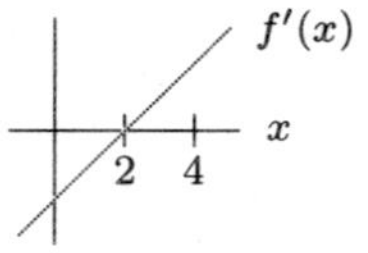

15

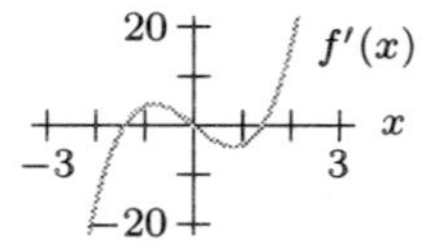

17

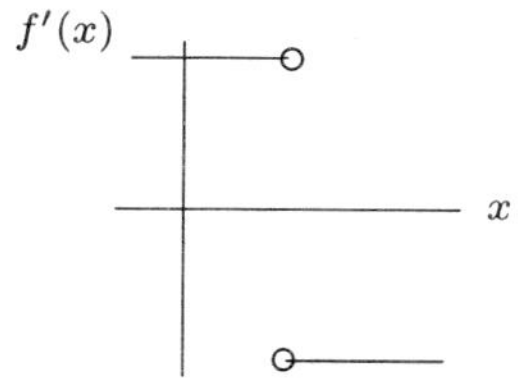

19

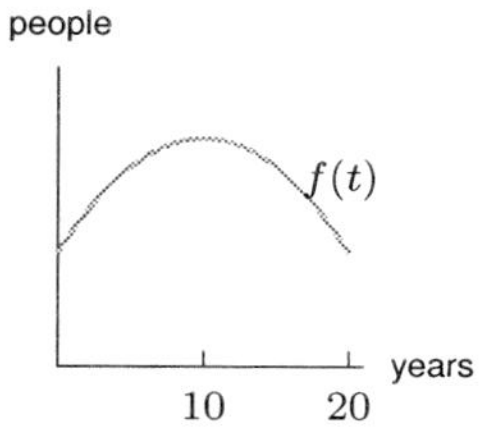

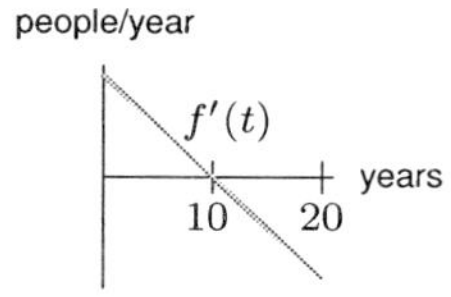

or

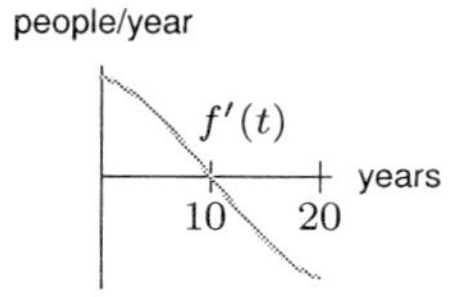

21 VIII

23 II

25

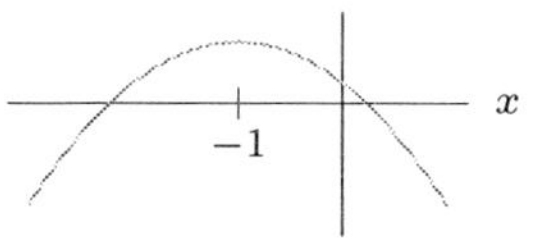

Other answers possible

27 (a) Graph II
(b) Graph I
(c) Graph III

29 $f'(2) = 4$
$f'(3) = 9$
$f'(4) = 16$
The pattern seems to be:
$f'(x) = x^2$.

31 (a) $t = 3$
(b) $t = 9$
(c) $t = 14$
(d)

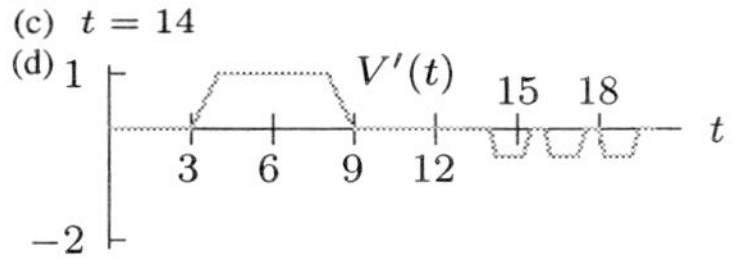

Section 2.3

1 dC/dW; dollars per pound

3 dP/dH; dollars per hour

5 (a) 12 pounds, 5 dollars
(b) Positive
(c) 12 pounds, 0.4 dollars/pound, extra pound costs about 40 cents

9 (a) Positive
(b) $^\circ$F/min

11 (a) Liters per centimeter
(b) About 0.042 liters per centimeter
(c) Cannot expand much more

13 (a) Investing the \$1000 at 5% would yield about \$1649 after 10 years
(b) Extra percentage point would yield an increase of about \$165; dollars/%

15 (a) Consuming 1800 Calories per day results in a weight of 155 pounds; Consuming 2000 Calories per day causes neither weight gain nor loss
(b) Pounds/(Calories/day)

17 (a) Positive
(b) Child weighs 45 pounds at 8 years
(c) lbs/year
(d) The child is growing at a rate of 4 lbs/year at 8 years of age
(e) Decrease

19 $f'(t) \approx 6$ where t is retirement age in years, $f(t)$ is age of onset in weeks

21 (a) kg/week
(b) Growing at 96 gm/wk at week 24

23 (a) Less
(b) Greater

25 About 3.4, about 2.6

27 (a) In 2008: Net sales 5.1 bn dollars; rate increase 0.22 bn dollars/yr
(b) About 5.98 billion dollars

29 (a) Dose for 140 lbs is 120 mg
Dose increases by 3mg/lb
(b) About 135 mg

31 (a) $f(0) = 80$; $f'(0) = 0.50$
(b) $f(10) \approx 85$

33 (a) 1.7 (liters/minute)/hour
(b) About 0.028 liter/minute
(c) $g'(2) = 1.7$

35 (a) $f'(a)$ is always positive
(c) $f'(100) = 2$: more
$f'(100) = 0.5$: less

37 (a) Fat
(b) Protein

39 (a) 2.0 kg/week
(b) 0.6 kg/week
(c) 0.3 kg/week

41 I-fat, II-protein

43 $f(4) = 200$ million users;
$f'(4) \approx 12.5$ million users/month;
Increasing at about 6.25%/month

45 0.50

Section 2.4

1 (a) Negative
(b) Negative
(c) Positive

3 $f'(x) > 0$
$f''(x) > 0$

5 $f'(x) < 0$
$f''(x) = 0$

7 $f'(x) > 0$
$f''(x) < 0$

9 (a)

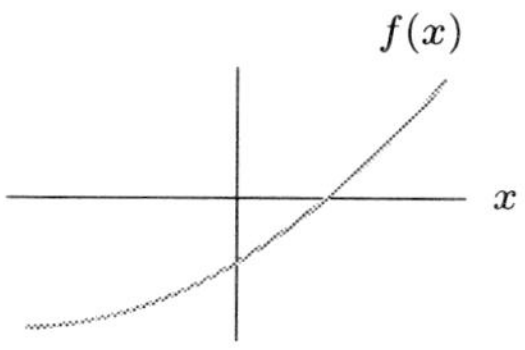

(b)

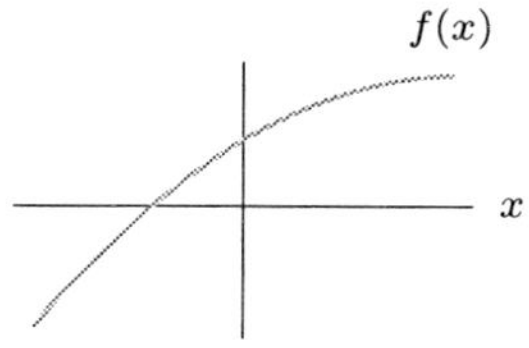

(c)

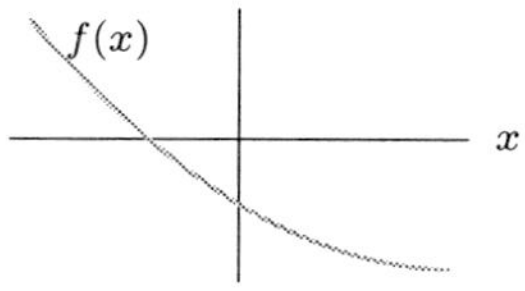

(d)

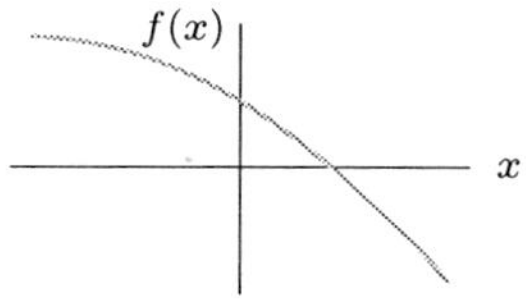

11 $s'(t)$: positive
$s''(t)$: positive or zero

13 A positive second derivative indicates a successful campaign
A negative second derivative indicates an unsuccessful campaign

15 (a) Positive; positive
(b) Number of cars increasing at 3.24 million cars per year in 1975

17 Derivative:
Pos. $-2.3 < t < -0.5$
Neg. $-0.5 < t < 4$
Second derivative:
Pos. $0.5 < t < 4$
Neg. $-2.3 < t < 0.5$

19

Point	f	f'	f''
A	$-$	0	$+$
B	$+$	0	$-$
C	$+$	$-$	$-$
D	$-$	$+$	$+$

21 (e)

23 22 only possible value

25 (a)

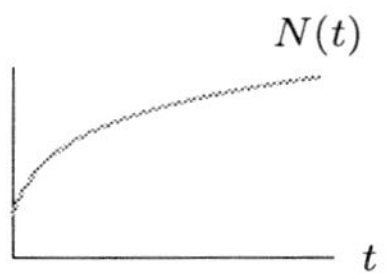

(b) dN/dt is positive.
d^2N/dt^2 is negative.

27 (a)

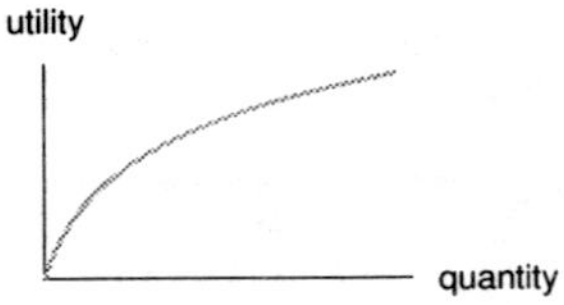

(b) Derivative of utility is positive
2nd derivative of utility is negative

Section 2.5

1 (a) Dollars/barrel
(b) 101 barrels cost about \$3 more than 100 barrels

3 At $q = 5$;
At $q = 40$

5 About \$16.67 (answers may vary)

7 $C'(2000) \approx \$0.37$/ton
The marginal cost is smallest on the interval $2500 \le q \le 3000$.

9 (a) About \$2408
(b) About \$2192

11 (a) About \$4348
(b) \$11 profit
(c) No, company will lose money

13 (a) \$1850 profit
(b) About \$0.40 increase; increase production
(c) About \$0.45 decrease; decrease production

15 (a) Fixed costs
(b) Decreases slowly, then increases

Chapter 2 Review

1 (a) (i) 8.4 m/sec
(ii) 8.04 m/sec
(iii) 8.004 m/sec
(b) 8 m/sec

3 (a) Negative
(b) $f'(1) = -3$

5 Positive: A and D
Negative: C and F
Most positive: A
Most negative: F

7 (a) (i) -1.00801 m/sec
(ii) -0.8504 m/sec
(iii) -0.834 m/sec
(b) -0.83 m/sec

9

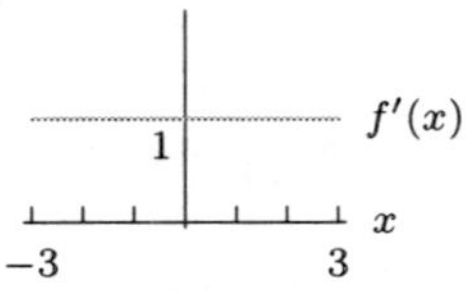

11

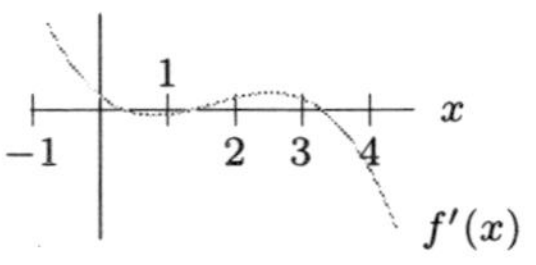

13

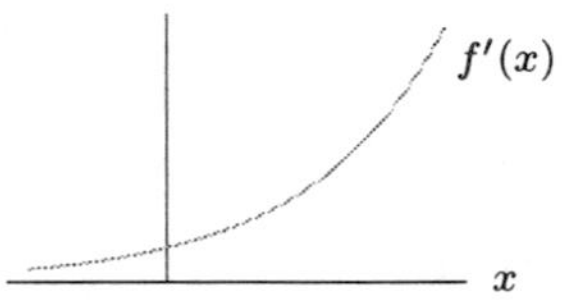

15 (a) $x_1 < x < x_3$
(b) $0 < x < x_1; x_3 < x < x_5$

21 $f(21) \approx 65$
$f(19) \approx 71$
$f(25) \approx 53$

23 Wind stronger at 15.1 km than at 15 km

25 About 1.338 billion people in 2009; growing at 8 million people per year

27 (a) $f'(t) > 0$: depth increasing
$f'(t) < 0$: depth decreasing
(b) Depth increasing 20 cm/min
(c) 12 meters/hr

31 (a) Negative
(b) Degrees/min

33 Dollars/year; negative

35 (a)

(b)

(c)

(d)

37 (a) Minutes/kilometer
(b) Minutes/kilometer^2

39 (a) x_4, x_5
(b) x_3, x_4
(c) x_3, x_4
(d) x_2, x_3
(e) x_1, x_2, x_5
(f) x_1, x_4, x_5

41 (a) t_3, t_4, t_5
(b) t_2, t_3
(c) t_1, t_2, t_5
(d) t_1, t_2, t_4
(e) t_3, t_4

43 (a)

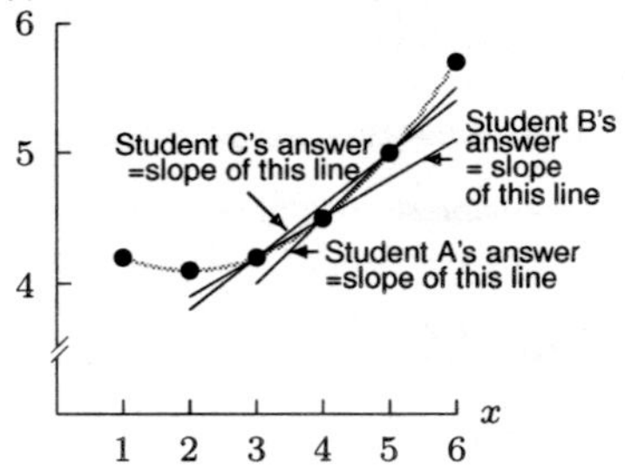

(b) Student C's
(c) $f'(x) = \dfrac{f(x+h) - f(x-h)}{2h}$

Ch. 2 Understanding

1 False

3 True

5 True

7 True

9 False

11 False

13 True

15 False

17 False

19 False

21 False

23 False

25 True

27 True

29 True

31 True

33 False

35 Possible answer:
$$f(x) = \begin{cases} 1 & x \le 2 \\ x & x > 2 \end{cases}$$

37 Possible answer $f(x) = 1/(x + 7\pi)$

39 Possible answer:
$f(x) = (x-1)/(x-2)$

41 $f(x) = x, g(x) = -2x$

43 $f(x) = e^x, g(x) = e^{-2x}$

45 False; $f(x) = x/(x^2+1)$

47 False; $y = x + 1$ at points $(1, 2)$ and $(2, 3)$

49 False; $y = 4x + 1$ starting at $x = 1$

51 False

53 True; $f(x) = 0$

55 True

57 True

59 False

61 True

63 False

65 False

67 (a) Follows
(b) Does not follow (although true)
(c) Follows
(d) Does not follow

Theory: Limits, Derivatives

1

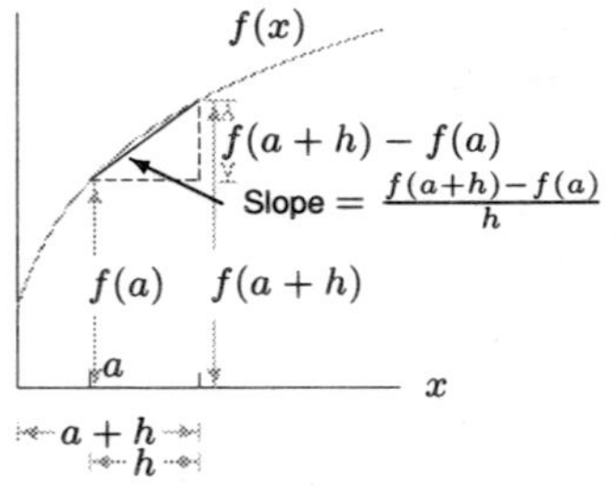

3 1

5 27

7 2.7

9 Yes

11 No; yes

13 Yes

15 Yes

17 No

19 Not continuous

21 Not continuous

23 Not continuous

Section 3.1

1 0

3 $12x^{11}$

5 $\frac{4}{3}x^{1/3}$

7 $12t^3 - 4t$

9 $-4x^{-5}$

11 $2x + 5$

13 $6x + 7$

15 $8.4q - 0.5$

17 $-5t^{-6}$

19 $-(7/2)r^{-9/2}$

21 $-(1/3)\theta^{-4/3}$

23 $15t^4 - \frac{5}{2}t^{-1/2} - 7t^{-2}$

25 $6t - 6/t^{3/2} + 2/t^3$

27 $(3/2)x^{1/2} + (1/2)x^{-1/2}$

29 $2kx$

31 $2aP + 3bP^2$

33 $b/(2\sqrt{t})$

35 $3ab^2$

37 (a) $P'(1)$: Positive;
$P'(3)$: Zero;
$P'(4)$: Negative

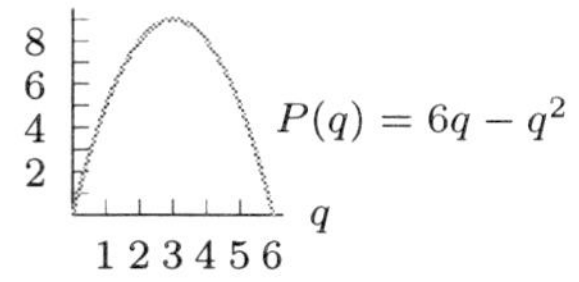

(b) $P'(1) = 4, P'(3) = 0, P'(4) = -2$

39 (a) $2t - 4$
(b) $f'(1) = -2, f'(2) = 0$

41 Height $= 625$ cm,
Changing (eroding) at -30 cm/year

43 $f(10) = 400$ tons;
$f'(10) = 60$ tons per year;
Relative rate $= 15\%$ per year

45 $f'(t) = 6t^2 - 8t + 3$
$f''(t) = 12t - 8$

47 (a) $f(100) = 11.11$ seconds
(b) $f'(100) = 0.05555$ seconds per foot

49 $y = 2x - 1$

51 $y = -2t + 16$

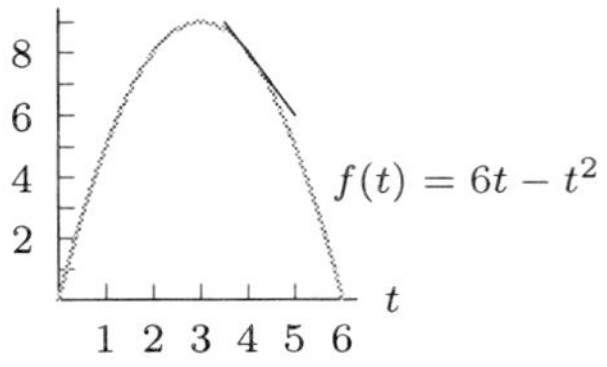

53 (a) $dA/dr = 2\pi r$
(b) Circumference of a circle

55 $100

57 (a) 770 bushels per acre
(b) 40 bushels per acre per pound of fertilizer
(c) Use more fertilizer

59 (a) $dC/dq = 0.24q^2 + 75$
(b) $C(50) = \$14{,}750$; $C'(50) = \$675$ per item

61 $f'(x) = 3x^2 - 12x - 15$,
$x = -1$ and $x = 5$

63 (a) $R(q) = bq + mq^2$
(b) $R'(q) = b + 2mq$

Section 3.2

1 $2e^x + 2x$

3 $10t + 4e^t$

5 $(\ln 2)2^x - 6x^{-4}$

7 $(\ln 2)2^x + 2(\ln 3)3^x$

9 $3 - 2(\ln 4)4^x$

11 $3e^{3t}$

13 $-4e^{-4t}$

15 $-30e^{-0.6t}$

17 $3000(\ln 1.02)(1.02)^t$

19 Ce^t

21 $Ae^x - 2Bx$

23 $3/q$

25 $2t + 5/t$

27 $2x + 4 - 3/x$

29 $f'(-1) \approx -0.736$
$f'(0) = -2$
$f'(1) \approx -5.437$

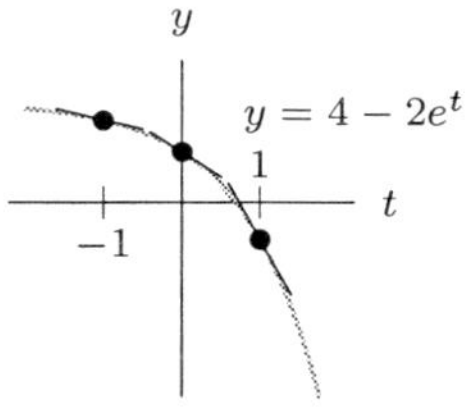

31 $y = -2t + 1$

33 (a) 13, 394 fish
(b) 8037 fish/month

35 $f(2) = 6065, f'(2) = -1516$

37 $f(5) = \$563.30$;
$f'(5) = \$70$ per week;
Relative rate $= 12.4\%$ per week

39 -444.3 people/year

41 $c = -1/\ln 2$

43 $C(50) \approx 1365, C'(50) \approx 18.27$

45 (a) 0.021 micrograms/year
(b) 779.4 years old in 1998

47 (a) $P = 1.166(1.015)^t$
(b) $\frac{dP}{dt} = 1.166(1.015)^t(\ln 1.015)$
$\frac{dP}{dt}|_{t=0} = 0.017$ billion people per year
$\frac{dP}{dt}|_{t=25} = 0.025$ billion people per year

49 (a) $y = x - 1$
(b) 0.1; 1
(c) Yes

Section 3.3

1 $56x(4x^2 + 1)^6$

3 $8q(q^2 + 1)^3$

5 $300t^2(t^3 + 1)^{99}$

7 $3s^2/(2\sqrt{s^3 + 1})$

9 $216q(3q^2 - 5)^2$

11 $25e^{5t+1}$

13 $(e^{\sqrt{s}})/(2\sqrt{s})$

15 $1/(x - 1)$

17 $e^{-x}/(1 - e^{-x})$

19 $25/(5t + 1)$

21 $3/(3t + 2)$

23 $5 + 1/(x + 2)$

25 $0.5/(x(1 + \ln x)^{0.5})$

27 $-x/\sqrt{1 - x^2}$

29 $10/(10t + 5)$

31 5

33 $2/t$

35 1.5

37 $2/t$

39 $y = 3x - 5$

41 $v(t) = 10e^{\frac{t}{2}}$

43 $f(10) = 31.640$ feet;
$f'(10) = 4.741$ ft/sec;
Relative rate $= 15\%$ per second

45 0

47 1/2

49 Approx 1

51 Approx 1.9

Section 3.4

1 $f'(x) = 12x - 1$

3 $e^x(x + 1)$

5 $5e^{x^2} + 10x^2e^{x^2}$

7 $\ln x + 1$

9 $30t + 11$

11 $t + 2t\ln t$

13 $-(5/t^2) - (12/t^3)$

15 $e^{-t^2}(1 - 2t^2)$

17 $2p/(2p + 1) + \ln(2p + 1)$

19 $2we^{w^2}(5w^2 + 8)$

21 $(3t^2 + 5)(t^2 - 7t + 2) + (t^3 + 5t)(2t - 7)$

23 $(1 - x)/e^x$

25 $-2/(1 + t)^2$

27 $(15 + 10y + y^2)/(5 + y)^2$

29 $(ak - bc)/(cx + k)^2$

31 $ae^{-bx} - abxe^{-bx}$

33 $(1 - 2\alpha)e^{-2\alpha}e^{\alpha e^{-2\alpha}}$

35 $y = 0$

37 60.65 mg, 30.33 mg/hr,
41.04 mg, -12.31 mg/hr

39 $(kc)/(k + r)^2$;
Approx change in P per unit increase in r

41 Revenue $R(10) \approx 22{,}466$.
$R'(10) \approx$ $449/dollar.

43 $1/t$

45 $(fg)'/(fg) = (f'/f) + (g'/g)$

Section 3.5

1 $5\cos x$

3 $2t - 5\sin t$

5 $2q + 2\sin q$

7 $3\cos(3x)$

9 $-8t\sin(t^2)$

11 $2x\cos(x^2)$

13 $-4\sin(4\theta)$

15 $2x\cos x - x^2\sin x$

17 $3\theta^2\cos\theta - \theta^3\sin\theta$

19 $(2t\cos t + t^2\sin t)/(\cos t)^2$

21 $y = -x + \pi$

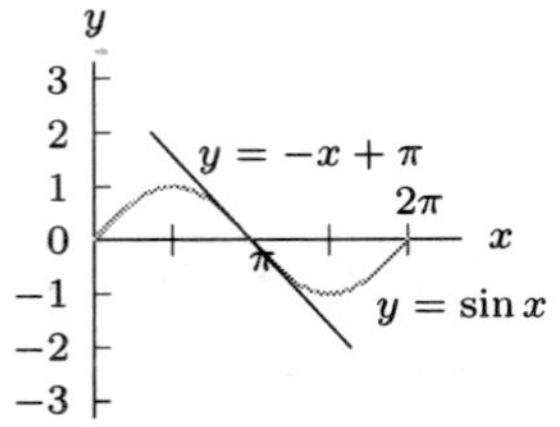

23 Decreasing, concave up

25 (a) max $2600; min $1400; April 1

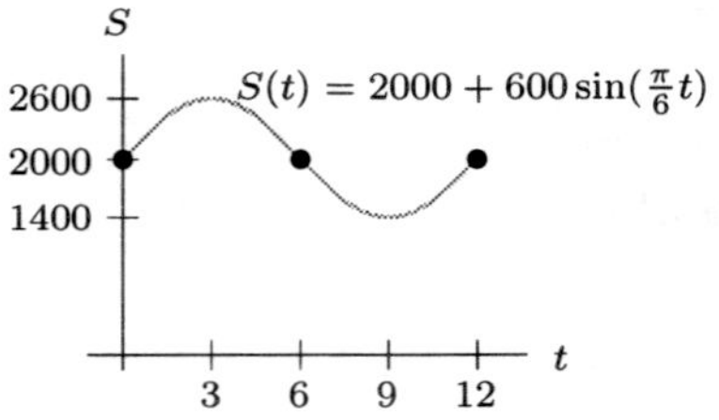

(b) $S(2) \approx 2519.62$; $S'(2) \approx 157.08$

27 (a) Falling, 0.38 m/hr
(b) Rising, 3.76 m/hr
(c) Rising, 0.75 m/hr
(d) Falling, 1.12 m/hr

29 (a) $H'(t) = (-10\pi/3)\sin((\pi/15)t)$
(b) $t = 0, 15, 30$ days; the percentage illuminated is not changing
(c) $H'(t)$ negative for $0 < t < 15$ and positive for $15 < t < 30$

Chapter 3 Review

1 $24t^3$

3 $2e^{2t}$

5 $0.08e^{0.08q}$

7 $e^{3x}(1+3x)$

9 $6(1+3t)e^{(1+3t)^2}$

11 $5 + 3.6e^{1.2t}$

13 $90(5x-1)^2$

15 $10e^x(1+e^x)^9$

17 $x(2\ln x + 1)$

19 $8t + 7\cos t$

21 $(-e^{-t} - 1)/(e^{-t} - t)$

23 $(50x - 25x^2)/(e^x)$

25 $2x\cos x - x^2\sin x$

27 e

29 $(e^x - 2 - e^{-x})/(1 - e^{-x})^2$

31 $xe^x + e^x$

33 $-3x^2\sin(x^3)$

35 $(t^2 + 6t + 13)/(t+3)^2$

37 $((\ln 3)3^x)/3 - (33x^{-3/2})/2$

39 $6y^2e^{(y^3)}e^{2e^{(y^3)}}$

41 $f'(0) = 0$, $f'(1) = 2$, $f'(2) = 4$, $f'(-1) = -2$

43 $y = -4 - x$

45 30% per year

47 49.7% per year

49 73.7 m/yr

51 (a)

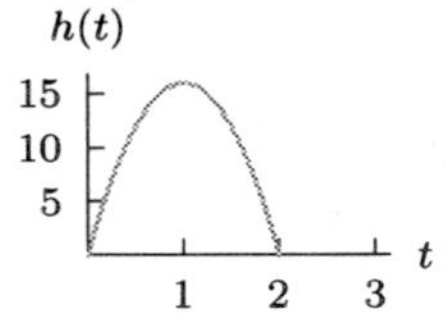

(b) $v(t) = 32 - 32t$

(c) $v(1) = 0$ ft/sec; $h(1) = 16$ ft

53 (a) $f'(0) = -1$
(b) $y = -x$

55 $596.73/yr

57 (a) $\pi/\sqrt{9.8l}$
(b) Decr

59 (a) $dH/dt = -60e^{-2t}$
(b) $dH/dt < 0$
(c) At $t = 0$

61 (a) $13,050
(b) $V'(t) = -4.063(0.85)^t$; thousands of dollars/year
(c) $V'(4) = -\$2121$ per year

63 (a) $P(d) = 4.07264 \cdot 10^{-10}d^{3/2}$
(b) Approximately 88 days
(c) $P'(d) = 6.10896 \cdot 10^{-10}d^{1/2}$ days/mile. Farther from the sun, time increases

65 $n = 4, a = 3/32$

67 (a) $H(4) = 1$
(b) $H'(4) = 30$
(c) $H(4) = 4$
(d) $H'(4) = 56$
(e) $H'(4) = -1$

69 $b = 1/40$ and $a = 169.36$

71 Approx 0.4

73 Approx 0.7

75 Approx -21.2

77 (a) $x < 1/2$
(b) $x > 1/4$
(c) $x < 0$ or $x > 0$

79 (a) $dy/dt = -(4.4\pi/6)\sin(\pi t/6)$ ft/hr
(b) Occurs at $t = 0, 6, 12, 18$, and 24 hrs

81 (c) $B'(10) \approx -3776.63$

83 If $a \neq e$, two solutions
If $a = e$, one solution

85 (a) 0.04 gal/mile; 0.06 gal/mile
(b) 25 mpg; 16.67 mpg
(d) 1.4 gallons
(e) 2.8 gal/hour; 1.8 gal/hour

87 $331.3 + 0.606T$ m/sec

Ch. 3 Understanding

1 True

3 True

5 True

7 False

9 True

11 False

13 False

15 False; $f(x) = |x|$

17 False; $\cos t + t^2$

19 False; $f(x) = 6, g(x) = 10$

21 False; $f(x) = 5x + 7, g(x) = x + 2$

23 False; $f(x) = x^2, g(x) = x^2 - 1$

25 False; $f(x) = e^{-x}, g(x) = x^2$

27 (a) Not a counterexample
(b) Not a counterexample
(c) Not a counterexample
(d) Counterexample

29 False

31 False

33 Possible answer

$$f(x) = \begin{cases} x & \text{if } 0 \le x < 2 \\ 19 & \text{if } x = 2 \end{cases}$$

Practice: Differentiation

1 $2t + 4t^3$

3 $15x^2 + 14x - 3$

5 $-2/x^3 + 5/(2\sqrt{x})$

7 $10e^{2x} - 2 \cdot 3^x(\ln 3)$

9 $2pe^{p^2} + 10p$

11 $2x\sqrt{x^2+1} + x^3/\sqrt{x^2+1}$

13 $16/(2t+1)$

15 $2^x(\ln 2) + 2x$

17 $6(2q+1)^2$

19 bke^{kt}

21 $2x\ln(2x+1) + 2x^2/(2x+1)$

23 $10\cos(2x)$

25 $15\cos(5t)$

27 $2e^x + 3\cos x$

29 $17 + 12x^{-1/2}$

31 $20x^3 - 2/x^3$

33 $1, x \neq -1$

35 $-3\cos(2-3x)$

37 $6/(5r+2)^2$

39 $ae^{ax}/(e^{ax}+b)$

41 $5(w^4 - 2w)^4(4w^3 - 2)$

43 $(\cos x - \sin x)/(\sin x + \cos x)$

45 $6/w^4 + 3/(2\sqrt{w})$

47 $(2t - ct^2)e^{-ct}$

49 $(\cos\theta)e^{\sin\theta}$

51 $3x^2/a + 2ax/b - c$

53 $2r(r+1)/(2r+1)^2$

55 $2e^t + 2te^t + 1/(2t^{3/2})$

57 $x^2 \ln x$

59 $6x\left(x^2+5\right)^2\left(3x^3-2\right)\left(6x^3+15x-2\right)$

61 $(2abr - ar^4)/(b+r^3)^2$

63 $20w/(a^2-w^2)^3$

Section 4.1

1 One

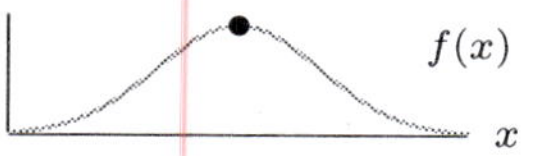

3 Three

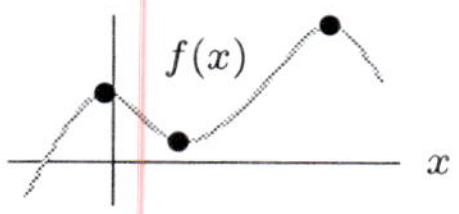

5 (a)

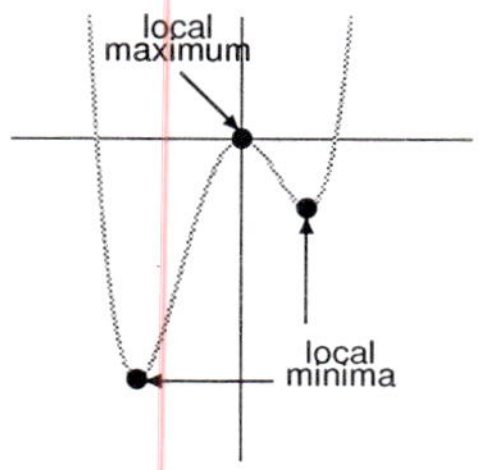

(b)

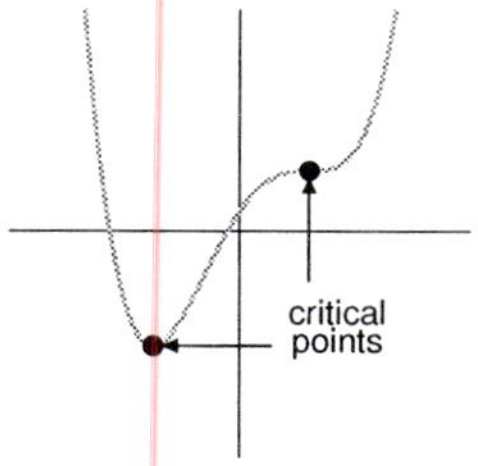

7 (a)

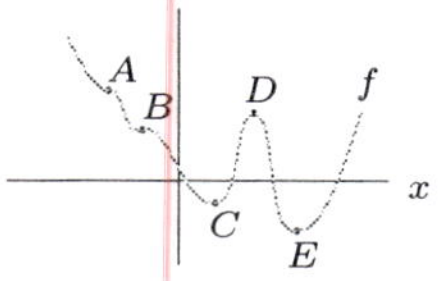

(b)

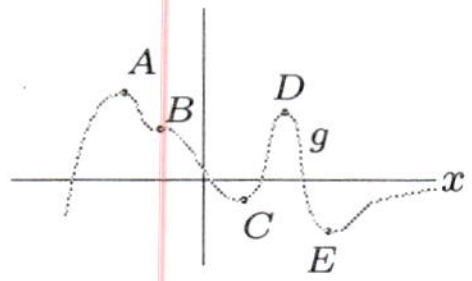

9 A: local max
B: local min
C: neither

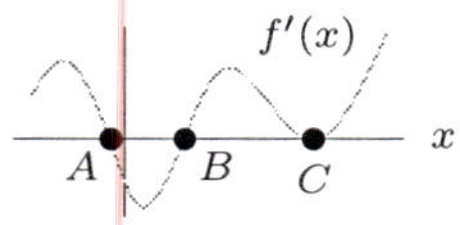

11 Increasing for all x, no critical point

13 Local min: $(2.3, -13.0)$

15 Alternately incr/decr

17

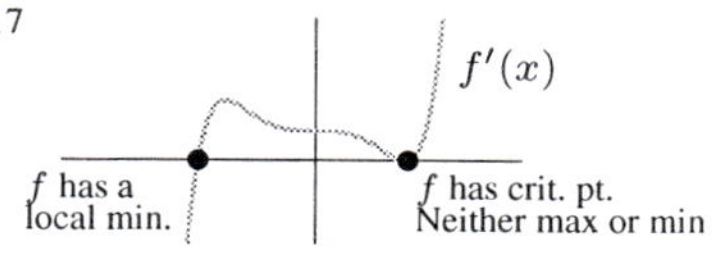

19 $t = 0.5\ln(V/U)$

21 (a) $x \approx 2.5$ (or any $2 < x < 3$)
$x \approx 6.5$ (or any $6 < x < 7$)
$x \approx 9.5$ (or any $9 < x < 10$)
(b) $x \approx 2.5$: local max;
$x \approx 6.5$: local min;
$x \approx 9.5$: local max

23 $x = 0$: not max/min
$x = 3/7$: local max
$x = 1$: local min

25 (a)

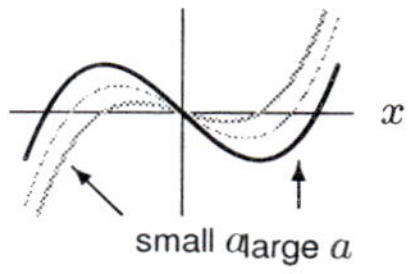

(b) 2 critical points move farther from origin

(c) $x = \pm\sqrt{a/3}$

27 $a = -6$; $b = 14$

29

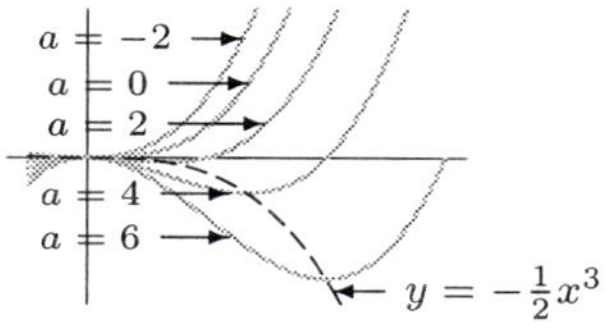

31 $a = -1/3$

33 (a) $x = 0, x = a^2/4$
(b) $a = \sqrt{20}$; Local minimum

35 Domain: All real numbers except $x = b$;
Critical points: $x = 0$, $x = 2b$

37 (a) $f(\theta) = 0$ at $\theta = 0$
(b) $f'(\theta) = 1 - \cos\theta > 0$
for $0 < \theta \le 1$.
Only zero is at origin

Section 4.2

1 Three

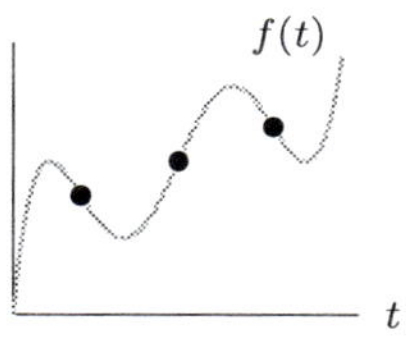

3 One

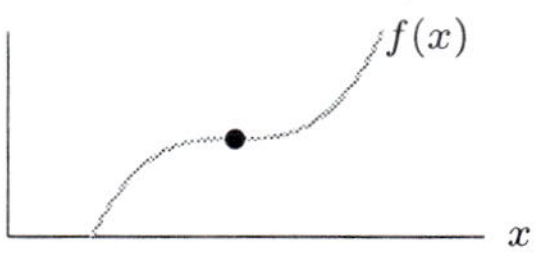

5 (a)

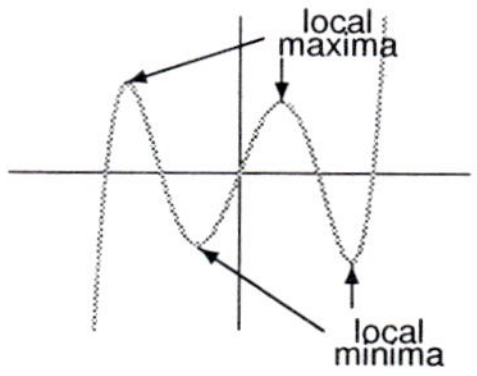

(b) 3

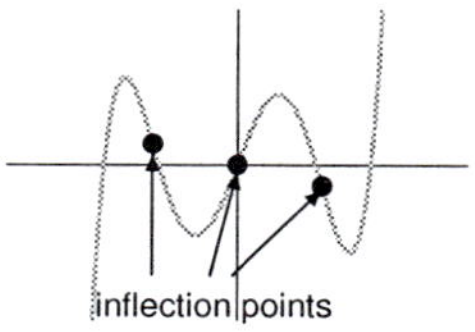

7

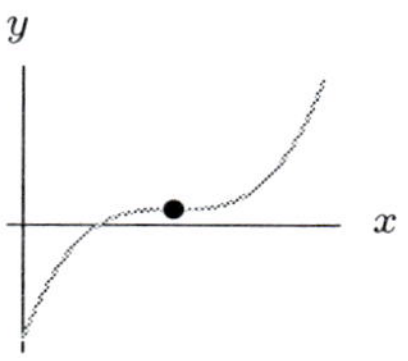

9 (a) 6 pm
(b) Noon; another between noon and 6 pm

11 Critical points: $x = \frac{5}{2}$, local minimum
Inflection points: None

13 Critical points:
$x = -3$ (local max) and $x = 2$ (local min)
Inflection point: $x = -1/2$

15 Critical points: $x = -1$, local min; $x = 0$ local max; $x = 1$, local min
Inflection points: $x = -1/\sqrt{3}$, $x = 1/\sqrt{3}$

17 Critical points:
$x = 0$ (local max) and $x = \pm 2$ (local minima)
Inflection points: $x = \pm 2/\sqrt{3}$

19 Critical points: $x = 0$, local max; $x = 4$, local min
Inflection point: $x = 3$

21 $x = -1, 1/2$

23

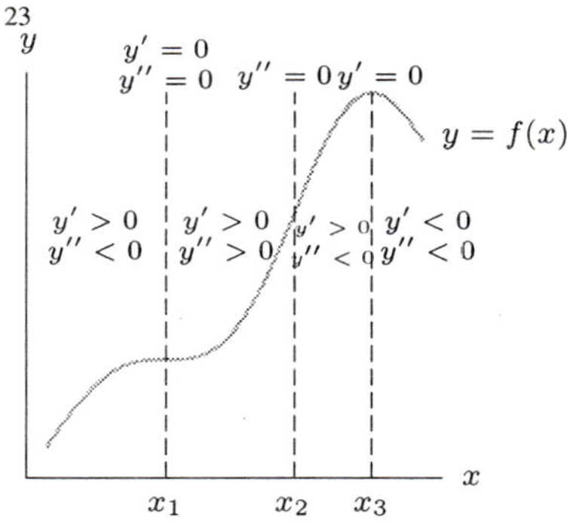

27 (a) Yes, at 2000 rabbits

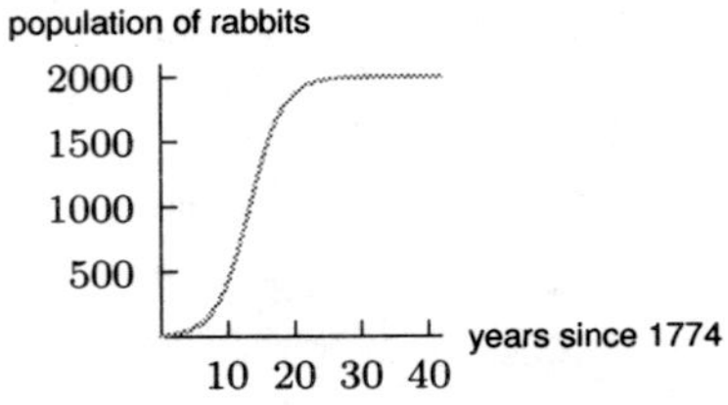

(b) 1787, 1000 rabbits
(c) 1787, 1000 rabbits

29

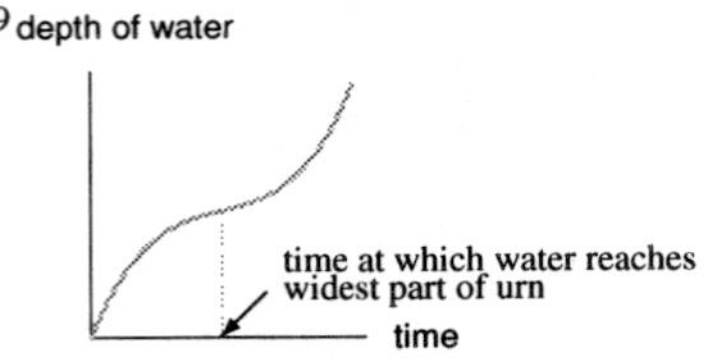

31

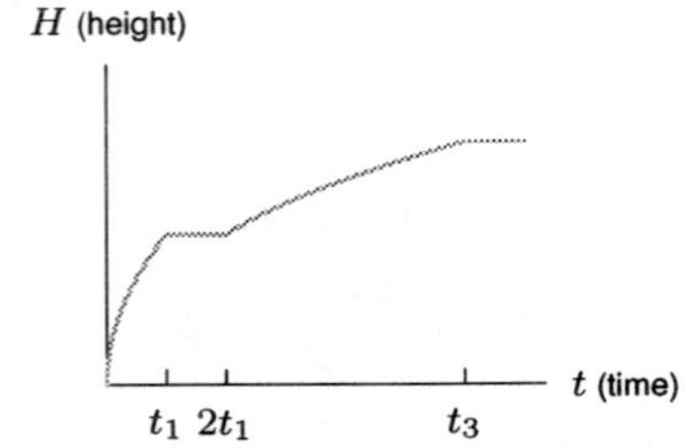

33 $y = x^3 - 3x^2 + 6$

35 (a)

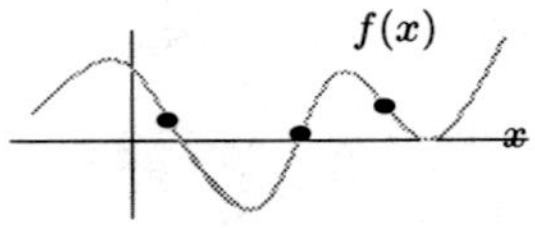

(b)

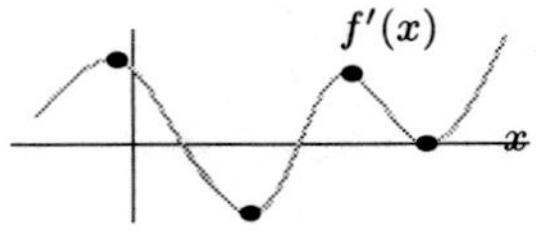

(c)

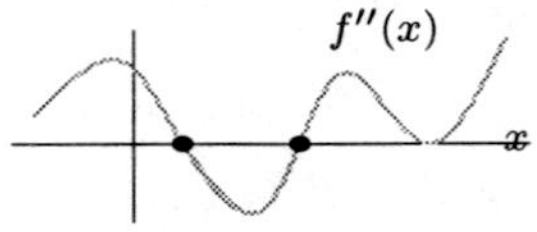

Section 4.3

1

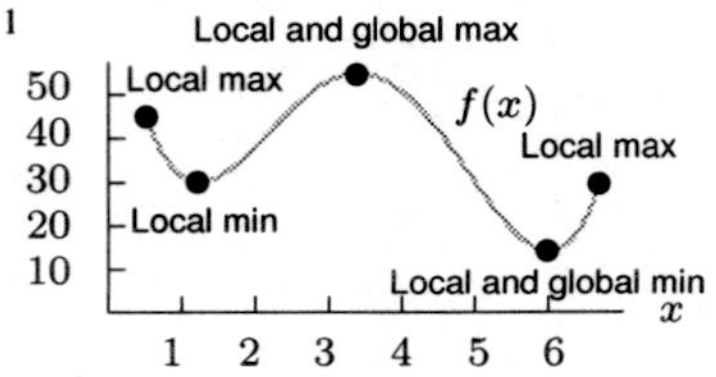

3 (a) (IV)
(b) (I)
(c) (III)
(d) (II)

5

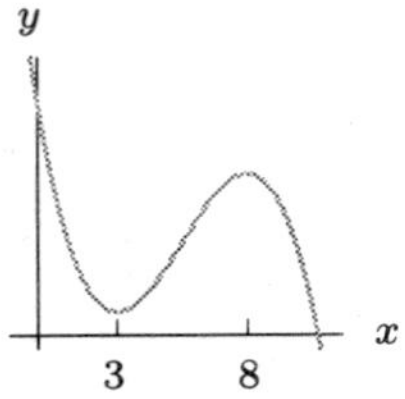

7

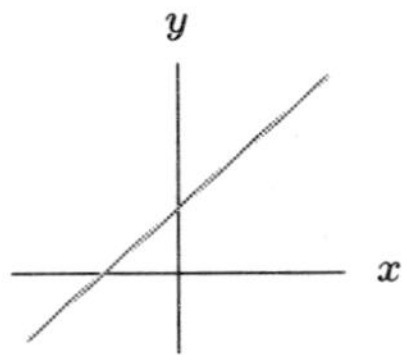

9

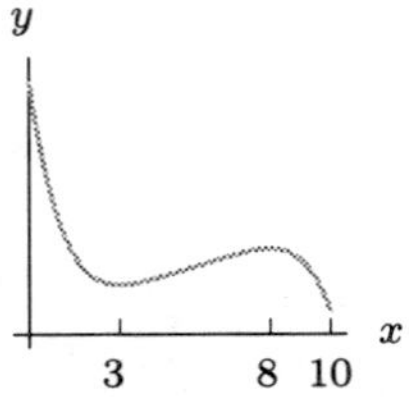

11

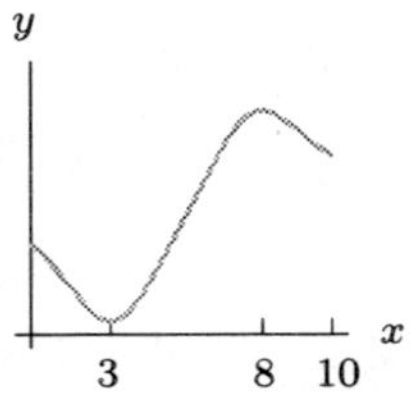

13 44.1 feet

15 (a) $f(0.91)$ is global minimum
$f(3.73)$ is global maximum
(b) $f(-3)$ is global minimum
$f(2)$ is global maximum

17 (a) $D = C$
(b) $D = C/2$

19 (a) $f'(x) = 6x^2 - 18x + 12$,
$f''(x) = 12x - 18$.
(b) $x = 1, 2$
(c) $x = 3/2$
(d) Local minimum: $x = 2$
Local maximum: $x = 1$
Global minimum: $x = -0.5$
Global maximum: $x = 3$
(e)

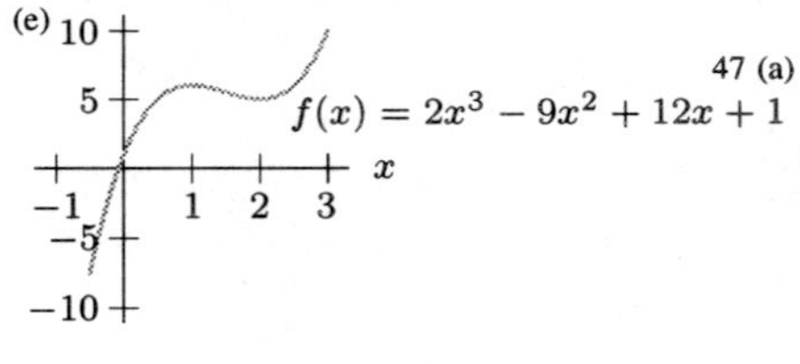

21 (a) $f'(x) = 1 + \cos x$,
$f''(x) = -\sin x$.
(b) $x = \pi$
(c) $x = 0, \pi, 2\pi$
(d) Global minimum: $x = 0$
Global maximum: $x = 2\pi$
(e)

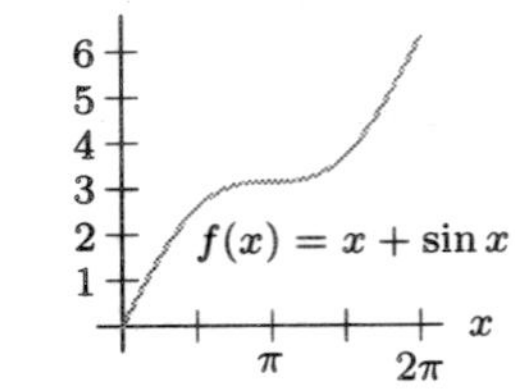

23 Global max $= -1$ at $x = 2$
No global min

25 Global max $= 1/e$ at $t = 1$
No global min

27 Global max $= 1/2$ at $t = 1$
Global min $= -1/2$ at $t = -1$

29 $x = -b/2a$,
Max if $a < 0$, min if $a > 0$

31 (b) Yes, at $x = 0$
(c) Max: $x = -2$, Min: $x = 2$
(d) $5 > g(0) > g(2)$

33 1250 square feet

35 (a) $x + 27/x$ meters
(b) 5.2 meters

37 (a) $y = \begin{cases} 200x & x \le 200 \\ 600x - x^2 & x > 200 \end{cases}$

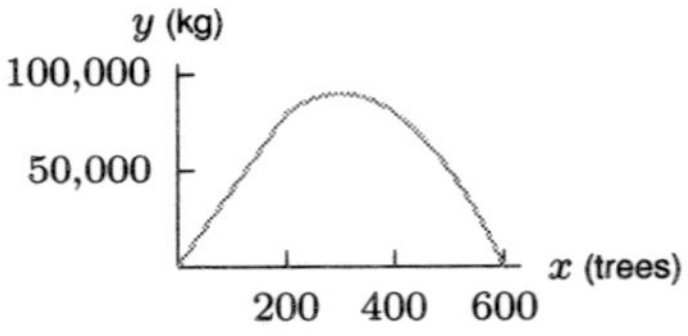

(b) 300 trees/km^2

39 (a) $H = 1/b$, $S = ae^{-1}/b$
(b) a: Increases
b: Decreases

41 (a) 10
(b) 9

43 (a) $k(\ln k - \ln S_0) - k + S_0 + I_0$
(b) Both

45 (a) 120 mm Hg, 80 mm Hg
(b) 0.8 sec
(c)

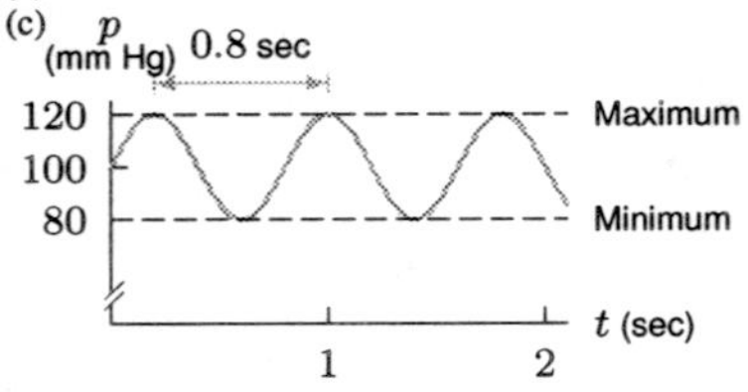

47 (a) $0 \le y \le a$

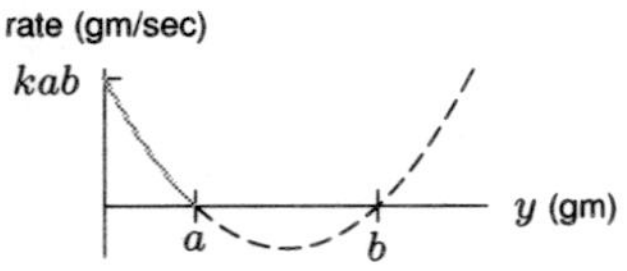

(b) $y = 0$

Section 4.4

1 $5.5 < q < 12.5$ positive;
$0 < q < 5.5$ and $q > 12.5$ negative;
Maximum at $q \approx 9.5$

3 (a) $q = 2500$
(b) \$3 per unit
(c) \$3000

5 (a) \$9
(b) $-\$3$
(c) $C'(78) = R'(78)$

7 (a) Increase production
(b) $q = 8000$

9 \$0.20/item

11 $q = 0$ or $q = 3000$

13 One; between 40 and 50

15 (a) MR; increase production
(b) MC; decrease production

17 Global maximum of \$6875 at $q = 75$

19 (a) Approximately \$1
(b) No
(c) About 400 items

21 (a) \$10; \$30,000; \$50,000
(b) $R(q) = 70q - 0.02q^2$
(c) 1750
(d) \$35
(e) \$61,250

23 \$14.

25 (a) $10{,}000 + 2q$
(b) $q = 37{,}820 - 5544p$
(c) $\pi = -0.00018q^2 + 4.822q - 10{,}000$
(d) 13, 394 items, \$22, 294

27 Maximum revenue = \$27,225
Minimum = \$0

29 (a) q/r months
(b) $(ra/q) + rb$ dollars
(c) $C = (ra/q) + rb + kq/2$ dollars
(d) $q = \sqrt{2ra/k}$

31 $L = [\beta pcK^{\alpha}/w]^{1/(1-\beta)}$

Section 4.5

1 (a) No
(b) Yes

3 (a) $a(1000) \approx \$1.60$ per unit
(b)

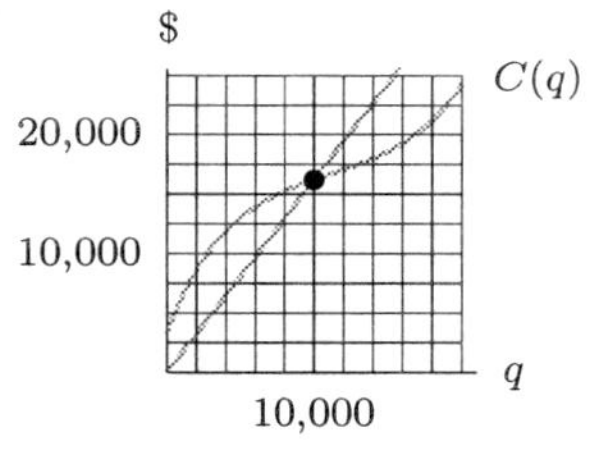

(c) 18,000 units

5 $MC = \$20$; $a(q) = \$25$

7 (a) $q = 6$

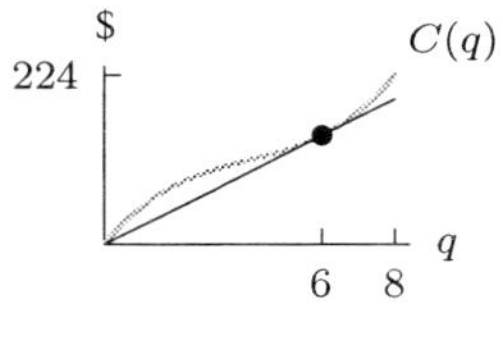

(b) $q = 6$

9 (a) $C(q) = 0.01q^3 - 0.6q^2 + 13q$
(b) \$1
(c) $q = 30$, $a(30) = 4$
(d) Marginal cost is 4

11 (a)

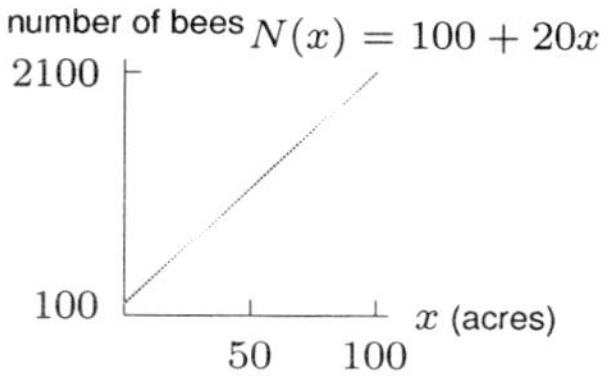

(b) (i) $N'(x) = 20$
(ii) $N(x)/x = (100/x) + 20$

13

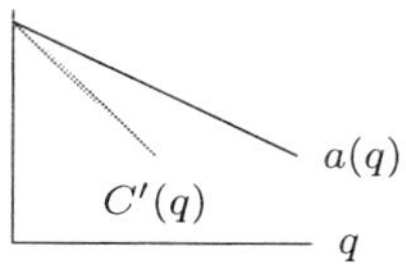

Section 4.6

1 (a) 1.5% decrease
(b) 1.5% increase

5 Elastic

7 High

9 Low

11 (a) $E \approx 0.470$, inelastic
(c) $P = 1.25$ and 1.50

13 (a) $q = 4960$
(b) $E = 0.016$, so demand is inelastic

Section 4.7

1 (a) 40 billion
(b)

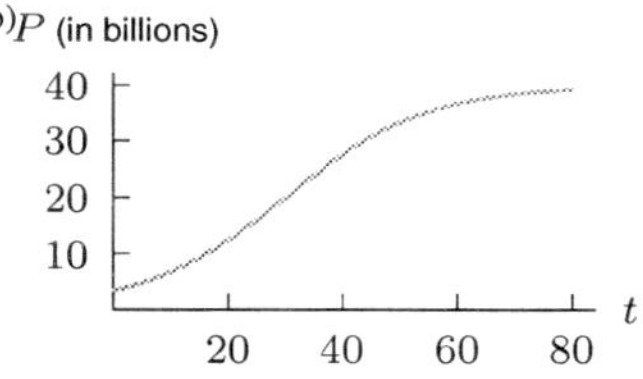

(c) 2020; 2095

3 (a)

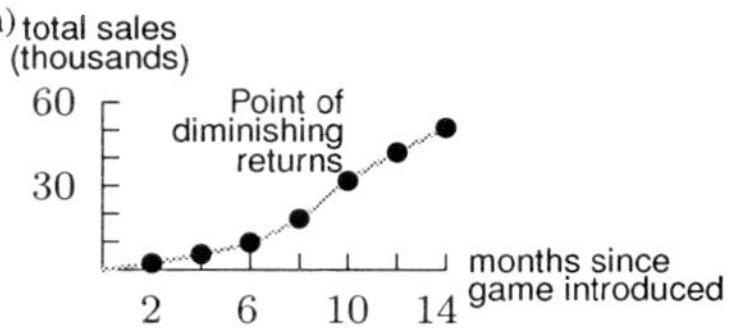

(b) 60,000

7 (a) 36 thousand; total number infected
(b) $t \approx 16$, $n \approx 18$ thousand
(c) Virus spreading fastest
(d) Number infected half total

9 (a) (b)

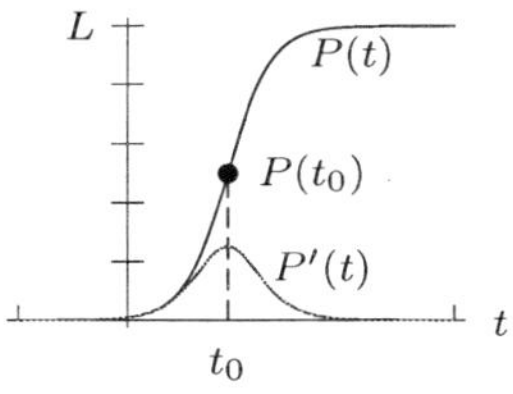

(c) P' maximum

11 (b) April 12, 2003
(d) $t = 19$; 1600 cases
(e) 1760 cases

15 (a)

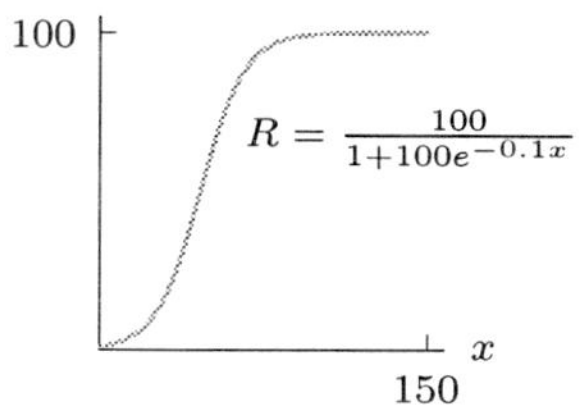

(b) About 46.05
(c) Between 32.12 and 54.52 mg

17 (b) $f'(10) < 11$
$f'(20) < 11$

19 10 mg to 18 mg

Section 4.8

1 (a)

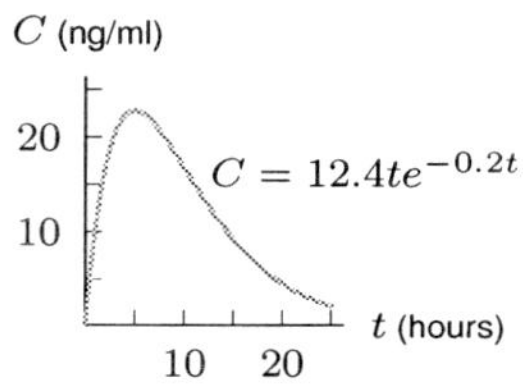

(b) 5 hrs, 22.8 ng/ml
(c) 1 to 14.4 hrs
(d) At least 20.8 hrs

9 $a = 49.3$; $b = 0.769$

13 (b) Min effective ≈ 0.2 Max safe ≈ 1.0
(c) Min effective ≈ 1.2 Max safe ≈ 2.0

Chapter 4 Review

1

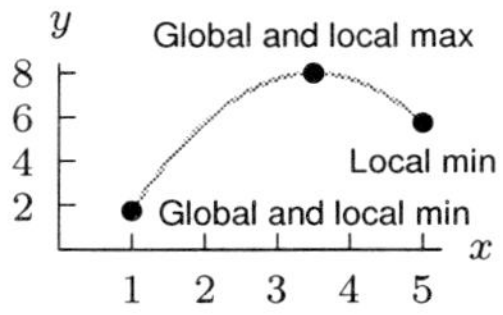

3 (a) $f(1)$ local minimum;
$f(0)$, $f(2)$ local maxima
(b) $f(1)$ global minimum
$f(2)$ global maximum

5 Global min $= 3$ at $z = 1/2$
No global max

7 Global max $= 1/2$ at $x = 1$
No global min

11 (a) Increasing for all x
(b) No maxima or minima

13 (a) Incr: $-1 < x < 0$ and $x > 1$
Decr: $x < -1$ and $0 < x < 1$
(b) Local max: $f(0)$
Local min: $f(-1)$ and $f(1)$

15 (a) cm/week

(b) Growing at 1.6 cm/wk in week 24

17 (a) Week 14
(b) Point of fastest growth

19 $a = 3e, b = -3$

21 (a) $5x^4 + 1$; positive
(b) One

25 (a) Polynomial; negative leading coefficient; degree 2
(b) Exponential.
(c) Logistic
(d) Logarithmic.
(e) Polynomial; positive leading coefficient; degree 2
(f) Exponential.
(g) Surge

27 (a) $0
(b) $96.56
(c) Raise the price by $5

29 (a) $\pi(q)$ max when $R(q) > C(q)$ and R and Q are farthest apart

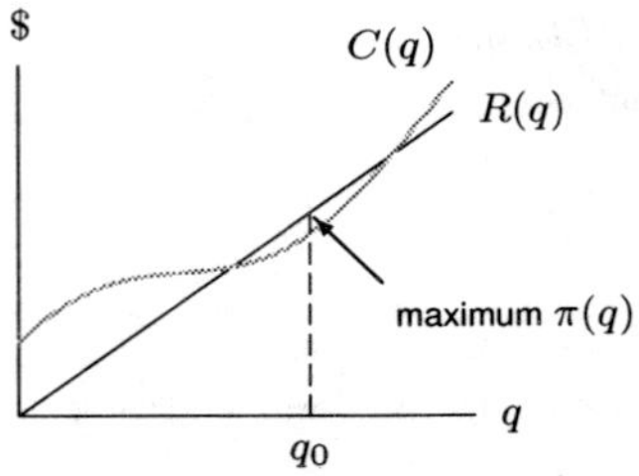

(b) $C'(q_0) = R'(q_0) = p$
(c)

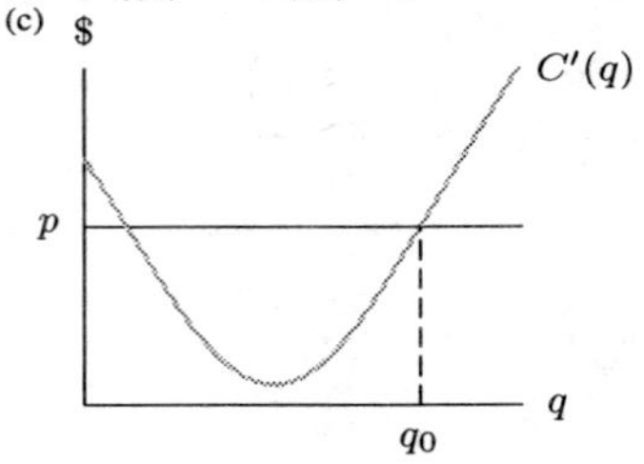

31

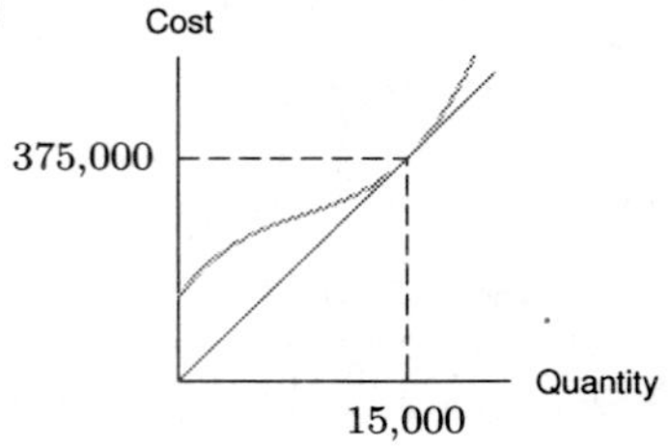

33 $C'(2)$

35 $(C(75) - C(50))/25$

37 $C(3)/3$

39 $E = 0.05$, demand is inelastic.

43 50 m by 50 m

45 (a) $1/5, 1$
(b) Local max: $p = 1/5$
Local min: $p = 1$
(c) Max: $256/3125$
Min: 0

47 (a) $E = 500e\left(\frac{2 - \cos\theta}{\sin\theta}\right) + 2000e$
$\left(\arctan\left(\frac{500}{2000}\right) \le \theta \le \pi/2\right)$
(b) $\theta = \pi/3$
(c) Independent of e, but dependent on $\overline{AB}/\overline{AL}$

49 (a) $k(1 - 2P/L) \cdot dP/dt$

51 Intercepts: $(0, 0)$, $(1.77, 0)$, $(2.51, 0)$
Critical Points: $(0, 0)$, $(1.25, 1)$, $(2.17, -1)$, $(2.80, 1)$
Inflection Points: $(0.81, 0.61)$, $(1.81, -0.13)$, $(2.52, 0.07)$

53 (a)

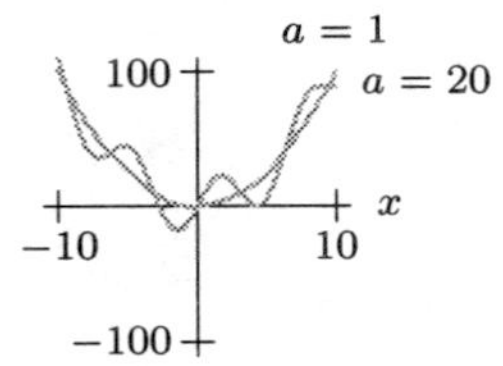

(b) $-2 < a < 2$

55 (a)

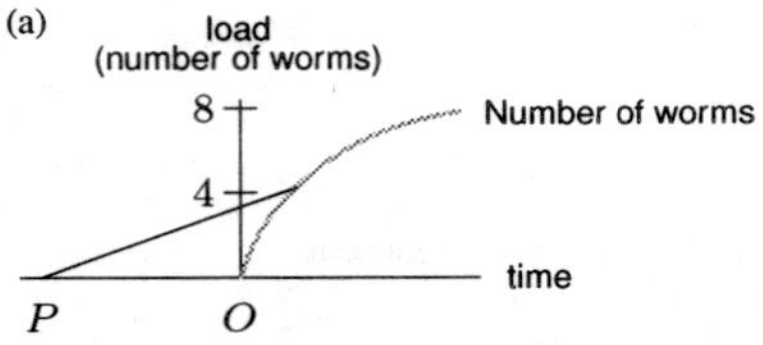

(b) 7 worms
(c) increases

57 Max slope $= 1/(3e)$ at $x = 1/3$

Ch. 4 Understanding

1 True

3 False

5 False

7 False

9 One possibility: $f(x) = ax^2, a \ne 0$

11 (a) True
(b) False
(c) True
(d) False
(e) True

13 False

15 True

17 True

19 True

21 False

23 False

25 $f(x) = x^2 + 1$

27 $f(x) = -x^2 - 1$

29 Impossible

Section 5.1

1 (a)

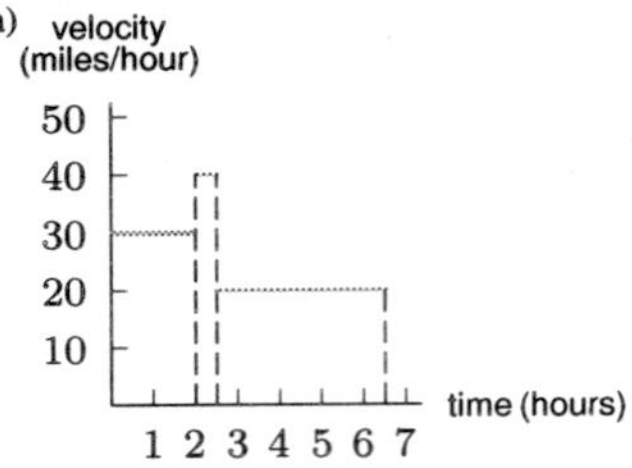

(b)

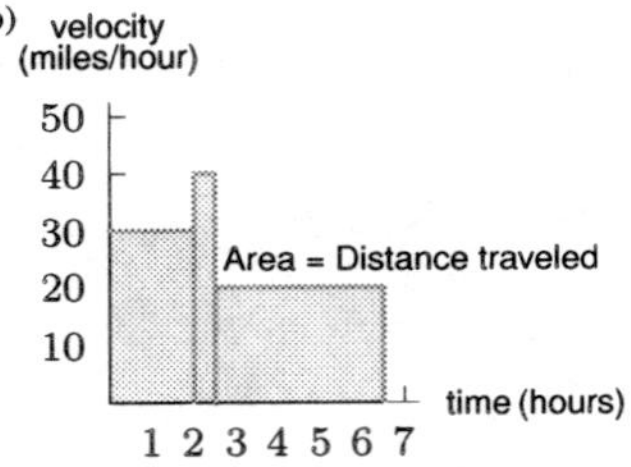

3 (a) Lower estimate $= 122$ ft
Upper estimate $= 298$ ft
(b)

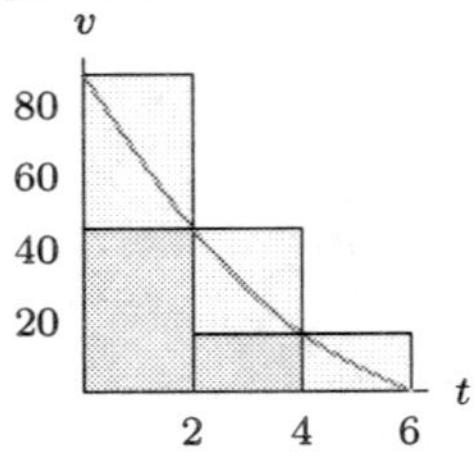

5 250 meters

7 ≈ 455 feet or 0.086 miles

9 (a) About 420 kg
(b) 336 and 504 kg

11 (a) 570 m^3
(b) Every 2 minutes

13 (a)

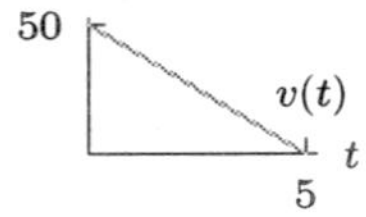

(b) 125 feet
(c) 4 times as far

15 (a) Car A
(b) Car A
(c) Car B

17 60 m (Other answers possible.)

19 (b) $6151

Section 5.2

1 27

3 1692.5

5 About 543

7 (a) 224

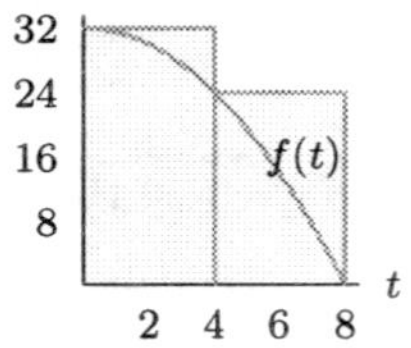

(b) 96

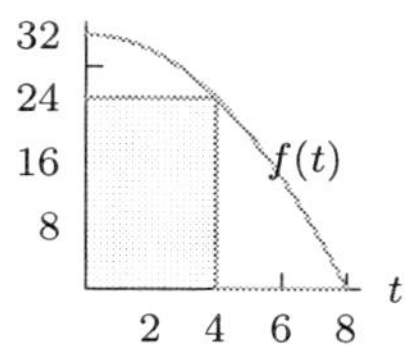

(c) 200

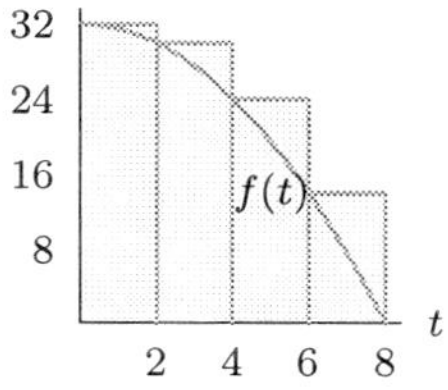

(d) 136

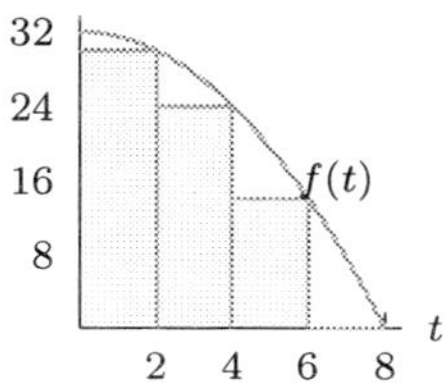

9 350

11 About 20

13 (a) 3.6

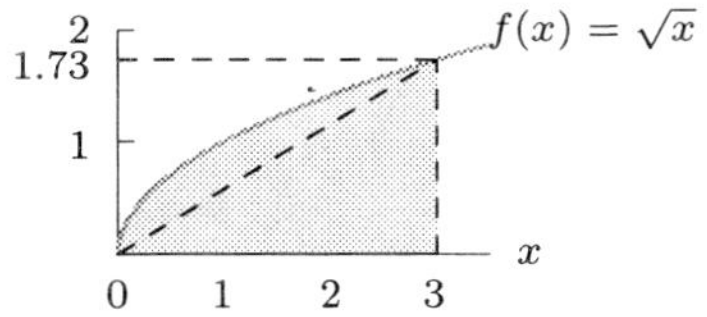

(b) 3.4641

15 (a) 2

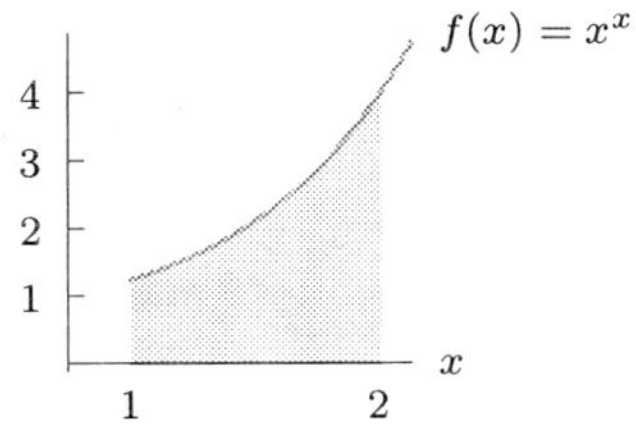

(b) 2.05045

17 (a) 78

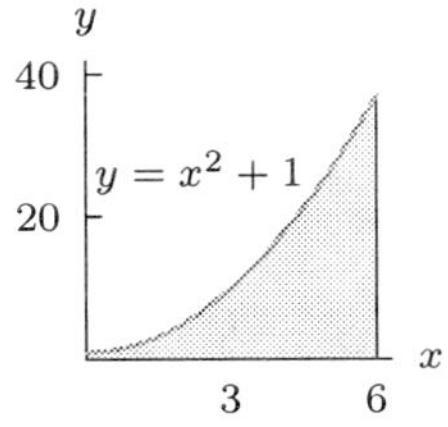

(b) 46; underestimate

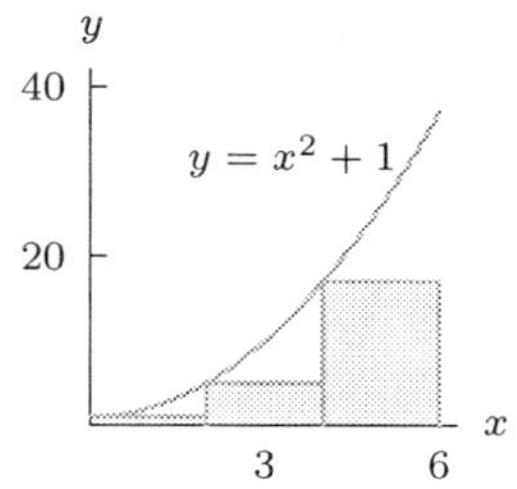

(c) 118; overestimate

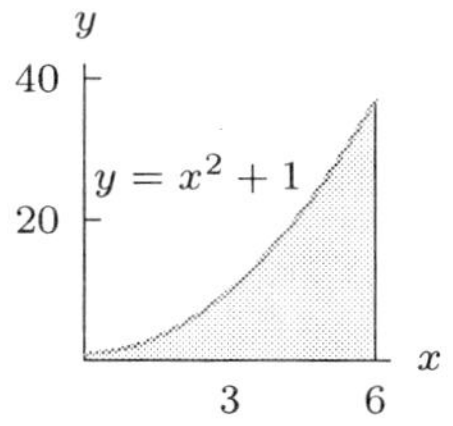

19 93.47

21 448.0

23 2.350

25 2.9

27 1.30

29 1.728, 1.816, 1.772, $n = 140$

31 (a) 4; 0, 4, 8, 12, 16; 25, 23, 22, 20, 17
(b) 360; 328
(c) 8; 0, 8, 16; 25, 22, 17
(d) 376; 312

Section 5.3

1 8

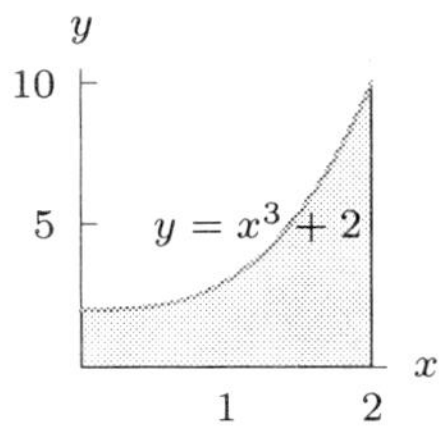

3 $\int_0^2 ((x + 5) - (2x + 1))\, dx = 6$

5 (a) 13
(b) 1

7 Positive

9 Positive

11 −40

13 (a) −4
(b) 0
(c) 8

15 II

17 III

19 V < IV < II < III < I
I, II, III positive
IV, V negative

21 (a) -2
(b) $-A/2$

23 About 3.34

25 0.80

27 14,052.5

29 14.688

31 13.457

33 2.762

Section 5.4

1 (a) Emissions 1970–2000, m. metric tons
(b) 772.8 million metric tons

3 Change in position; meters

5 Change in world pop; bn people

7 Total amount $= \int_0^{60} f(t)\, dt$.

9 1417 antibodies

11 29,089.813 megawatts

13 3.4 ft

15 (a) Concave up
(b) 3.1 kg

17 (a) 2023, at crossing of curves
(b) 2023, at highest point of curve

19 2627 acres

21 Tree B is taller after 5 years.
Tree A is taller after 10 years.

23 (a) Boys: black curve; girls: colored curve
(b) About 43 cm
(c) Boys: about 23 cm; girls: about 18 cm
(d) About 13 cm taller

25 More fat

27 (a) About 750 liters
(b) $\int_0^3 60g(t)\, dt$
(c) About 150 liters

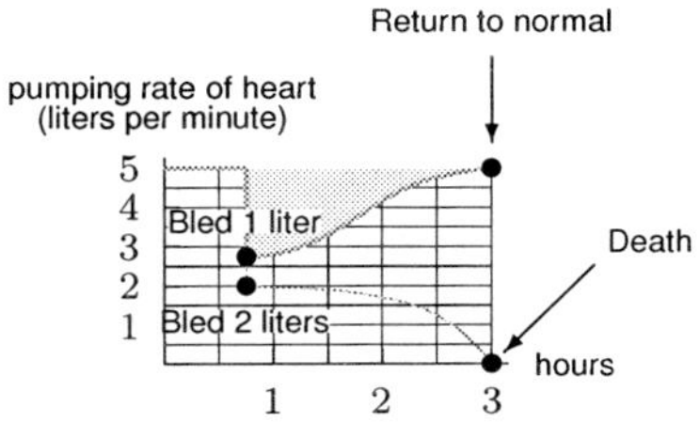

29 15 cm to the left

31 25 cm to the right

33 65 km from home
3 hours
90 km

35 Product B has a greater peak concentration
Product A peaks sooner
Product B has a greater overall bioavailability
Product A should be used

37

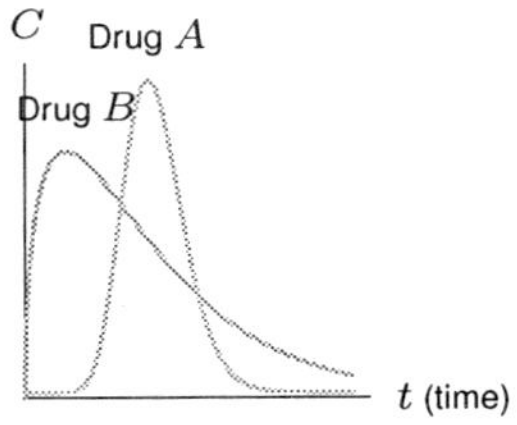

39 About $13,800

41 (a) $\int_0^T 49(1 - (0.8187)^t)\,dt$ (meters)
(b) $T \approx 107$ seconds

Section 5.5

1 Dollars; cost of increasing production from 800 tons to 900 tons

3 (a)

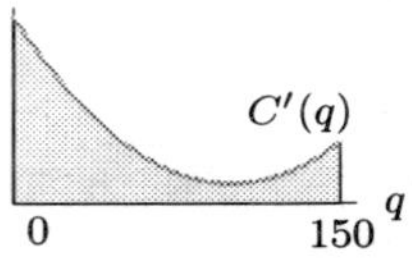

(b) $22,775
(c) $C'(150) = 18.5$
(d) $C(151) \approx \$22,793.50$

5 7.65 million people

7 (a) $18,650
(b) $C'(400) = 28$

9 (a)

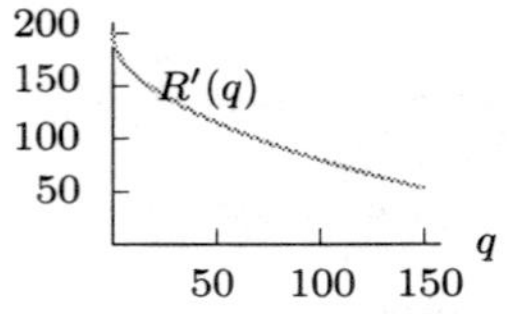

(b) $12,000
(c) Marginal revenue is $80/unit
Total revenue is $12,080

11 (a) 7.54 inches after 8 hours
(b) 1.41 inches/hour

Chapter 5 Review

1 (a) 408
(b) 390

3 1096

5 Meters per second

7 Foot-pounds

9 1.44

11 1.15

13 −0.083

15 84

17 0.0833

19 About 0.1667

21 (a) 430 ft
(b) (ii)

23 (a)

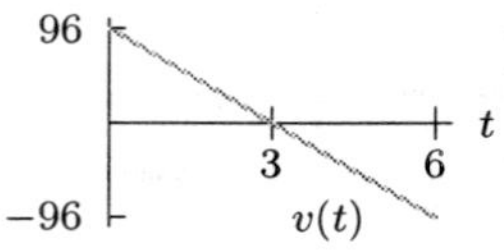

(b) 3 sec, 144 feet
(c) 80 feet

25 (a) Species B for both
(b) Species A

27 (a) Inflection point
(b) About 50 cm

29 (a) Negative
(b) Positive
(c) Negative
(d) Positive

33 (a) $\int_0^5 (10 + 8t - t^2)\,dt$
(b) 100
(c) 108.33

35 (a) Forward for $t < 3$, backward for $t > 3$
(b) Farthest forward: $t = 3$
Farthest backward:
no upper bound.

37 $300,000

39 (a)

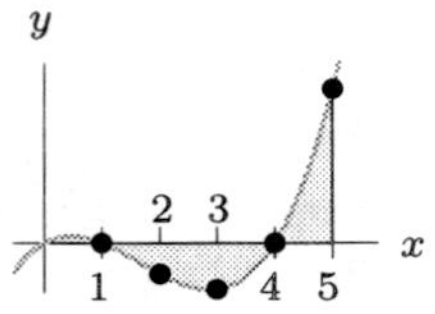

(b) I_1 is largest.
I_4 is smallest.
I_1 is the only positive value.

41 9 years

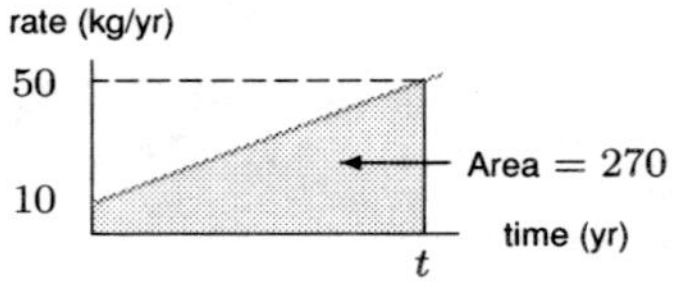

43 $F(x)$ is decreasing for $x < -2$ and $x > 2$;
$F(x)$ is increasing for $-2 < x < 2$

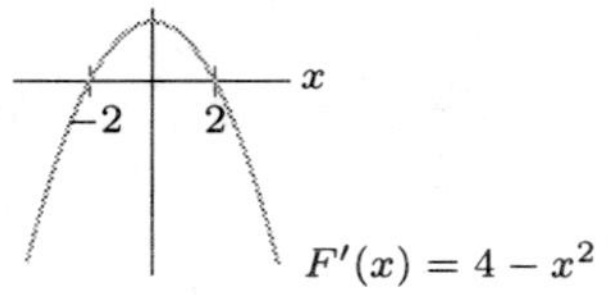

Ch. 5 Understanding

1 False

3 True

5 False

7 False

9 True

11 True

13 False

15 False

17 False

19 True

21 True

23 False

25 False

27

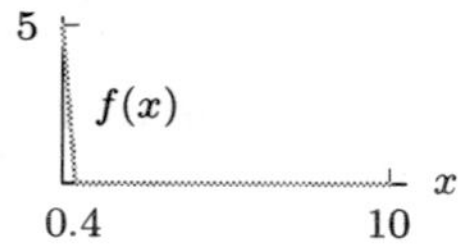

Theory: Second Fund. Thm.

1 x^3

3 xe^x

5 (a) 0
(b) F increases
(c) $F(1) \approx 1.4$, $F(2) \approx 4.3$, $F(3) \approx 10.1$

9 9

11 $8c$

Section 6.1

1 1.7

3 2

5 −3

7 About 8.5

9 (a) $26,667 per year
(b) Less 25–65; more 65–85

11 $6080

13 (a) 120 mm Hg
(b) 80 mm Hg
(c) 100 mm Hg
(d) Less

15 (a) 0.375 thousand/hour
(b) 1.75 thousands

17 (a) 9.9 hours
(b) 14.4 hours
(c) 12.0 hours

19 (a) Second half

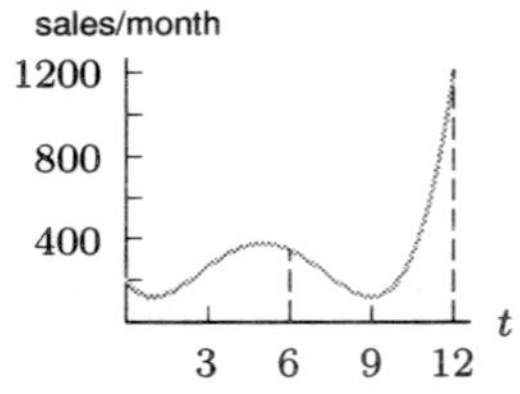

(b) $1531.20, $1963.20
(c) $3494.40
(d) $291.20/month

21 $(c) < (a) < (b)$

Section 6.2

1 (a) $p^* = \$30$, $q^* = 6000$
(b) Consumer surplus = $210,000;
Producer surplus ≈ $70,000

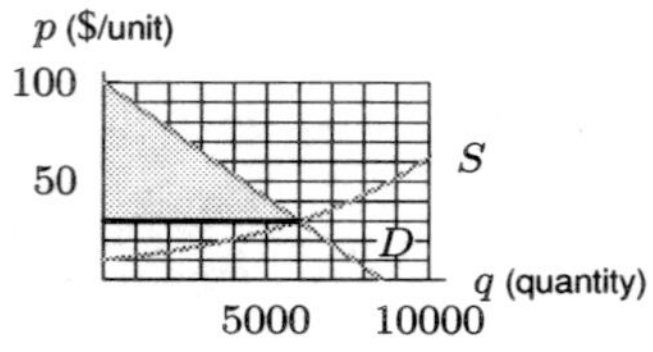

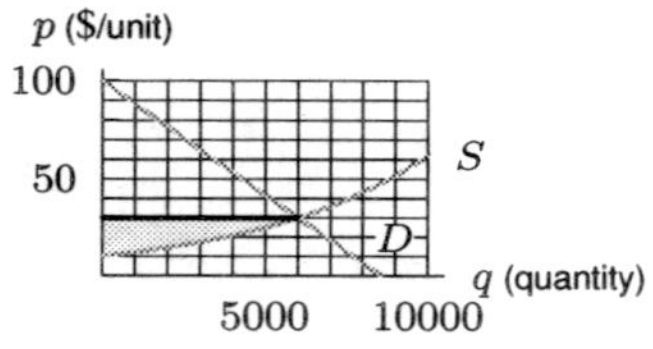

3 250

5 200

7 (a) Equilibrium price is $30
Equilibrium quantity is 125 units
(b) $3500
$2000

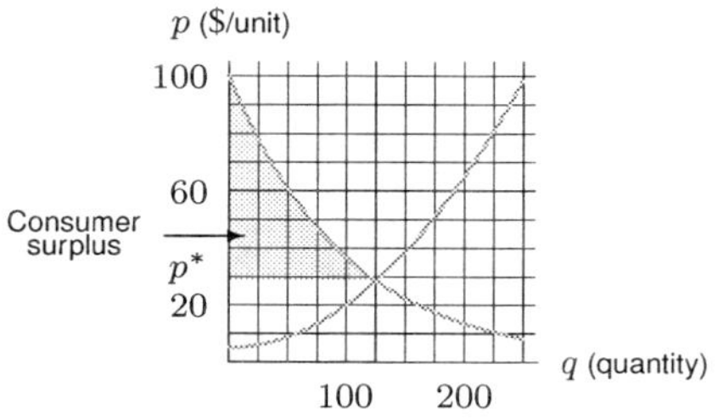

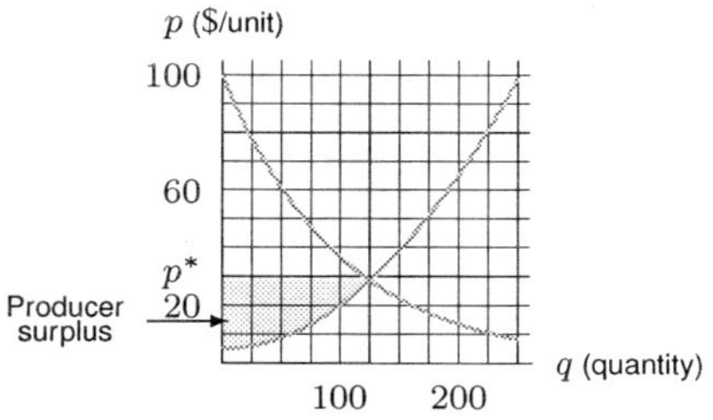

(c) $5500

9 (a) Supplied: $11, demanded: $7
(b) $p^* \approx \$9$
$q^* \approx 400$

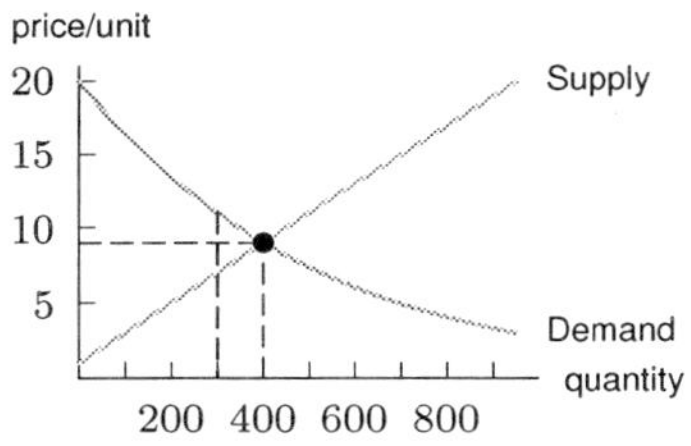

(c) $1907
(d) $1600

11 (a) Less
(b) Can't tell
(c) Less

13 (a) No, yes

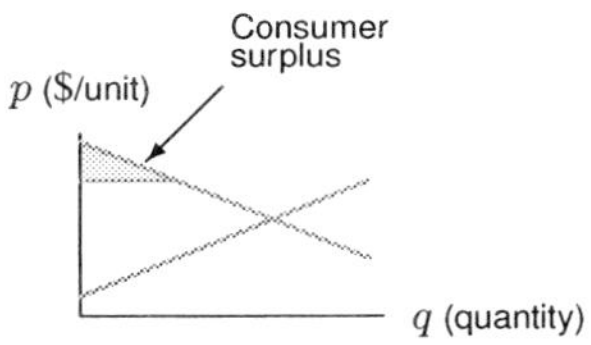

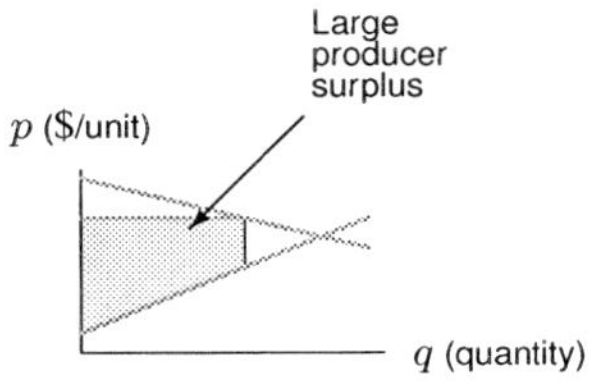

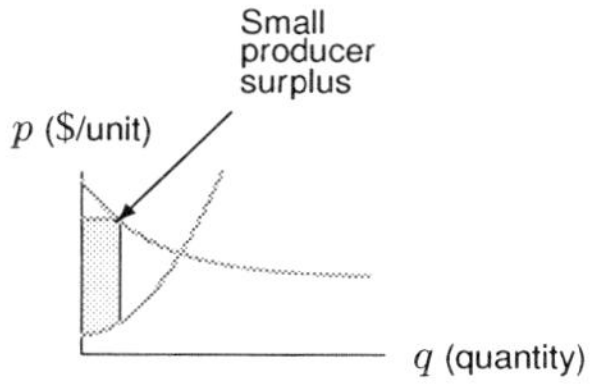

(b) Yes, no

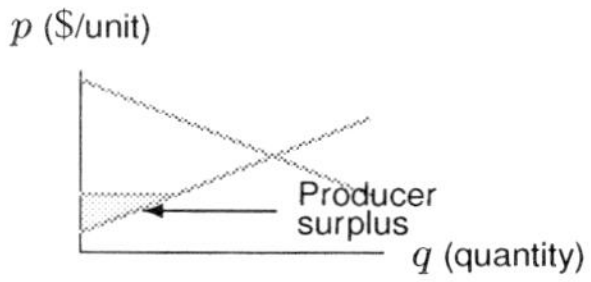

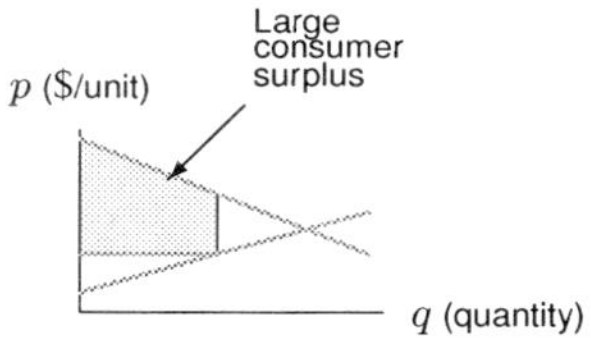

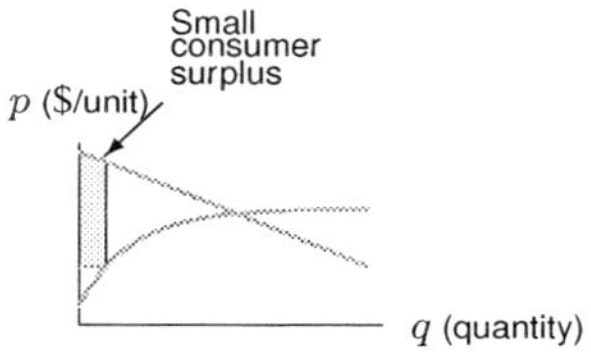

Section 6.3

1 Future value = $72,980.16
Present value = $29,671.52

3 (a) $P = \$47{,}216.32$
$F = \$77{,}846.55$
(b) $60,000; $17,846.55

5 (a) (i) $18,846.59
(ii) $16,484.00
(b) (i) $21,249.47
(ii) $24,591.24

7 (a) $65,022
(b) ≈ 2.27 years

9 (a) $417,635.11
(b) $228,174.64

11 $41,508

13 (a) $4.6 billion; $8.6 billion
(b) $54.7 billion
(c) $77.6 billion

15 About 1.75 years

17 In 10 years

Section 6.4

3 11%

5 Absolute:
2003–2004: 3.1 m
2006–2007: 2.0 m
Relative:
2003–2004: 10.3%
2006–2007: 5.3%

7 (a) $P = 4000 - 100t$
(b) $P = 4000(0.95)^t$
Case (a)

9 22% increase

11 No change

13 Increasing $0 \le t \le 10$

15 Decreasing: $0 < t < 5$
Increasing: $5 < t < 10$

17 (a) $P = 5.78(1.018)^t$
(b) 0.10404 mil/yr
0.10591 mil/yr
(c) 1.8%, 1.8%

19 Decreases by about 9.52%

Chapter 6 Review

1 (a) 20
(b) 10/3

3 (a) 0.79

5 (a) $Q(10) \approx 2.7$,
$Q(20) \approx 1.8$
(b) 2.21
(c) 2.18

7 About 17

9 (a) $(1/5)\int_0^5 f(x)\,dx$
(b) $(1/5)(\int_0^2 f(x)\,dx - \int_2^5 f(x)\,dx)$

11 (a)

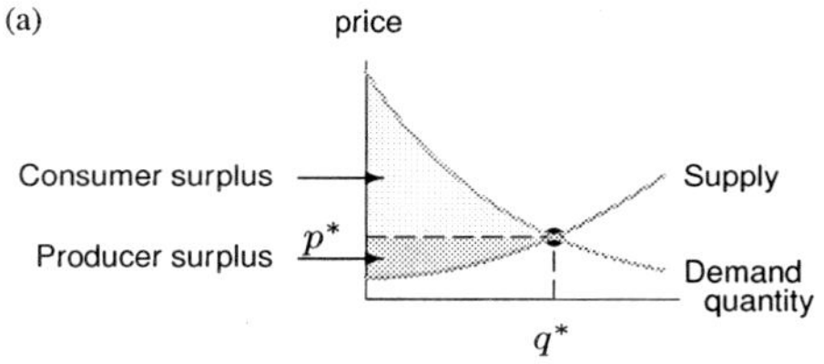

(b)

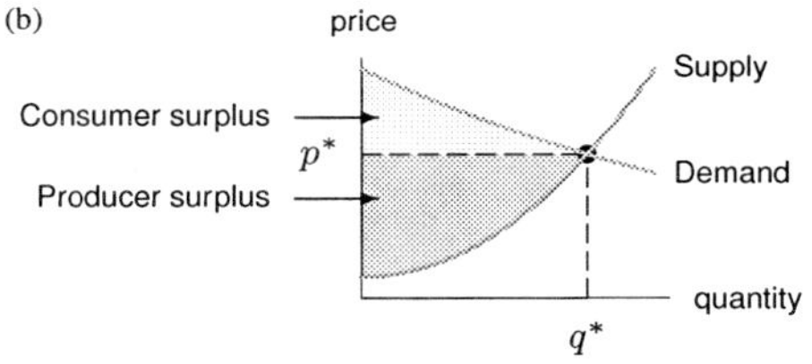

13 (a) Consumer: 87 units
Suppliers: 100 units
Down
(b) $p^* = \$48$
$q^* = 91$ units
(c) Consumer surplus: $2096
Producer surplus: $1143

15 (a) $5820 per year
(b) $36,787.94

17 $635.37 per year

19 (a) $0 \le t \le 10$, 10.5%
(b) $10 \le t \le 15$, 2.5%
(c) 7.8% increase

21 In 45 years; 2035

Ch. 6 Understanding

1 False

3 True

5 False

7 False

9 True

11 True

13 False

15 False

17 False

19 True

21 True

23 False

25 False

27

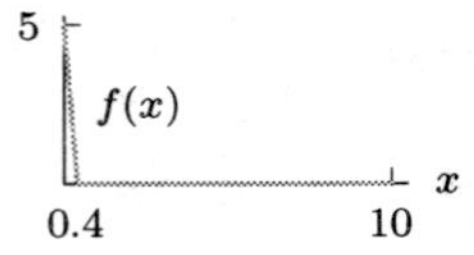

Section 7.1

1 $5x$

3 $t^3/3 + t^2/2$

5 $x^5/5$

7 $3x^4/2 + 4x$

9 $y^3 - y^4/4$

11 $x^3 + 5x$

13 $(x^3/3) - 3x^2 + 17x$

15 $5x^2/2 - 2x^{3/2}/3$

17 $\ln|z|$

19 $-1/2z^2$

21 $F(x) = x^7/7 + x^{-5}/35 + C$

23 $-e^{-3t}/3$

25 $G(t) = 5t + \sin t + C$

27 $F(x) = 3x$ (only possibility)

29 $F(x) = 2x + 2x^2 + (5/3)x^3$ (only possibility)

31 $F(x) = (2/3)x^{3/2}$ (only possibility)

33 $(5/2)x^2 + 7x + C$

35 $2x^3 + C$

37 $(x+1)^3/3 + C$

39 $t^3/3 + 5t^2/2 + t + C$

41 $t^4/4 + 2t^3 + C$

43 $2w^{3/2} + C$

45 $3\ln|t| + \dfrac{2}{t} + C$

47 $x^2/2 + 2x^{1/2} + C$

49 $e^x + 5x + C$

51 $e^{3r}/3 + C$

53 $-\cos t + C$

55 $25e^{4x} + C$

57 $3\sin x + 7\cos x + C$

59 $\frac{1}{2}\sin(x^2+4) + C$

61 $10x - 4\cos(2x) + C$

63 $F(x) = 3x^2 - 5x + 5$

65 $F(x) = -4\cos(2x) + 9$

67 $q^3 + 2q^2 + 6q + 200$

Section 7.2

1 $\frac{1}{5}(x^3+1)^5 + C$

3 $\frac{1}{4}(x+10)^4 + C$

5 $e^{q^2+1} + C$

7 $(1/2)e^{t^2} + C.$

9 $(1/33)(t^3-3)^{11} + C$

11 $(1/9)(x^2-4)^{9/2} + C$

13 $-2\sqrt{4-x} + C$

15 $4\sin(x^3) + C$

17 $(1/5)x^5 + 2x^3 + 9x + C$

19 $\cos(3-t) + C$

21 $-(2/9)(\cos 3t)^{3/2} + C$

23 $(1/7)\sin^7\theta + C$

25 $(\sin x)^3/3 + C$

27 $-\frac{1}{8}\cos(4x^2) + C$

29 $\frac{1}{6}e^{3x^2} + C$

31 $\frac{1}{10}\ln(5q^2+8) + C$

33 $(1/2)\ln(y^2+4) + C$

35 $2e^{\sqrt{y}} + C$

37 $2\sqrt{x+e^x} + C$

39 $(1/2)\ln(x^2+2x+19) + C$

41 (a) Yes; $-0.5\cos(x^2) + C$
(b) No
(c) No
(d) Yes; $-1/(2(1+x^2)) + C$
(e) No
(f) Yes; $-\ln|2+\cos x| + C$

43 (a) $x^4 + 2x^2 + C$
(b) $(x^2+1)^2 + C$
(c) Both correct but differ by a constant

Section 7.3

1 10

3 6

5 1/2

7 125

9 52

11 38/3

13 $(\ln 2)/2 \approx 0.35$

15 $5 - 5e^{-0.2} \approx 0.906$

17 $\sin 1 - \sin(-1) = 2\sin 1$

19 $20(e^{0.15} - 1)$

21 (a) $(\ln 2)/2$
(b) $(\ln 2)/2$

23 $\ln 10$

25 2

27 $\int_0^2 (6x^2+1)\,dx = 18$

29 2.32

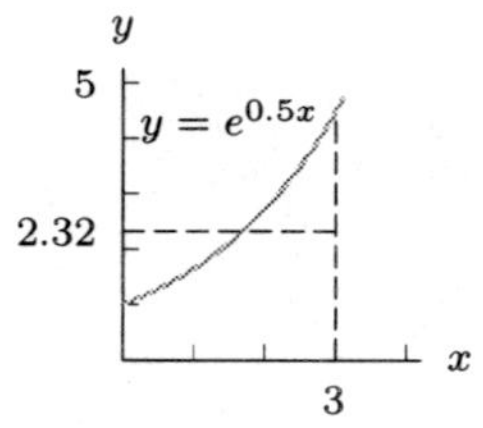

31 11

33 $(300)^{1/3} = 6.694$

35 (a) 6.9 billion, 7.8 billion
(b) 6.5 billion

37 $\int_1^\infty (1/x^2)\,dx$

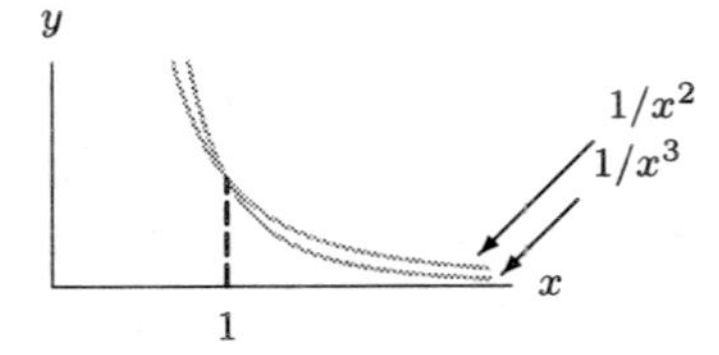

39 (a) 0.4988, 0.49998, 0.4999996, 0.499999998
(b) $(1 - e^{-2b})/2$
(c) Converges to 0.5

41 (a) $\int_0^\infty 1000te^{-0.5t}\,dt$
(b)

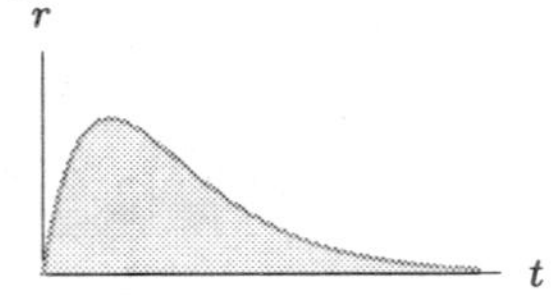

43 1/30

45 (a) No
(b) Yes
(c) Yes

Section 7.4

1 $\frac{1}{5}te^{5t} - \frac{1}{25}e^{5t} + C$

3 $(1/4)(2z+1)e^{2z} + C$

5 $(x^4/4)\ln x - (x^4/16) + C$

7 $(2/3)y(y+3)^{3/2} - (4/15)(y+3)^{5/2} + C$

9 $-(z+1)e^{-z} + C$

11 $-2y(5-y)^{1/2} - (4/3)(5-y)^{3/2} + C$

13 $-t\cos t + \sin t + C$

15 $\frac{1}{5}t^2e^{5t} - \frac{2}{25}te^{5t} + \frac{2}{125}e^{5t} + C$

17 $5\ln 5 - 4 \approx 4.047$

19 $(9/2)\ln 3 - 2 \approx 2.944$

21 (a) Substitution
(b) Substitution
(c) Substitution
(d) Substitution
(e) Parts
(f) Parts

23 $1 - 3e^{-2}$

25 $4\ln 4 - 3\ln 3 - 1$

27 45.71 (ng/ml)-hours

29 Integrate by parts choosing $u = x^n$, $v' = e^x$

Section 7.5

1 3.5, 2, 1.5, 2, 2.5

3 $F(0) = 0$
$F(1) = 1$
$F(2) = 1.5$
$F(3) = 1$
$F(4) = 0$
$F(5) = -1$
$F(6) = -1.5$

5
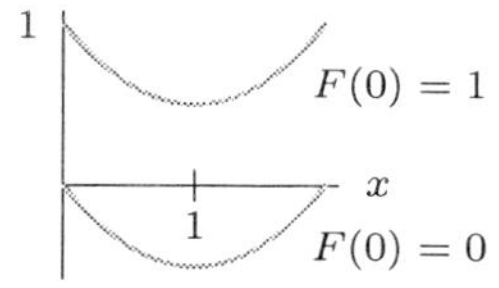

7
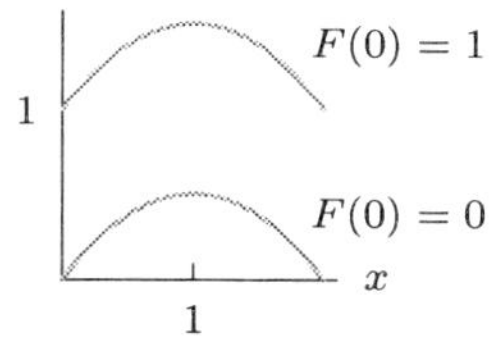

9 (a) Increasing for $x < -2$, $x > 2$
Decreasing for $-2 < x < 2$
Local maximum at $x = -2$
Local minimum at $x = 2$
(b)
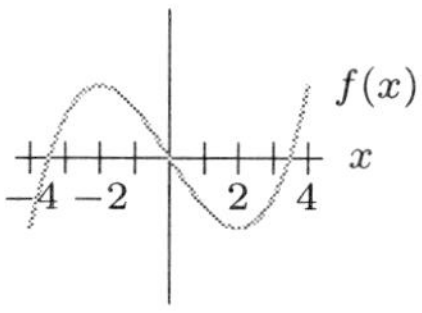

11 (a) $f(x)$ increasing when $2 < x < 5$
$f(x)$ decreasing when $x < 2$ or $x > 5$
f has a local minimum at $x = 2$
f has a local maximum at $x = 5$
(b)
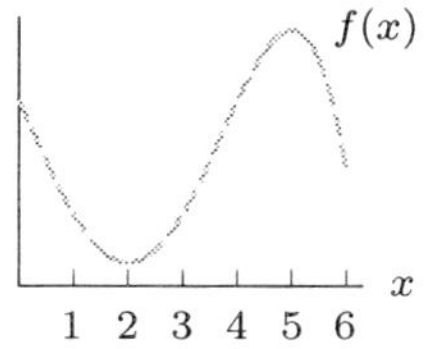

13
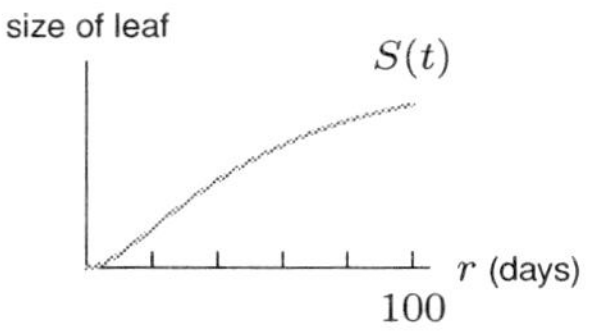

15 (a) (I) volume; (II) flow rate
(b) (I) is an antiderivative of (II)

17 Min: $(1.5, -20)$, max: $(4.67, 5)$

19 Critical points: $(0, 5)$, $(2, 21)$, $(4, 13)$, $(5, 15)$

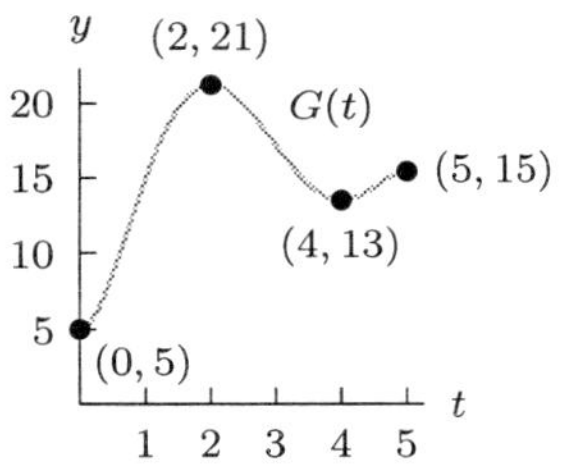

21 (a) $x = -1, x = 1, x = 3$
(b) Local min at $x = -1$, $x = 3$;
local max at $x = 1$
(c)
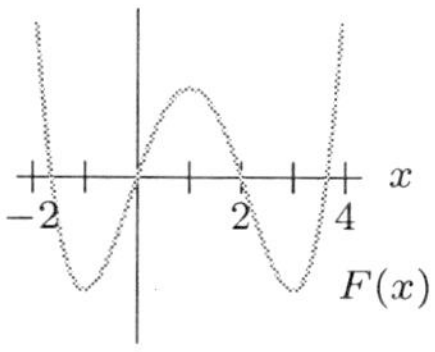

23 $f(3) - f(2)$,
$[f(4) - f(2)]/2$,
$f(4) - f(3)$

25
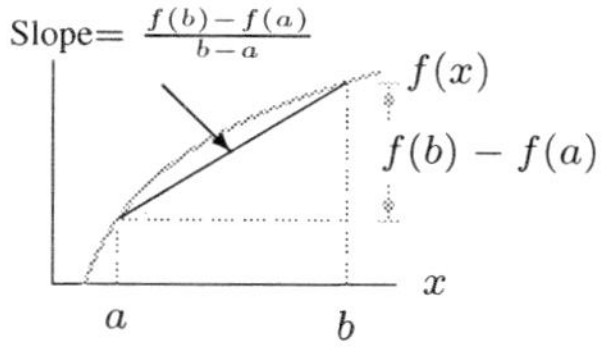

27
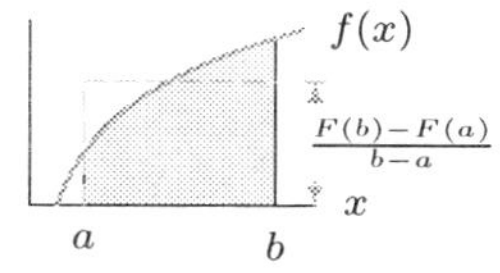

Chapter 7 Review

1 $2t^3/3 + 3t^4/4 + 4t^5/5$

3 $2x^3 - 4x^2 + 3x$

5 $P(r) = \pi r^2 + C$

7 $-1/t$

9 $G(x) = (x + 1)^4/4 + C$

11 $2t^2 + 7t + C$

13 $(x^4/4) - (x^2/2) + C$

15 $-5/t - 3/t^2 + C$

17 $4t^2 + 3t + C$

19 $2x^4 + \ln |x| + C$

21 $2 \ln |x| - \pi \cos x + C$

23 34

25 $1 - \cos 1 \approx 0.460$

27 $1/2$

29 $F(x) = x^2$
(only possibility)

31 $2\sqrt{x^2 + 1} + C$

33 $-(1/2)e^{-x^2} + C$

35 $-1/(3(3x + 1)) + C$

37 $\frac{1}{55}(5x - 7)^{11} + C$

39 $\frac{1}{3}(x^2 + 1)^{3/2} + C$

41 $0.5 \sin(t^2) + C$

43 $\ln 3$

45 (a) $\int_0^5 462e^{0.019t}\,dt$
(b) About 2423 quadrillion BTUs

47
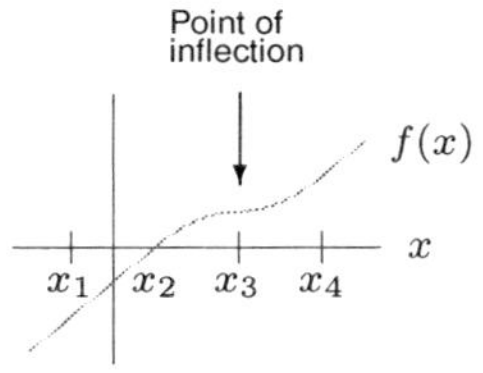

49 $F(1) = -1, F(3) = 7, F(4) = 5$

51 (a)
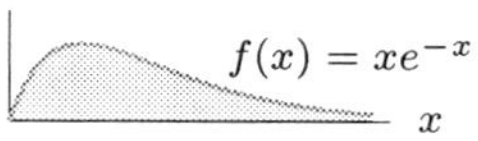

(b) $0.9596, 0.9995,$
0.99999996
(c) Converges to 1

53 250 mg

55 $-y \cos y + \sin y + C$

Ch. 7 Understanding

1 False

3 True

5 False

7 False

9 True

11 True

13 False

15 False

17 False

19 True

21 True

23 False

25 False

27
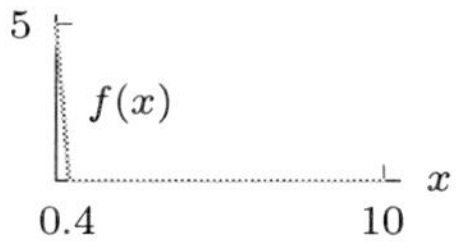

Practice: Integration

1 $q^3/3 + 5q^2/2 + 2q + C$

3 $(x^3/3) + x + C$

5 $4x^{3/2} + C$

7 $(x^4/4) + 2x^2 + 8x + C$

9 $w^5/5 - 3w^4 + 2w^3 - 10w + C$

11 $2q^{1/2} + C$

13 $p^3/3 + 5 \ln |p| + C$

15 $q^4/4 + 4q^2 + 15q + C$

17 $-5 \cos x + 3 \sin x + C$

19 $5 \ln |w| + C$

21 $q^2/2 - 1/(2q^2) + C$

23 $3p^2q^5 + C$

25 $2.5e^{2q} + C$

27 $Ax^4/4 + Bx^2/2 + C$

29 $x^3/3 + 8x + e^x + C$

31 $t^3/3 - 3t^2 + 5t + C$

33 $Aq^2/2 + Bq + C$

35 $\frac{1}{2}e^{2t} + 5t + C$

37 $3\sin(4x) + C$

39 $(1/18)(y^2 + 5)^9 + C$

41 $-(A/B)\cos(Bt) + C$

43 $\ln(2 + e^x) + C$

45 $2\sqrt{1 + \sin x} + C$

47 $xe^x - e^x + C$

Section 8.1

1 (a) 0.25
(b) 0.7
(c) 0.15

3 (a) 0.4375
(b) 0.49
(c) 0.2475

7 (a) 0.19
(b) Tenth; both same
(c) 0.02, 0.38, 0.21

9 1/15

11 1/5

13
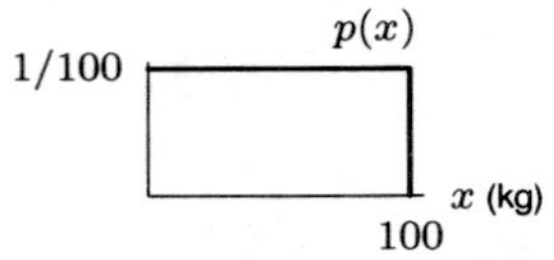

15
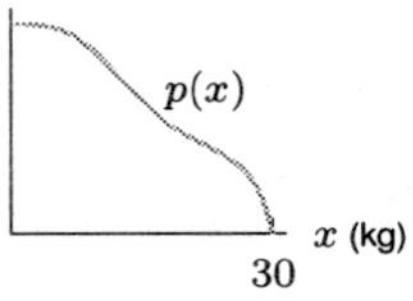

Section 8.2

1 (a)

t	0	1	2	3	4	...
$P(t)$	0	0.60	0.975	1	1	...

(b)
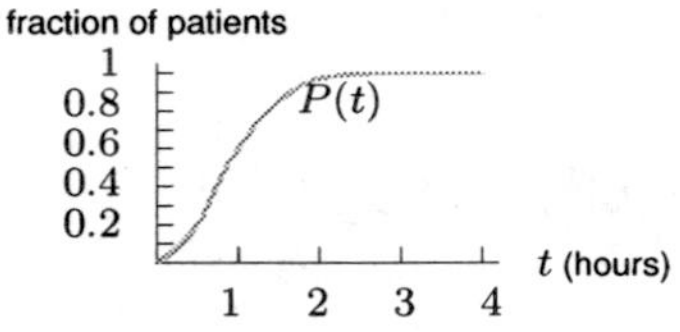

3 (a)
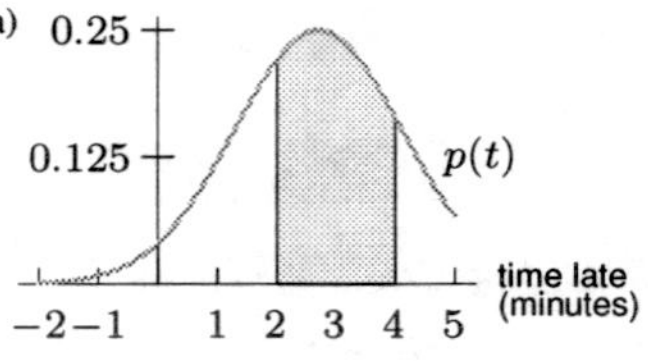

(b)
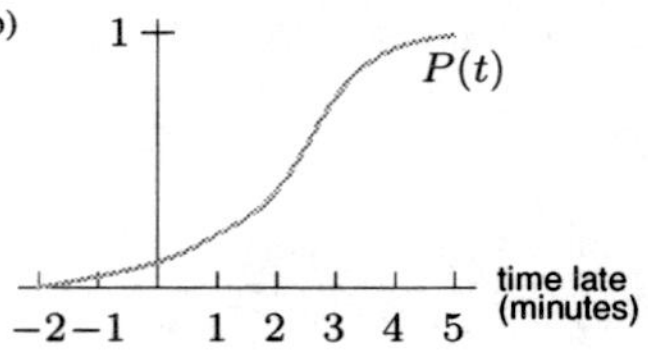

7 (a) 2/3
(b) 1/3
(c) Possibly many work just to pass
(d)
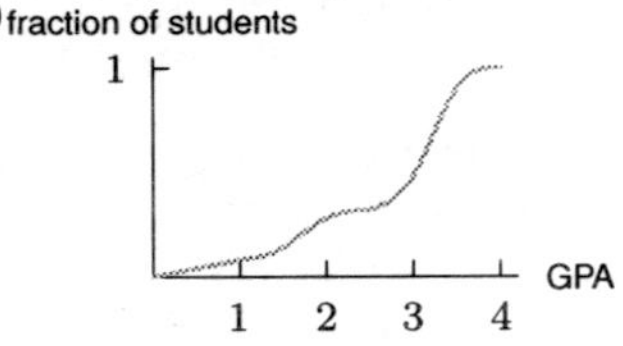

9 (a) First: January 30
Last: August 28
(b) 65%
(c) 25%
(d) 10%
(e) April 10 – April 30

11 (a) 25%
(b) 32.5%
(c) 0

13 Fraction dead at time t

15 (a) $-e^{-2} + 1 \approx 0.865$
(b) $-(\ln 0.05)/2 \approx 1.5$ km

Section 8.3

1 5.35 tons

3 (a) 0.2685
(b) 0.067

5 2.48 weeks

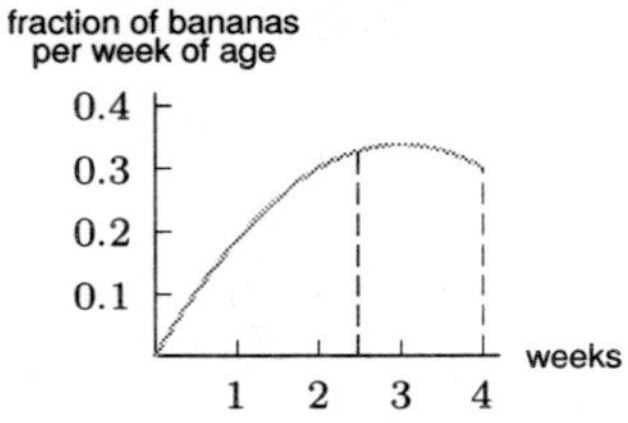

7 (a) 0.684 : 1
(b) 1.6 hours
(c) 1.682 hours

9 (a) 17.6%; 6%
(b) About $45,200
(c) False

Chapter 8 Review

1 (a)-(II), (b)-(I), (c)-(III)

3
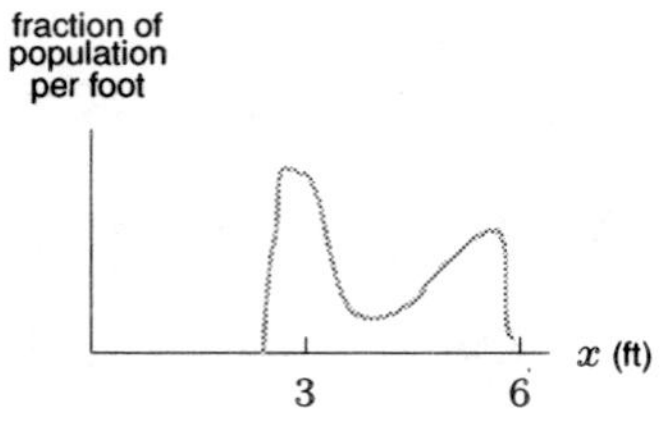

5
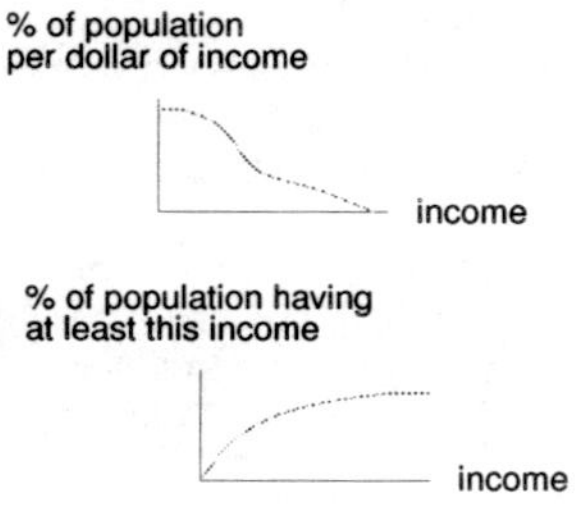

7 0.04

9 0.008

11 (a) Twelfth
(b) 1/4
(c) 7/16

13 (b) About 3/4

15 (a) $f(r) = 0.2\,(0 < r < 5)$
$f(r) = 0\,(5 \le r)$
(b) $F(r) = 0.2r$
$(0 \le r \le 5)$
$F(r) = 1\,(5 < r)$
(c) $G(v) = 0.124r^{1/3}$
$(0 \le v \le 523.6)$
$G(v) = 1\,(523.6 < v)$
(d) $g(v) = 0.0413v^{-2/3}\,(0 < v < 523.6)$
$g(v) = 1\,(523.6 \le v)$

17 (a) 22.1%
(b) 33.0%
(c) 30.1%
(d) $C(h) = 1 - e^{-0.4h}$

19 14.6 days

21 False

23 True

25 True

Ch. 8 Understanding

1 True

3 False

5 False

7 False

9 True

11 True

13 False

15 False

17 False

19 True

21 False

23 False

25 False

27 True

29 False

Section 9.1

3 $f(20, p)$: 2.65, 2.59, 2.51, 2.43
$f(100, p)$: 5.79, 5.77, 5.60, 5.53
$f(I, 3.00)$: 2.65, 4.14, 5.11, 5.35, 5.79
$f(I, 4.00)$ 2.51, 3.94, 4.97, 5.19, 5.60

7 $P = 0.052pf(I, p)/I$

9 Decreasing function of p
Increasing function of a

11 (a) 81°F
(b) 30%

13 Incr of A and r
Decr of t

15
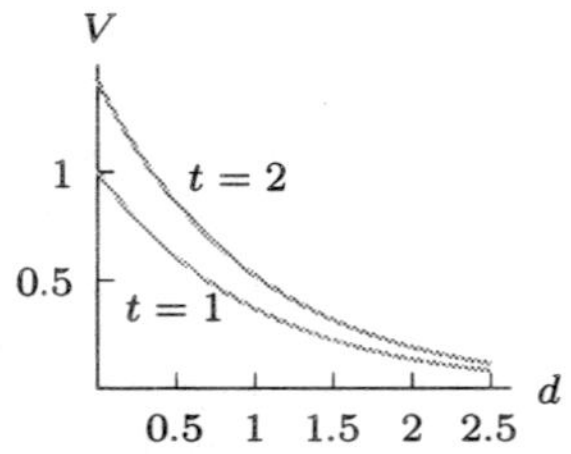

17 (a) 260 cal/m^3

(b)

T (°C) \ w (gm/m^3)	0.1	0.2	0.3	0.4
0	150	290	425	590
10	110	240	330	450
20	100	180	260	350
30	70	150	220	300
40	65	140	200	270

Section 9.2

1 Decreasing function of x
Increasing function of y

3 Heat coming in window

5

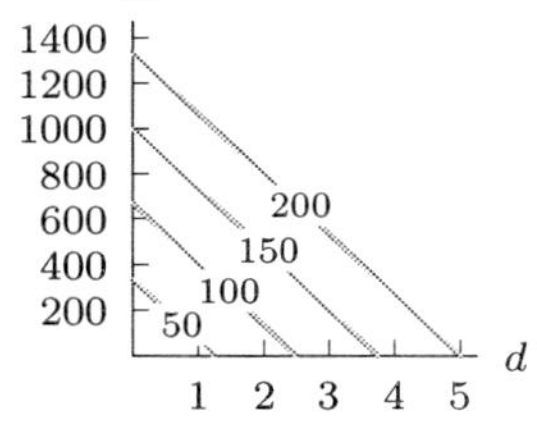

7

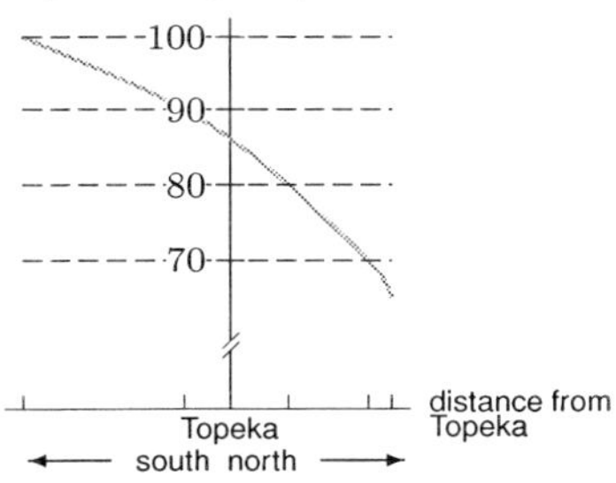

9

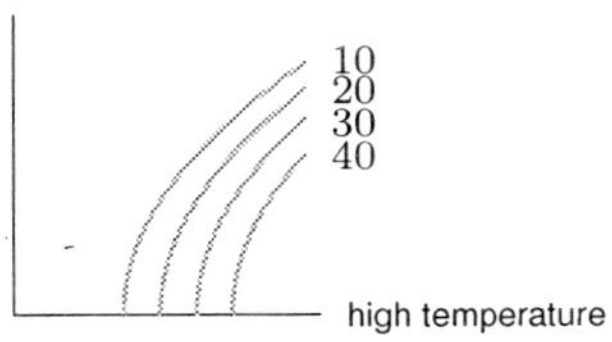

11 Contours evenly spaced

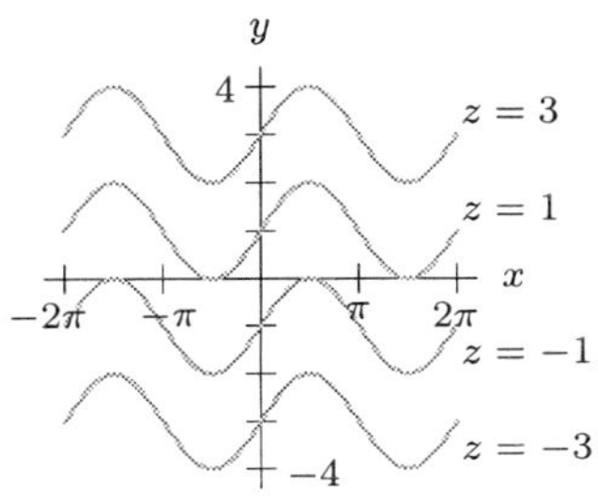

15 Contours evenly spaced

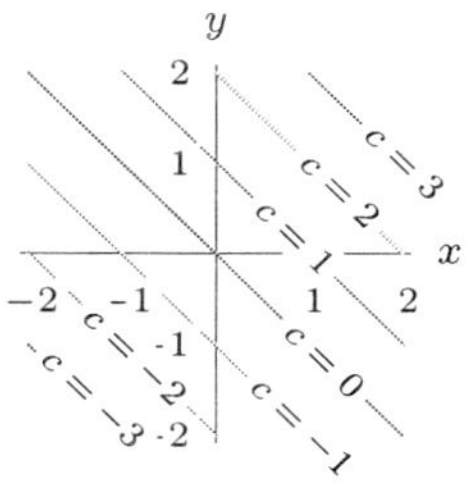

17 Contours evenly spaced

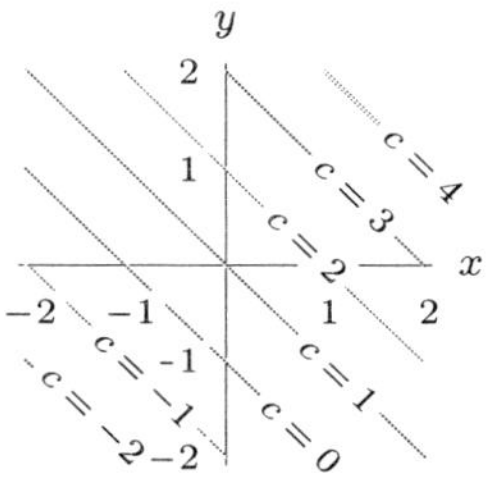

19 Contours evenly spaced

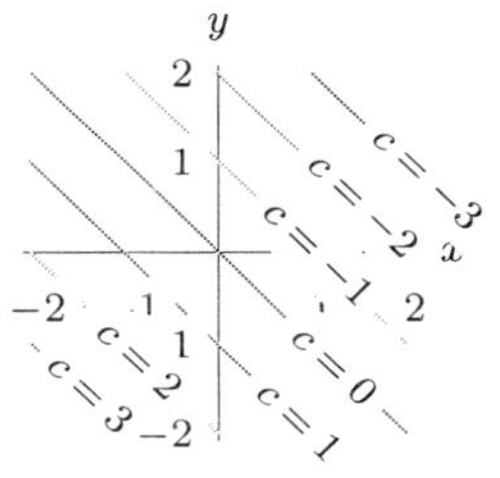

21 (a) False
(b) True
(c) False
(d) True

23 (a) 510.17 thousand pages/day
(b) 773.27 thousand pages/day
(c) 673.17 thousand pages/day
(d) 1020.34 thousand pages/day

25 (a) III
(b) II
(c) I
(d) IV

27 (a) 4 hours
(b) 40%
(c) Contours approx horizontal
(d) Increasing
(e) Increasing

Section 9.3

1 (a) Positive
(b) Negative
(c) Positive
(d) Zero

3 $\frac{\partial I}{\partial H}\Big|_{(10,100)} \approx 0.4$
$\frac{\partial I}{\partial T}\Big|_{(10,100)} \approx 1$

5 (a) f_c is negative
f_t is positive

7 (a) Concentration/distance
Rate of change of concentration with distance
$\partial c/\partial x < 0$

(b) Concentration/time
Rate of change of concentration with time
For small t, $\partial c/\partial t > 0$
For large t, $\partial c/\partial t < 0$

9 $z_x(1,0) \approx 2$
$z_x(0,1) \approx 0$
$z_y(0,1) \approx 10$

11 (a) Negative

13 (a) 3.3 % / in
(b) -5% / °F

15 $\partial f/\partial P_1 < 0$
$\partial f/\partial P_2 > 0$

19 14 %

21 5.67

23 (a)

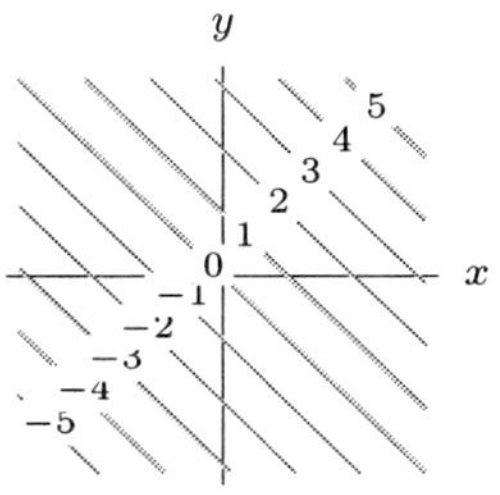

(b)

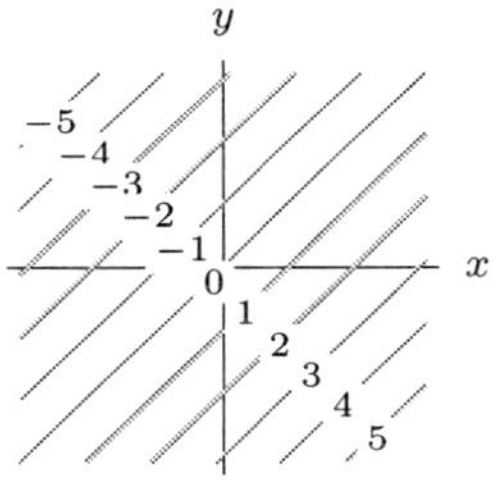

(c)

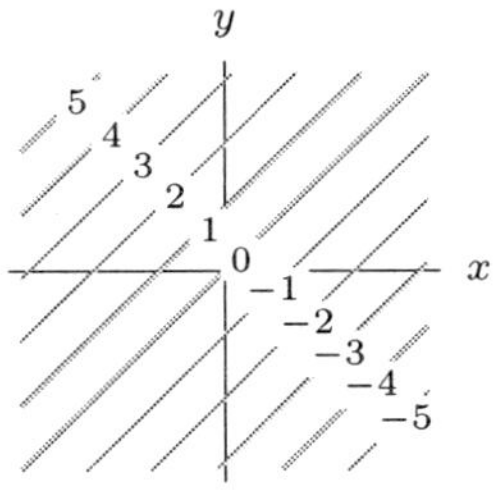

(d)

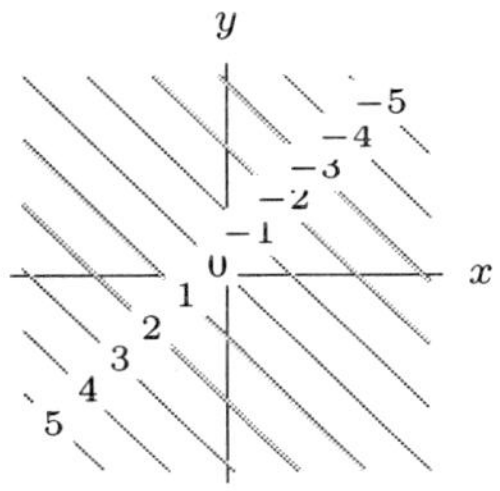

Section 9.4

1 $4x$; $6y$

3 $f_x = 2x + 2y$
$f_y = 2x + 3y^2$

5 $2u + 5v$; $5u + 2v$

7 $5a^2 - 9ap^2$

9 $f_x = 10xy^3 + 8y^2 - 6x$,
$f_y = 15x^2y^2 + 16xy$

11 $100te^{rt}$

13 $(1/2)v^2$

15 15; 5; 30

17 Pre^{rt}; Pte^{rt}; e^{rt}

19 80; 30; 313 tons
2.9 tons per worker
2.6 tons per \$25,000

21 $f_{xx} = 2, f_{xy} = 2,$
$f_{yy} = 2, f_{yx} = 2$

23 $f_{xx} = 0, f_{xy} = -2/y^2,$
$f_{yy} = 4x/y^3, f_{yx} = -2/y^2$

25 $f_{xx} = y^2e^{xy}$,
$f_{xy} = (xy + 1)e^{xy}$,
$f_{yy} = x^2e^{xy}$,
$f_{yx} = (xy + 1)e^{xy}$

27 $V_{rr} = 2\pi h, V_{hh} = 0,$
$V_{rh} = V_{hr} = 2\pi r$

29 $B_{xx} = 0, B_{tt} = 20xe^{-2t},$
$B_{xt} = B_{tx} = -10e^{-2t}$

31 $f_{rr} = 100t^2e^{rt}, f_{tt} = 100r^2e^{rt},$
$f_{tr} = f_{rt} = 100(rt + 1)e^{rt}$

35 (a) $-(c+1)B/(c+r)^2$
(b) Negative

Section 9.5

1 $f(0,0) \approx 12.5$ is a local and global maximum
$f(13,30) \approx 4.5$ is a local minimum
$f(37,18) \approx 2.5$ is a local and global minimum
$f(32,34) \approx 10.5$ is a local maximum

3 Max: 11 at $(5.1, 4.9)$
Min: -1 at $(1, 3.9)$

5 Max: 1 at $(\pi/2, 0)$; $(\pi/2, 2\pi)$
Min: -1 at $(\pi/2, \pi)$

7 $(-3, 6)$, neither

9 $(4, 2)$, neither

11 Saddle pts: $(1, -1), (-1, 1)$
local max $(-1, -1)$
local min $(1, 1)$

13 Local maximum at $(-1, 0)$,
Saddle points at $(1, 0)$ and $(-1, 4)$,
Local minimum at $(1, 4)$

15 Local max: $(1, 5)$

17 $A = 10, B = 4, C = -2$

19 (b) $p_1 = p_2 = 25$
Max revenue is 4375

21 $q_1 = 300, q_2 = 225$.

Section 9.6

1 $f(66.7, 33.3) = 13{,}333$

3 $f(9.26, 2.32) = 201.9$

5 Min $= -\sqrt{2}$, max $= \sqrt{2}$

7 Min $= -22$, max $= 22$

9 Min $= -2$, max $= 2$

11 $x = 6$; $y = 6$; $f(6,6) = 400$

13 (a) $C = 127x_1 + 92x_2$
(b) $x_1^{0.6}x_2^{0.4} = 500$

15 (b) $L = 40, K = 30$

17 (a) Reduce K by 1/2 unit,
increase L by 1 unit.

19 (a) $P(x,y)$; $C(x,y) = 50{,}000$
(b) $C(x,y)$; $P(x,y) = 2000$

21 (a) $P(K, L)$
(b) $C(K, L) = 600{,}000$
(c) Tons/dollar
(d) Extra dollar produces approximately extra 3.17 tons

23 (a) Quantity of fuel, $x_1 + x_2$
(b) Terminal velocity (as function of x_1 and x_2) $= v_0$
(c) Liters per meter/sec
(d) 51 meters/sec requires about 8 more liters than 50 meters/sec

25 1820.04; about 209

Chapter 9 Review

1 Decreasing function of x
Increasing function of y

3 Lines with slope 3/5, evenly spaced

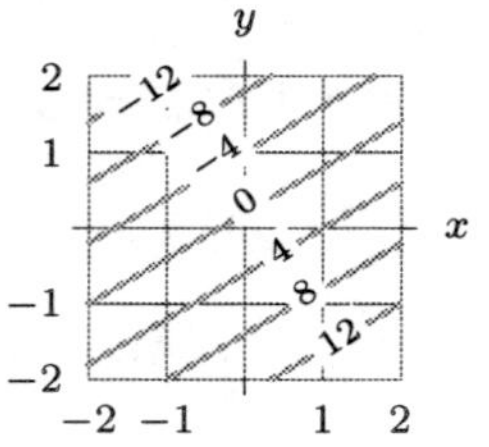

5 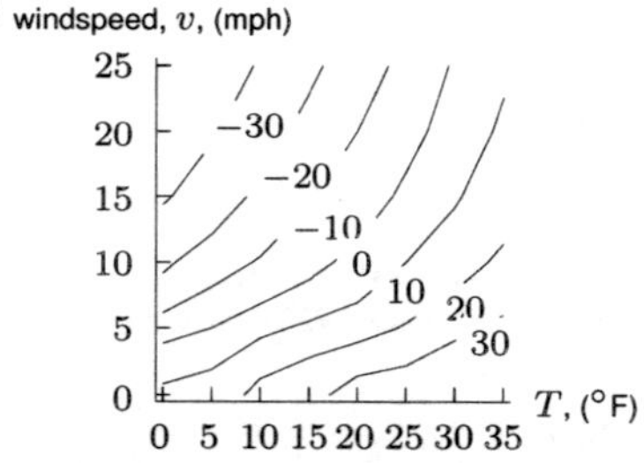

9 x-axis: price
y-axis: advertising

11 Answers in °C:

(a)

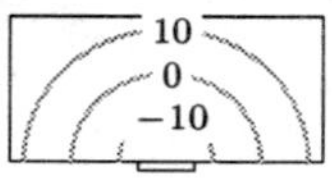

(b)

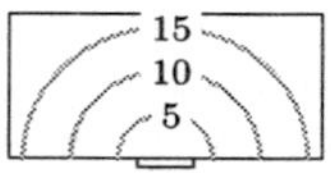

(c)

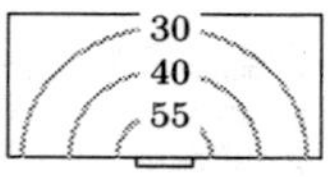

(d)

13 Table (a) matches (II)
Table (b) matches (III)
Table (c) matches (IV)
Table (d) matches (I)

15 Positive, Negative, 10, 2, -4

17 $f_w(10, 25) \approx -0.4°$F/mph

19 $f_w(5, 20) \approx -0.8$

21 (a) Both negative
(b) Both negative

23 (A) 0.06, -0.06
(B) 0, -0.05
(C) 0, 0

25 $P_a = 2a - 2b^2, P_b = -4ab$

27 $\dfrac{\partial f}{\partial x} = 5e^{-2t}, \dfrac{\partial f}{\partial t} = -10xe^{-2t}$

29 $f_x = x/\sqrt{x^2 + y^2}$
$f_y = y/\sqrt{x^2 + y^2}$

31 (a) $Q_K = 18.75K^{-0.25}L^{0.25}$,
$Q_L = 6.25K^{0.75}L^{-0.75}$
(b) $Q = 1704.33$
$Q_K = 21.3$
$Q_L = 4.26$

33 (a) 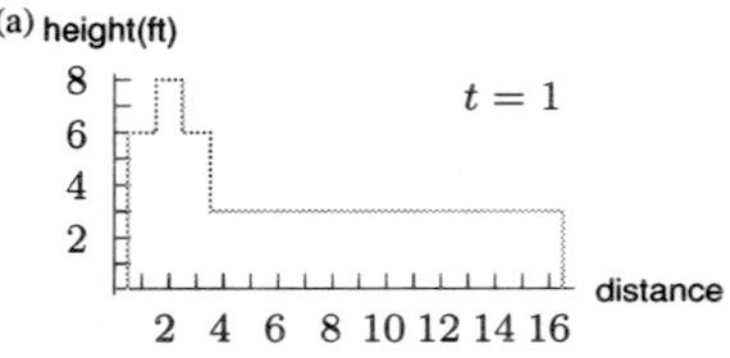

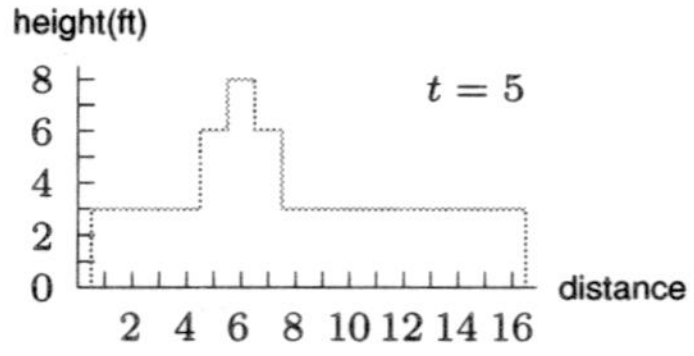

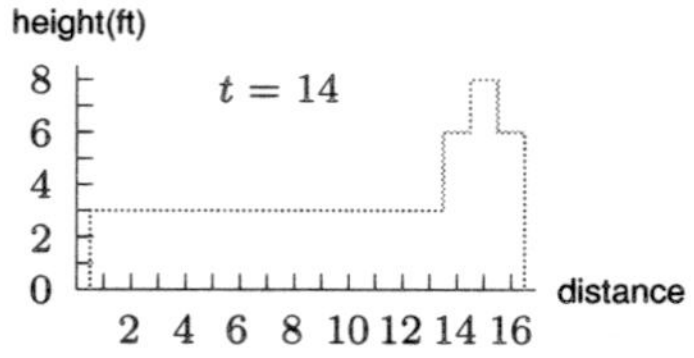

(b) 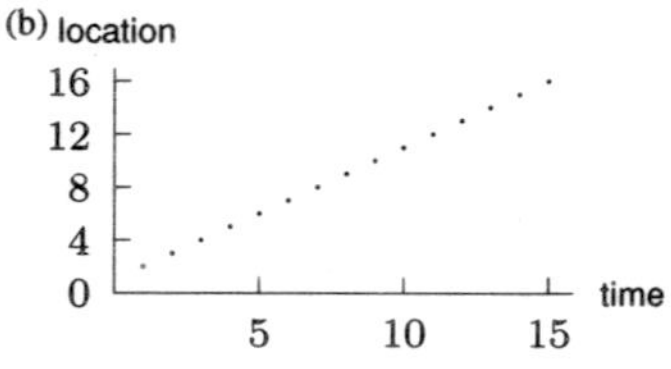

(c) The "wave" at a sports arena

35 $(1,0), (-1,0)$; $f(1,0)$ is a local minimum

37 $p_1 = 110, p_2 = 115$

39 43

41 (a) $Q = x_1^{0.3}x_2^{0.7}$
(b) $10x_1 + 25x_2 = 50{,}000$

43 (a) 2599
(b) 129; $\$4{,}712{,}958$
(c) $\$8572.54$ per car

45 (a) 475 units
(b) 505 units

Ch. 9 Understanding

1 Could not be true
3 Might be true
5 True
7 True
9 False
11 False
13 True
15 True
17 False
19 False
21 True
23 True
25 False
27 True
29 True
31 True
33 False
35 False
37 True
39 False
41 False
43 False
45 True
47 False
49 True
51 True
53 True
55 False
57 True
59 True
61 True
63 False
65 False

Theory: Least Squares

1 $y = 2/3 - (1/2)x$
3 $y = 2/3 - (1/2)x$
5 $y = x - 1/3$
7 (a) (i) Power function
(ii) Linear function
(b) $\ln N = 1.20 + 0.32 \ln A$
Agrees with biological rule

Section 10.1

1 (a) (III)
(b) (IV)
(c) (I)
(d) (II)

3 $dP/dt = kP, k > 0$
5 $dQ/dt = kQ, k < 0$
7 $dP/dt = -0.08P - 30$
9 $dA/dt = -1$
11 (a) $dA/dt = -0.17A$
(b) -17 mg
13 (a) Increasing, decreasing
(b) $W = 4$
15 $dN/dt = B + kN$

Section 10.2

1 $y = t^2 + C$
5 F
7 E
9 A
11 D
13 8, 8.5, 9.5, 11
15 4, 4, 4, 4
17 74, 78.8, 84.56 million
19 $k = -0.03$ and C is any number, or $C = 0$ and k is any number
21 $k = 5$

Section 10.3

1 (a)

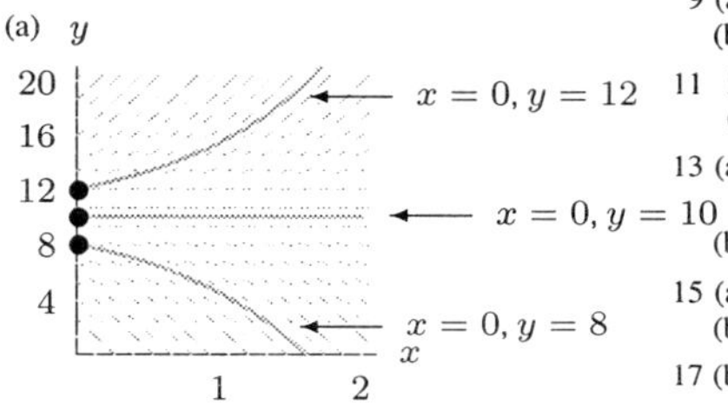

3 (a)

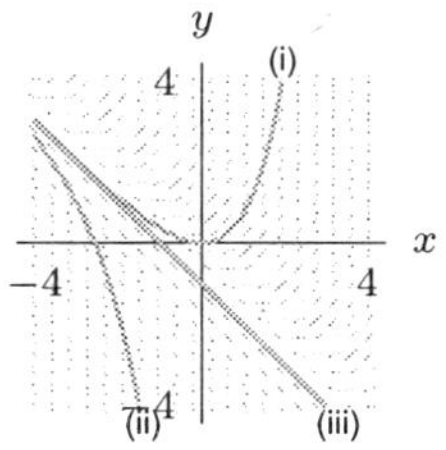

(b) $y = -x - 1$
5 (a) Slopes $= 2, 0, -1, -4$
(b)

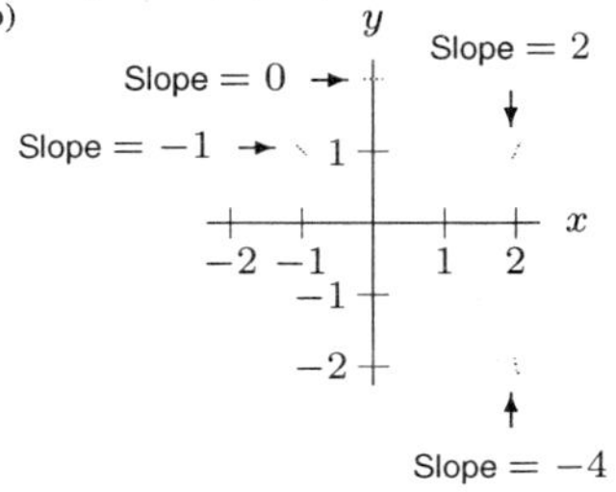

7 (a) II
(b) VI
(c) IV
(d) I
(e) III
(f) V

9 As x increases, $y \to \infty$.
11 As $x \to \infty$, y oscillates within a certain range
13 $y \to \infty$ as $x \to \infty$

Section 10.4

1 $w = 30e^{3r}$
3 $P = 20e^{0.02t}$
5 $Q = 50e^{(1/5)t}$
7 (a) $dM/dt = rM$
(b) $M = 1000e^{rt}$
(c)

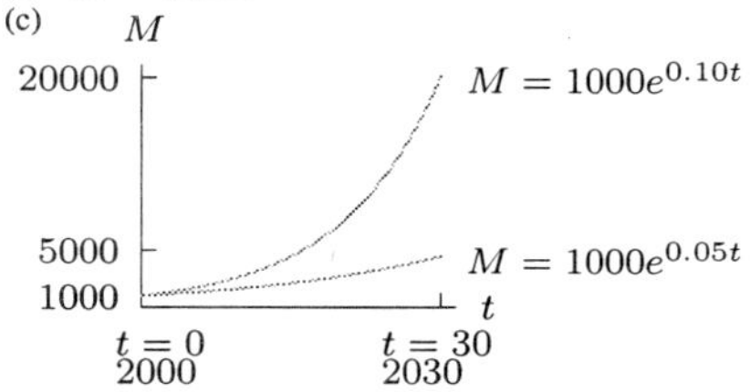

9 (a) $dB/dt = 0.10B + 1000$
(b) $B = 10{,}000e^{0.1t} - 10{,}000$
11 Michigan: 72 years
Ontario: 18 years
13 (a) $dQ/dt = -0.5365Q$
$Q = Q_0e^{-0.5365t}$
(b) 4 mg
15 (a) $dy/dt = ky$
(b) 0.2486 grams
17 (b) 2001

Section 10.5

1 $H = 75 - 75e^{3t}$
3 $P = 104e^t - 4$
5 $Q = 400 - 350e^{0.3t}$
7 $B = 25 + 75e^{2-2t}$
11 (a) $dV/dt = 0.02V - 80{,}000$
(b) $V = \$4{,}000{,}000$
(c) $V = 4{,}000{,}000 + Ce^{0.02t}$
(d) $\$2{,}728{,}751$
13 (a) $dB/dt = 0.05B + 1200$
(b) $B = 24{,}000(e^{0.05t} - 1)$
(c) $\$6816.61$
15 (a) $y = ce^{-t} + 100$
(b)

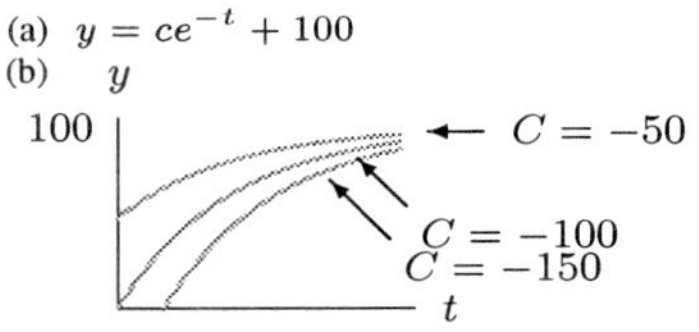

(c) $y = 100 - 100e^{-t}$
17 (a) $dy/dt = -k(y - a)$
(b) $y = (1 - a)e^{-kt} + a$
(c) a: fraction remembered in the long run
k: rate material is forgotten
19 (a) $y = 500$
(b) $y = 500 + Ce^{0.5t}$

(c)

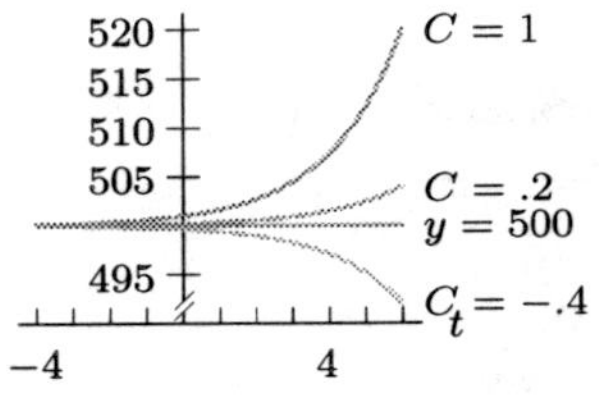

(d) Unstable

21 (a) $H = 200 - 180e^{-kt}$
(b) $k \approx 0.027$ (if t is in minutes)

23 (a) $dT/dt = -k(T - 68)$
(b) $T = 68 + 22.3e^{-0.06t}$; 3:45 am.

Section 10.6

1 (a) $x \to \infty$ exponentially
$y \to 0$ exponentially
(b) Predator-prey

3 (a) $x \to \infty$ exponentially
$y \to 0$ exponentially
(b) y is helped by the presence of x

5 $dx/dt = x - xy$,
$dy/dt = y - xy$

7 $dx/dt = -x - xy$,
$dy/dt = -y - xy$

11 Symmetric about the line $r = w$; solutions closed curves

13 Robins:
Max ≈ 2500
Min ≈ 500
When robins are at a max, the worm population is about 1 million

17 (a) $dw/dt = 0$
$dr/dt = 1.2$
(b) $w \approx 2.2, r \approx 1.1$
(c) At $t = 0.2$:
$w \approx 2.2, r \approx 1.3$
At $t = 0.3$
$w \approx 2.1, r \approx 1.4$

19 (a)

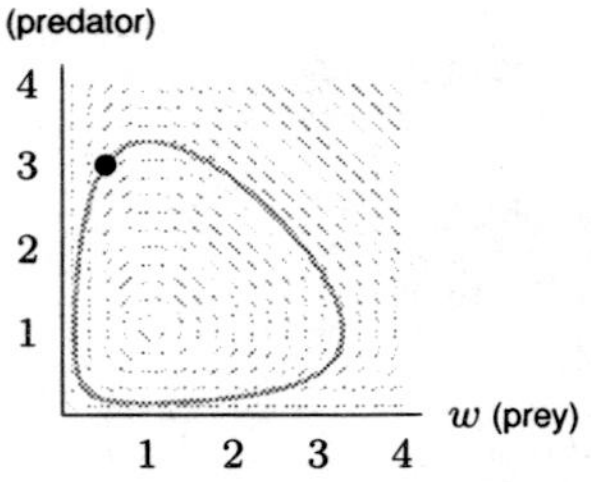

(b) Down and left
(c) $r = 3.3, w = 1$
(d) $w = 3.3, r = 1$

21 (a)

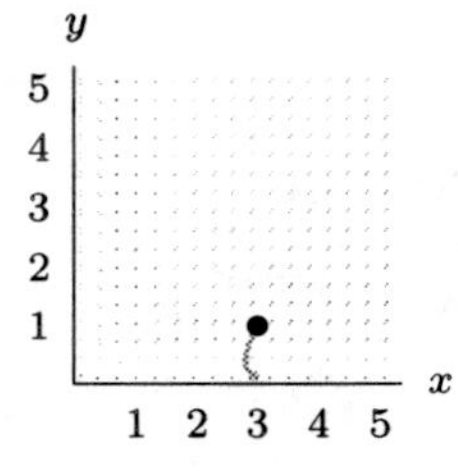

(b)

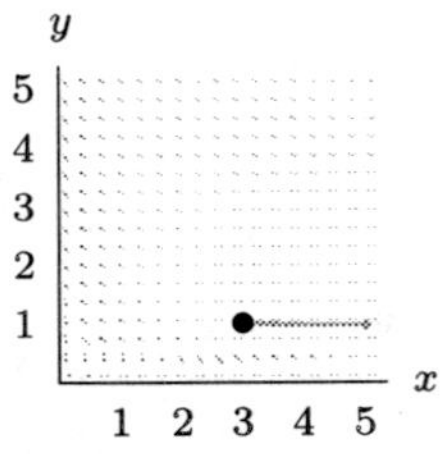

(c)

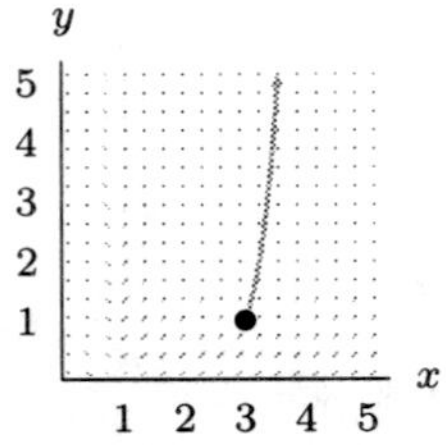

(d)

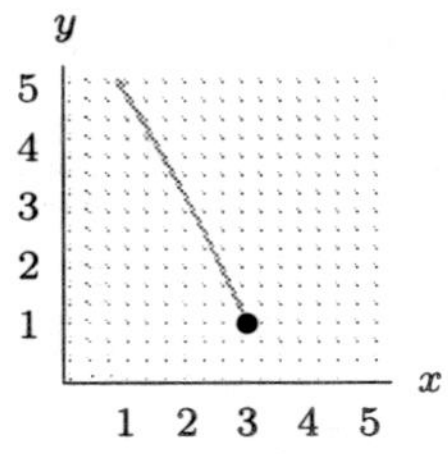

Section 10.7

5 (a) $I_0 = 1, S_0 = 349$
(b) Increases; spreads

7 About 300 boys;
$t \approx 6$ days

9 5

11 (a) b/a

Chapter 10 Review

1 (a) (III)
(b) (V)
(c) (I)
(d) (II)
(e) (IV)

3 Yes

5 (a) I is $y' = 1 + y$;
II is $y' = 1 + x$
(b) I

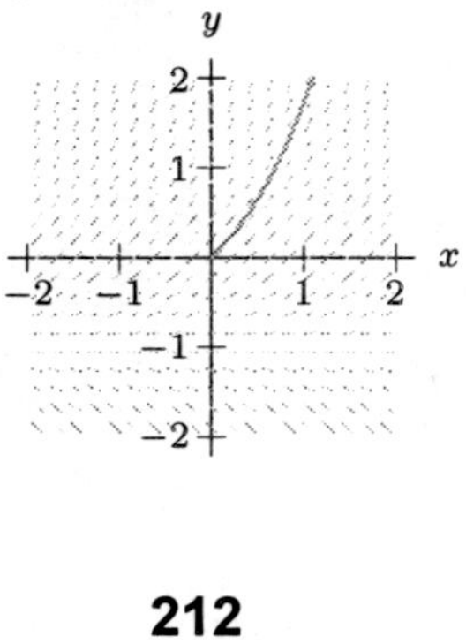

II

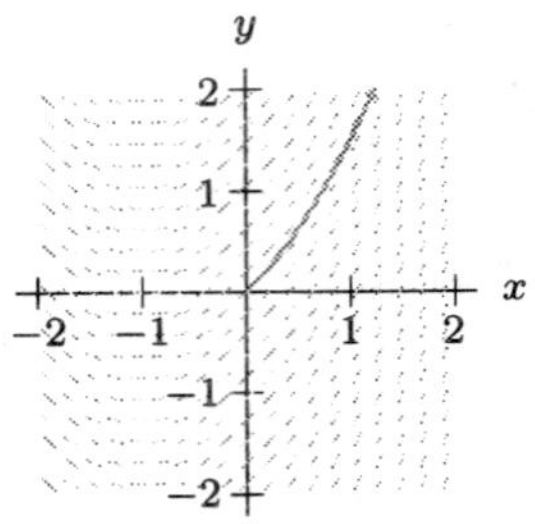

(c) I: $y' = -1$, unstable;
II: None

7 III: $y' = (1 + y)(2 - y)$

9 (a) $dB/dt = 0.07B$
(b) $B = B_0e^{0.07t}$
(c) $B = 5000e^{0.07t}$
(d) $B(10) \approx \$10{,}068.76$

11 $P = (1/2)t^2 + C$

13 $y = (5/2)t^2 + C$

15 $A = Ce^{-0.07t}$

17 $P = Ce^{-2t} + 5$

19 $y = Ce^{0.2x} + 40$

21 $dS/dt = -k(S - 65), k > 0$
$S = 65 - 25e^{-kt}$

23 (a) $k \approx 0.000121$
(b) 779.4 years

25 (a)

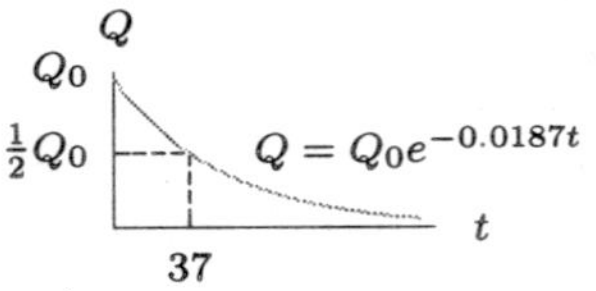

(b) $dQ/dt = -0.0187Q$
(c) 3 days

27 (b) $dQ/dt = -0.347Q + 2.5$
(c) $Q = 7.2$ mg

29 (a) $dW/dt = (1/3500)(I - 20W)$
(b) $W = I/20 + (W_0 - I/20)\,e^{-(1/175)t}$
(c)

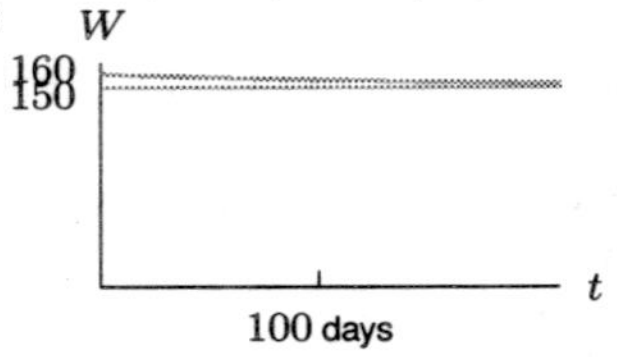

31 (a) $y = 1, y = 8, y = 16$
(d) Stable: $y = 1$
Unstable: $y = 8, y = 16$

33 Initially $x = 0$; y decreases, x increases. Then x increases, y increases. Finally y increases, x decreases

35 Populations oscillate

Ch. 10 Understanding

1 True

3 True

5 True

7 False

9 True

11 False

13 False

15 False; $f(x) = |x|$

17 False; $\cos t + t^2$

19 False; $f(x) = 6, g(x) = 10$

21 False; $f(x) = 5x + 7, g(x) = x + 2$

23 False; $f(x) = x^2, g(x) = x^2 - 1$

25 False; $f(x) = e^{-x}, g(x) = x^2$

27 (a) Not a counterexample
(b) Not a counterexample
(c) Not a counterexample
(d) Counterexample

29 False

31 False

33 Possible answer
$$f(x) = \begin{cases} x & \text{if } 0 \le x < 2 \\ 19 & \text{if } x = 2 \end{cases}$$

Theory: Separation of Vars

1 $P = e^{-2t}$

3 $P = \sqrt{2t+1}$

5 $u = 1/(1 - (1/2)t)$

7 $R = 1 - 0.9e^{1-y}$

9 $z = -\ln(1 - t^2/2)$

11 $y = -2/(t^2 + 2t - 4)$

13 (a) Yes (b) No (c) Yes
(d) No (e) Yes (f) Yes
(g) No (h) Yes (i) No
(j) Yes (k) Yes (l) No

15 $Q = b - Ae^{-t}$

17 $R = -(b/a) + Ae^{at}$

19 $y = -1/\left(k(t + t^3/3) + C\right)$

21 (a)

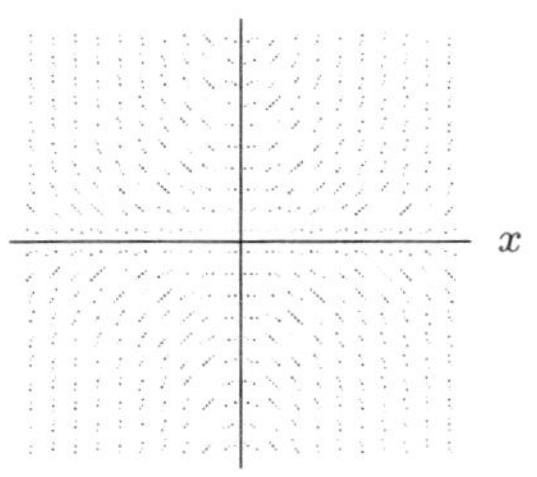

(b)

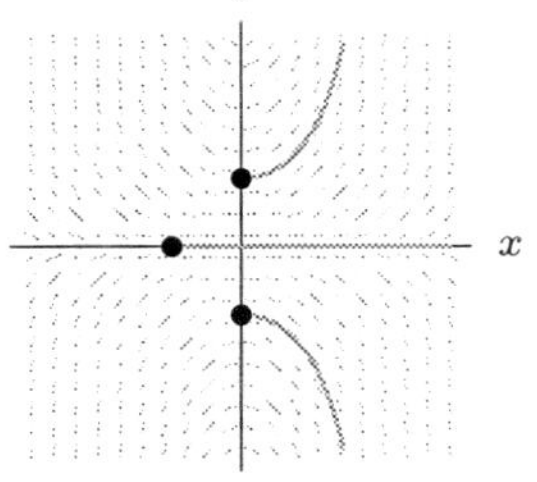

(c) $y(x) = Ae^{x^2/2}$

Section 11.1

1 21

3 3,985,805

5 555.10

7 96.154

9 39,375

11 1039.482

13 400

15 30.51, 37.75, 39.47, 39.87, Yes

17 (a) \$3037.75, \$2537.75
(b) \$6167.78, \$5667.78

19 (a) 39 mg
(b) 41.496 mg
(c) 41.667 mg

21 (a) 40 mg
(b) 0.57 mg/kg, Yes
(c) (i) Greater than 100 kg
(ii) Less than 13.3 kg

Section 11.2

1 Balance = \$48,377.01, \$20,000. from deposits, \$28,377.01 from interest

3 \$33,035.37

5 \$42, 567.82

7 17.54 payments

9 \$44,407.33, \$360,183.21

11 0.12%

13 \$1081.11

15 (a) \$1250
(b) 12.50

17 (a) \$400 billion
(b) \$900 billion

19 (a) $N + (1 - r)N$ dollars
(b) $N + (1 - r)N + (1 - r)^2N$ dollars
(c) N/r dollars

Section 11.3

1 230.159 billion barrels

3 (a) 98 mg
(b) 121.5 mg
(c) 125 mg

5 (a) 400 mg
(b) 400(0.30) = 120 mg

7 (a) 39.99 mg, 9.99 mg
(b) 40 mg, 10 mg

9 (a) 4.10 mg
(b) 7.54 mg
(c) 14.46 mg
(d) 35.24 mg
(e) 69.87 mg

11 24.5 years

13 Until 2046

15 About 34 years

17 Lasts forever

Chapter 11 Review

1 2046

3 Does not exist

5 200

7 1.9961

9 (a) (i) \$16.43 million
(ii) \$24.01 million
(b) \$16.87 million

11 \$27,979.34

13 \$400 million

15 (a) $N(k/(1 - k))$
(b) $5.667N$

17 (a) $0.232323\ldots = 0.23 + 0.23(0.01) + 0.23(0.01)^2 + \cdots$
(b) $0.23/(1 - 0.01) = (23)/(99)$

19 \$25,503

Ch. 11 Understanding

1 False

3 True

5 False

7 False

9 False

11 False

13 False

15 False

17 True

19 False

21 False

23 False

25 False

27 False

29 True

31 False

33 False

35 False

37 False

39 True

41 True

43 True

45 True

Appendix A

1 (a)

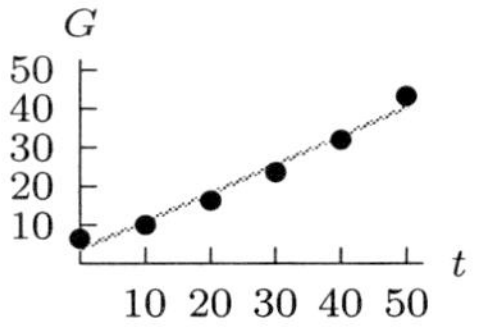

(b) Increasing at a rate of \$734 billion/year
(c) \$43.9 trillion, \$54.9 trillion, More confidence in the 2005 prediction

3 (a)

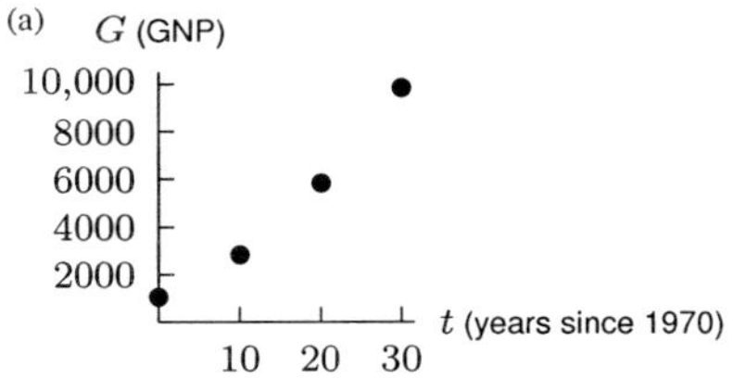

Yes

(b)

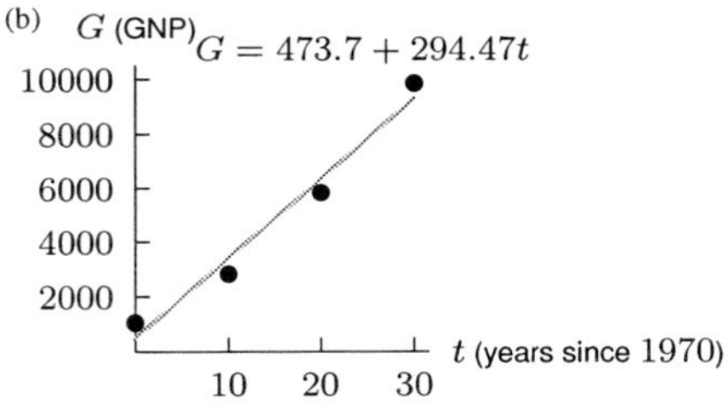

$G = 294.47t + 473.7$

(c) For 1985: 4891
For 2020: 15,197
More confidence in 1985

5 (a) $S = 0.08v + 1.77$
(b)

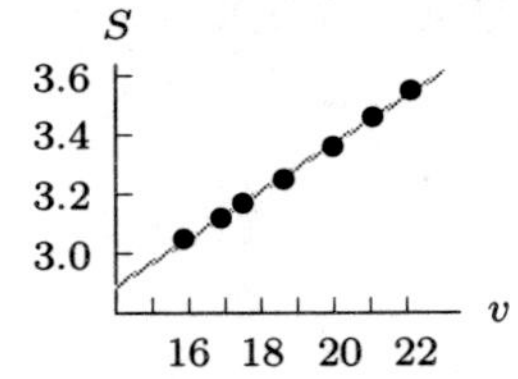

Yes

(c) At $v = 18$ ft/sec, $S = 3.21$
At $v = 10$ ft/sec, $S = 2.57$
$v = 18$ better

7 (a) $0.0026 = 0.26\%$
(b) For 1900, 272.27
For 1980, 335.1

9 (a)

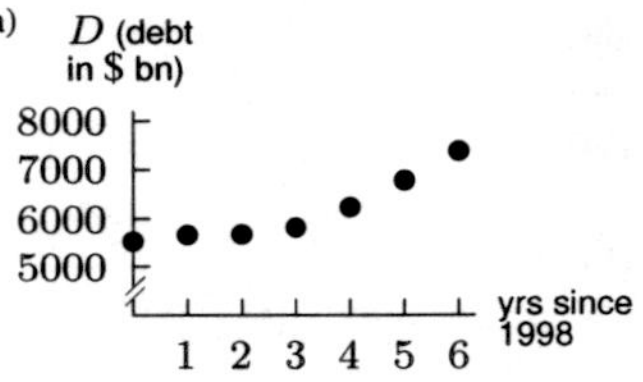

(b) Exponential
(c) $D = 5309(1.048)^t$, answers may vary

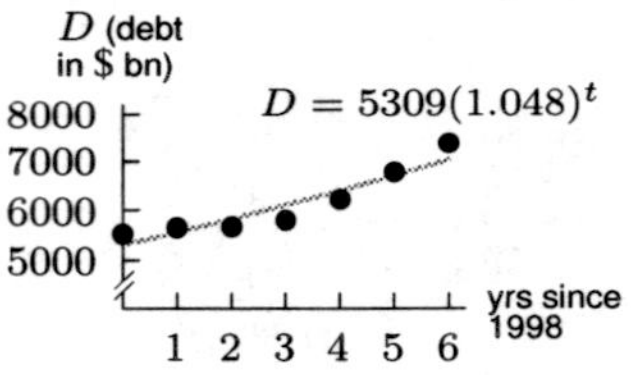

(d) About 4.8%
(e) No

11 (a) $r = 1$
(b) $r = 0.7$
(c) $r = 0$
(d) $r = -0.98$
(e) $r = -0.25$
(f) $r = -0.5$

13 (a) Exponential

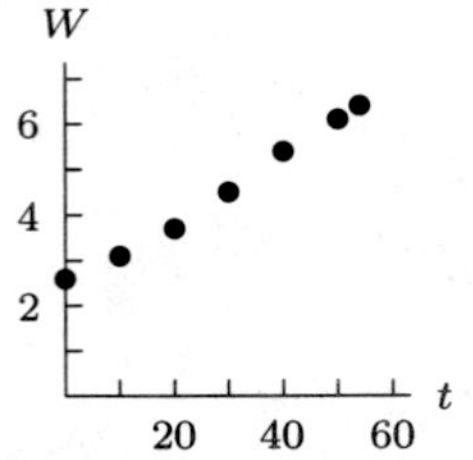

(b) $2.6(1.0165)^t$; answers may vary
(c) 1.65%
(d) At year 2020, 8.175 billion
At year 2050, 13.357 billion
2020 prediction more accurate

15 (a)

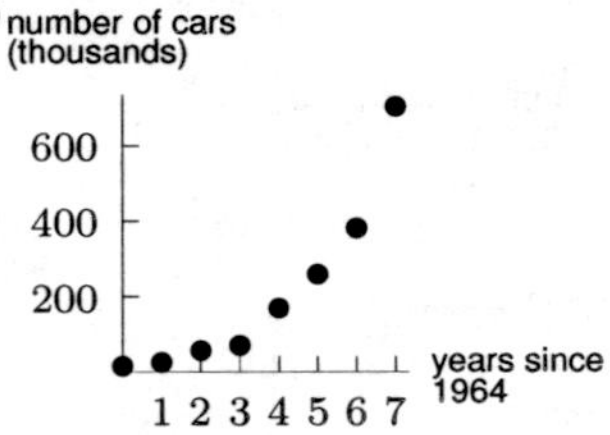

(b) Exponential
(c) $C = 15.9 \cdot (1.725)^t$, answers may vary

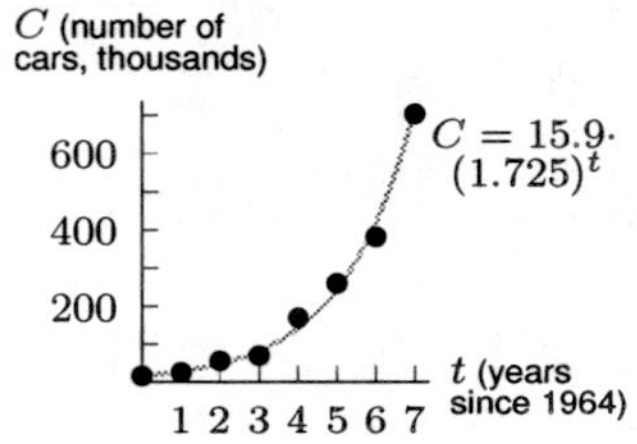

(d) About 73%
(e) No

17 (a) Negative
(b) $f(t) = -0.03t^2 + 1.01t + 13.82$

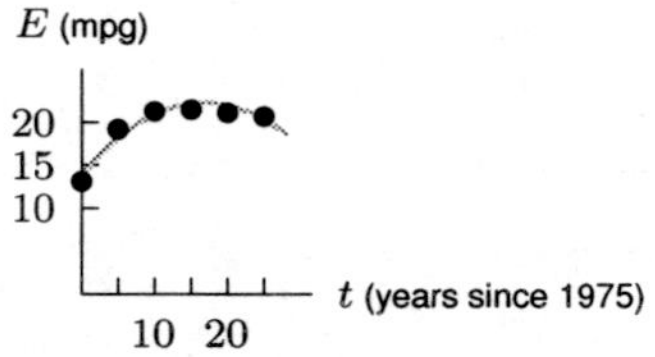

19 (a) Exponential
(b) $S = 29.96(1.30)^t$, answers may vary
Increasing 30%/yr
(c) 21,141 megawatts
(d)

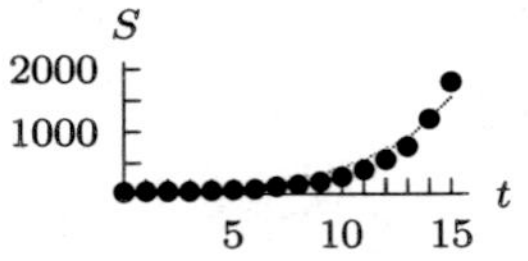

21 (a) Linear
(b) $C = 320 + 1.5t$
Increasing at 1.5 ppm/yr
(c) 387.5 ppm
(d)

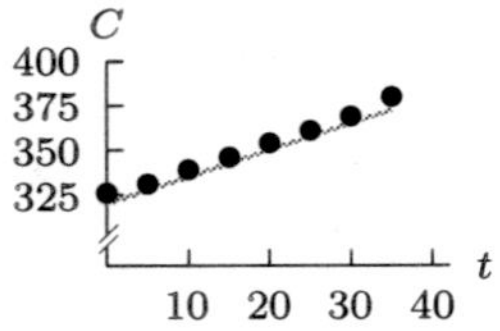

Appendix B

1 27%

3 (a) $160,356.77
(b) $165,510.22
(c) $165,891.05
(d) $165,989.48
(e) $166,005.85

5 (a) 1.0408107
1.0408108
1.0408108
4% compounded continuously
$\approx 4.08108\%$
(b) $e^{0.04} \approx 1.048108$

7 4.88%

9 (a) (i) 5.126978...%
(ii) 5.127096...%
(iii) 5.127108...%
(b) 5.127%
(c) $e^{0.05} = 1.05127109\ldots$

11 (a) = V, (b) = III, (c) = IV,
(d) = I, (e) = II

13 (a) 13,900 cruzados
(b) 24.52%

GRAPHS OF FAMILIES OF FUNCTIONS

Linear function: $y = b + mx$

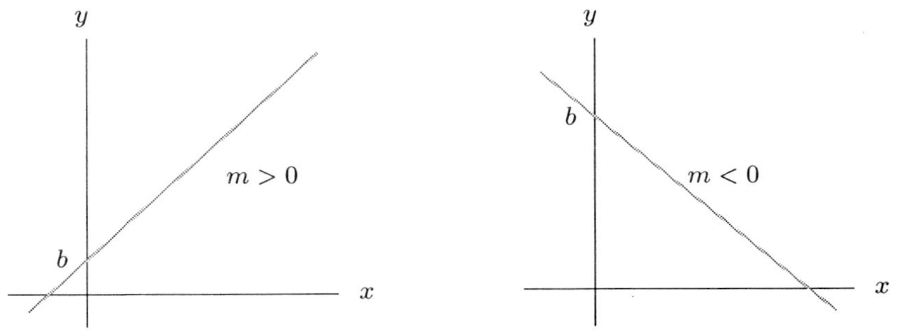

Power function: $y = x^p$

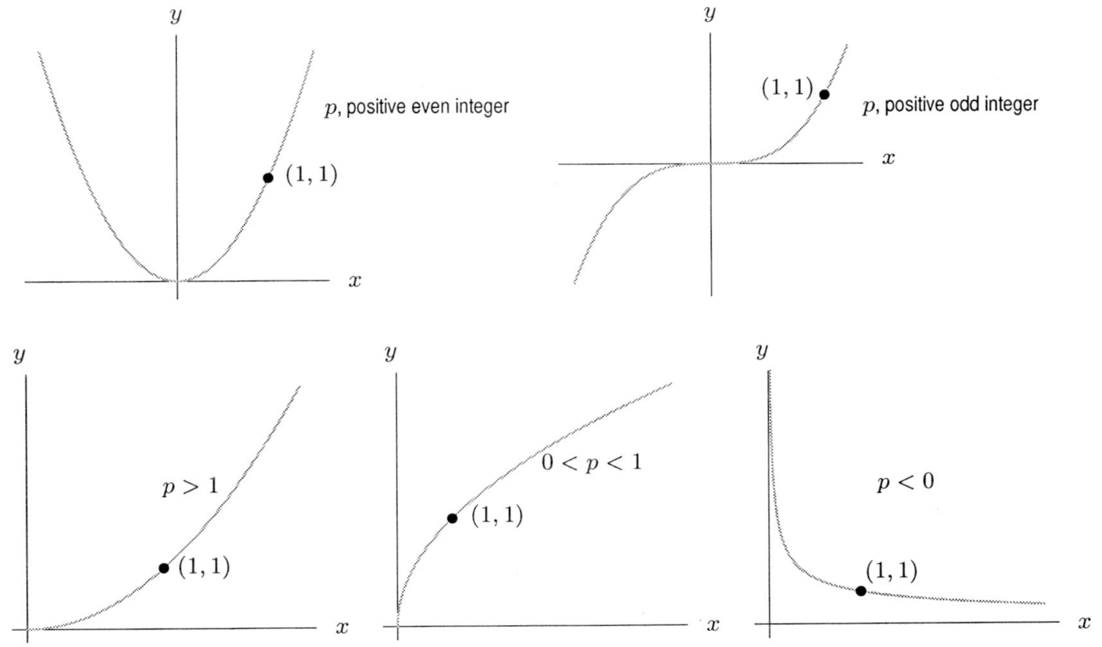

Exponential function: $y = P_0 a^x$

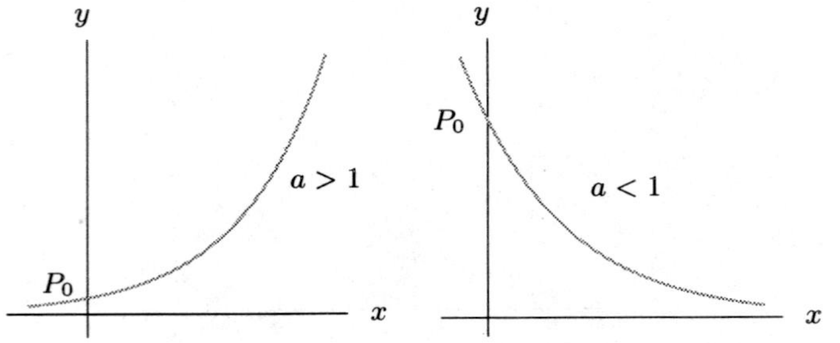

Logarithm function: $y = \ln x$

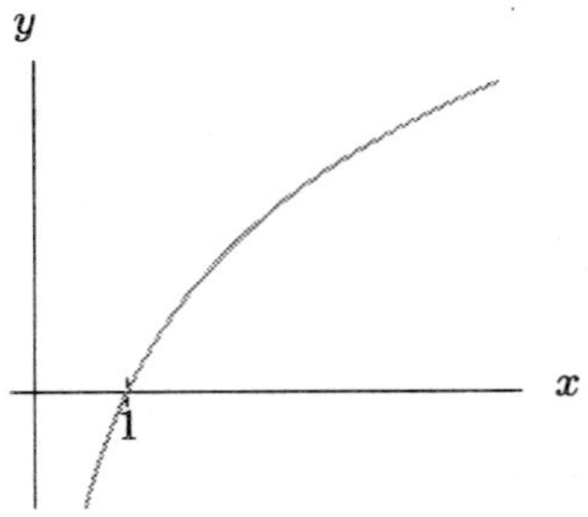

Periodic functions

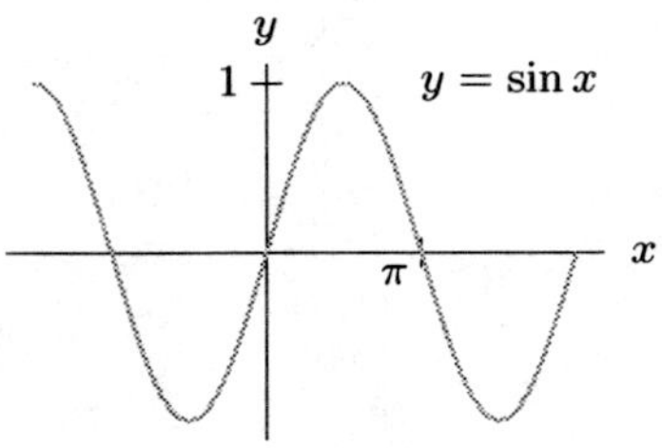

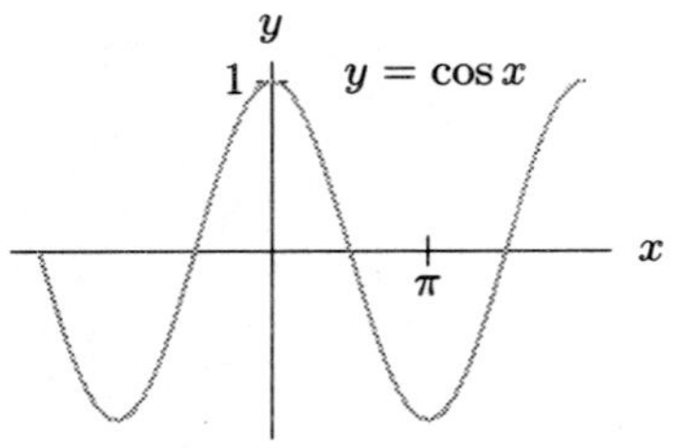

Logistic function: $y = \dfrac{L}{1 + Ce^{-kx}}$

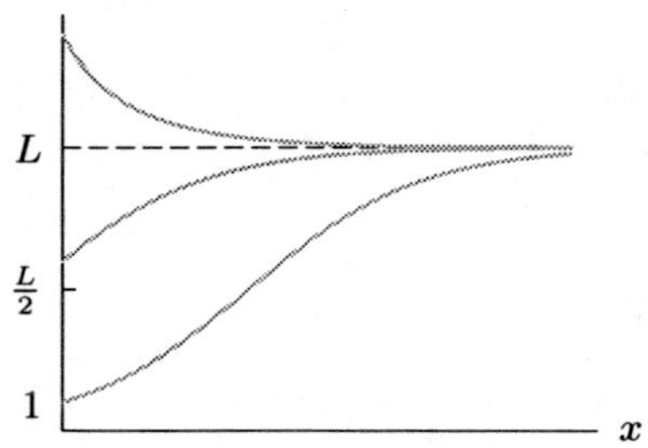

Surge function: $y = axe^{-bx}$

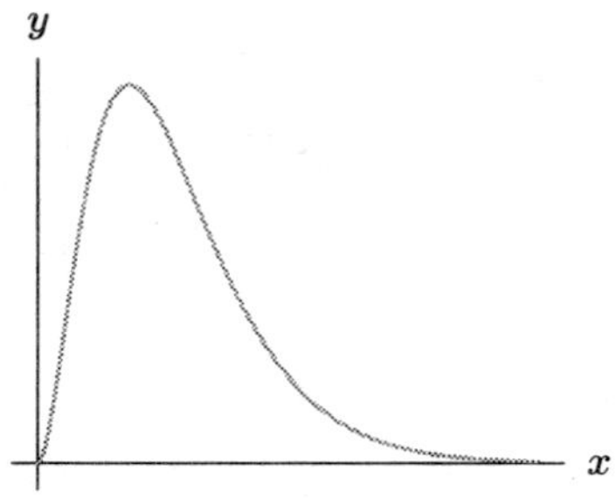

FORMULA SUMMARY: ALGEBRA

Lines

Slope of line through (x_1, y_1) and (x_2, y_2):

$$m = \frac{y_2 - y_1}{x_2 - x_1}$$

Point-slope equation of line through (x_1, y_1) with slope m:

$$y - y_1 = m(x - x_1)$$

Slope-intercept equation of line with slope m and y-intercept b:

$$y = b + mx$$

Definition of Zero, Negative, and Fractional Exponents

$$a^0 = 1, \quad a^{-1} = \frac{1}{a}, \quad \text{and, in general, } a^{-x} = \frac{1}{a^x}$$

$$a^{1/2} = \sqrt{a}, \quad a^{1/3} = \sqrt[3]{a}, \quad \text{and, in general, } a^{1/n} = \sqrt[n]{a}.$$

$$\text{Also, } a^{m/n} = \sqrt[n]{a^m} = (\sqrt[n]{a})^m.$$

Rules of Exponents

1. $a^x \cdot a^t = a^{x+t}$ For example, $2^4 \cdot 2^3 = (2 \cdot 2 \cdot 2 \cdot 2) \cdot (2 \cdot 2 \cdot 2) = 2^7$.
2. $\dfrac{a^x}{a^t} = a^{x-t}$ For example, $\dfrac{2^4}{2^3} = \dfrac{2 \cdot 2 \cdot 2 \cdot 2}{2 \cdot 2 \cdot 2} = 2^1$.
3. $(a^x)^t = a^{xt}$ For example, $(2^3)^2 = 2^3 \cdot 2^3 = 2^6$.

Definition of Natural Log

$y = \ln x$ means $e^y = x$; *for example:* $\ln 1 = 0$, since $e^0 = 1$.

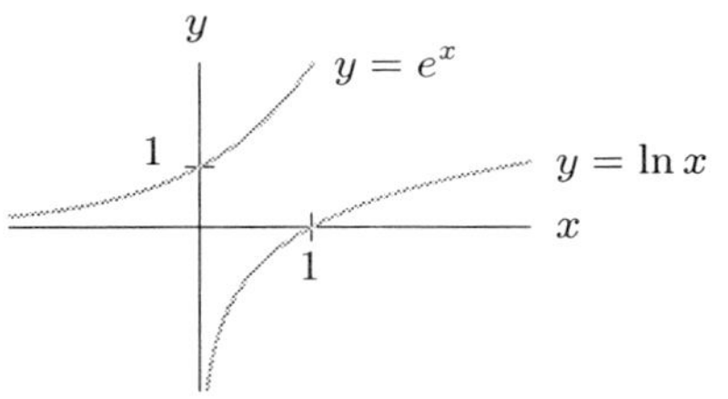

Rules of Natural Logarithms

$$\ln(AB) = \ln A + \ln B$$

$$\ln\left(\frac{A}{B}\right) = \ln A - \ln B$$

$$\ln A^p = p \ln A$$

Identities

$$\ln e^x = x$$

$$e^{\ln x} = x$$

FORMULA SUMMARY: CALCULUS

Differentiation Formulas

1. $(f(x) \pm g(x))' = f'(x) \pm g'(x)$

2. $(kf(x))' = kf'(x)$

3. $(f(x)g(x))' = f(x)g'(x) + g(x)f'(x)$

4. $\left(\dfrac{f(x)}{g(x)}\right)' = \dfrac{g(x)f'(x) - f(x)g'(x)}{(g(x))^2}$

5. $(f(g(x)))' = f'(g(x)) \cdot g'(x)$

6. $\dfrac{d}{dx}(x^n) = nx^{n-1}$

7. $\dfrac{d}{dx}(e^x) = e^x$

8. $\dfrac{d}{dx}(a^x) = a^x \ln a \quad (a > 0)$

9. $\dfrac{d}{dx}(\ln x) = \dfrac{1}{x}$

10. $\dfrac{d}{dx}(\sin x) = \cos x$

11. $\dfrac{d}{dx}(\cos x) = -\sin x$

Integration Formulas

1. $\displaystyle\int x^n\,dx = \frac{1}{n+1}x^{n+1} + C, \quad n \neq -1$

2. $\displaystyle\int \frac{1}{x}\,dx = \ln|x| + C$

3. $\displaystyle\int e^x\,dx = e^x + C$

4. $\displaystyle\int e^{kx}\,dx = \frac{1}{k}e^{kx} + C$

5. $\displaystyle\int \sin x\,dx = -\cos x + C$

6. $\displaystyle\int \cos x\,dx = \sin x + C$